American Book Company

Meeting Standards,
Exceeding Expectations

Dear Educator,

Thank you for your interest in American Book Company's state-specific test preparation resources. We commend you for your interest in pursuing your students' success. Feel free to contact us with any questions about our books, software, or the ordering process.

Our Products Feature	Your Students Will Improve
Multiple-choice and open-ended diagnostic tests	Confidence and mastery of subjects
Step-by-step instruction	Concept development
Frequent practice exercises	Critical thinking
Chapter reviews	Test-taking skills
Multiple-choice practice tests	Problem-solving skills

American Book Company's writers and curriculum specialists have over 100 years of combined teaching experience, working with students from kindergarten through middle, high school, and adult education.

Our company specializes in effective test preparation books and software for high stakes graduation and grade promotion exams across the country.

How to Use This Book

Each book:

*contains a chart of standards which correlates all test questions and chapters to the state exam's standards and benchmarks as published by the state department of education. This chart is found in the front of all preview copies and in the front of all answer keys.

*begins with a full-length pretest (diagnostic test). This test not only adheres to your specific state standards, but also mirrors your state exam in weights and measures to help you assess each individual student's strengths and weaknesses.

*offers an evaluation chart. Depending on which questions the students miss, this chart points to which chapters individual students or the entire class need to review to be prepared for the exam.

*provides comprehensive review of all tested standards within the chapters. Each chapter includes engaging instruction, practice exercises, and chapter reviews to assess students' progress.

*finishes with two full-length practice tests for students to get comfortable with the exam and to assess their progress and mastery of the tested standards and benchmarks.

While we cannot <u>guarantee</u> success, our products are designed to provide students with the concept and skill development they need for the graduation test or grade promotion exam in their own state. We look forward to hearing from you soon.

Sincerely,

The American Book Company Team

PO Box 2638 ★ Woodstock, GA 30188-1383 ★ Phone: 1-888-264-5877 ★ Fax: 1-866-827-3240

ISTEP+ GQE MATHEMATICS
CHART of STANDARDS

ISTEP Standard Algebra 1	Chapter #	Diagnostic Exam	Progress 1	Progress 2	ISTEP Standard Algebra 1	Chapter #	Diagnostic Exam	Progress 1	Progress 2
Operations With Real Numbers	Chapter				Pairs of Linear Equations and Inequalities				
A1.1.1	4				A1.5.1	15	87	96	
A1.1.2	4	1,16,41	1,	1,12	A1.5.2	14		82	
A1.1.3	5	15		3	A1.5.3	15	48	6	7,10
A1.1.4	4	2,40	2,	2	A1.5.4	15	36,40,80		
A1.1.5	19	55	13	86	A1.5.5	15			
Linear Equations and Inequalities					A1.5.6	8	13,	51	90
					Polynomials				
A1.2.1	6,7	3,	3,	33					
A1.2.2	12		14		A1.6.1	9	8	12	
A1.2.3	5	43	1	5	A1.6.2	9			
A1.2.4	7	5,6			A1.6.3	10		5	
A1.2.5	7		53	93	A1.6.4	9	59	5	3,29
A1.2.6	8	39	54	54,63	A1.6.5	9	86		87
Relations and Functions					A1.6.6	10			
					A1.6.7	10	88	85	79
A1.3.1	6,7				A1.6.8	10			
A1.3.2	15	34	46	19	Algebraic Fractions				
A1.3.3	4			13					
A1.3.4	16	57	38,60	21	A1.7.1	10	94,106	78	
Graphing Linear Equations and Inequalities					A1.7.2	10			
					Quadratic, Cubic and Radical Equations				
A1.4.1	13	20	47						
A1.4.2	13	21,23	19,25	77	A1.8.1	13	48		
A1.4.3	13			97	A1.8.2	11	53	50	52
A1.4.4	13	12	82	14	A1.8.3	11	38	28	28
A1.4.5	13		37		A1.8.4	11	45	24	45
A1.4.6	13	68	7	56	A1.8.6	11	97	81	22,85
					A1.8.7	8		77	
					A1.8.8	11		9	37

Algebra 1 standards account for 30% of the questions on each test. The other items are distributed over the seven standards adopted for 7th and 8th grade.listed on the following page.

Chart of Standards Continued

	ISTEP Standard	Chapters #	Diagnostic Test	Practice Test 1	Practice Test 2
1	Number Sense	4	14,15,18,32,40,82,83	11,34,35,39,72,83,	1,35,36,62,72,78,95
2	Computation	1,2,3,	44,46,72,74,14,87,104	2,8,41,42,95,	2,38,41,42,65,65,71,88
3	Algebra and Functions	5,6,7,8,13,14	9,10,16,17,19,49,51,52,60,73,79,80,85,90,102	10,14,15,17,19,43,44,45,47,55,67,69,74,86,89,	4,8,11,12,14,24,43,44,46,59,63,66,63,66,74,89
4	Geometry	16,17,22	20,21,22,23,24,25,27,33,54,78,89	18,20,29,30,80,84	16,17,25,26,47,50,62,83,84,91
5	Measurement	18	28,29,30,31,48,56,58,71,75,76,93,96,98,99,100,101,103	21,22,26,27,49,52,73,88	18,20,23,48,49,51,64,75
6	Data Analysis and Probability	19,20	7,11,34,35,36,37,50,61,62,63,64,70,77,81,84	16,33,56,5758,59,64,70,71,75,76,79,93,94	27,29,30,32,34,53,55,67,68,69,70,73,76,80,81
7	Problem Solving	21	42,65,66,67,68,69,95,107	7,32,61,6263,65,66,90,91,	56,57,58,61,92,94

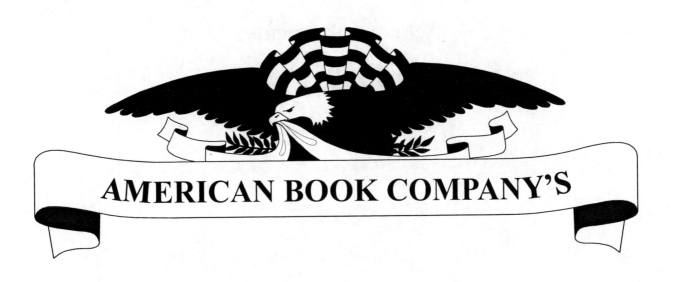

AMERICAN BOOK COMPANY'S

PASSING THE
ISTEP+ GQE
IN
MATHEMATICS

Colleen Pintozzi

AMERICAN BOOK COMPANY
P O BOX 2638
WOODSTOCK, GEORGIA 30188-1383
TOLL FREE: 1 (888) 264-5877   PHONE: 770-928-2834   FAX: 770-928-7483
Web site: www.americanbookcompany.com

Acknowledgements

In preparing this book, we would like to acknowledge the writing contributions of Laurie Sabbarese, Erica Day, Alan Fuqua and the editorial assistance of Tracy Jones and Andy Marzka.

TABLE OF CONTENTS

PASSING THE
ISTEP GQE IN MATHEMATICS

Preface

   PASSING THE ISTEP✦ GQE IN MATHEMATICS will help you review and learn important concepts and skills related to mathematics that will be assessed on the ISTEP✦ GQE Mathematics Test. Some of this material will be a review of skills you have already learned. In addition to a review of Algebra I standards you will also be able to review Numbers and Operations, Algebra, Measurement and Geometry, and Data Analysis and Probability. To help identify which areas are of greater challenge for you, begin by taking the Diagnostic Test at the beginning of this book. Once you have taken the test, complete the evaluation chart with your instructor in order to help you identify the chapters which require your careful attention. When you have finished your review of all of the material your teacher assigns, take the progress tests to evaluate your understanding of the ISTEP✦ GQE mathematics standards. **The materials in this book are based on the curriculum standards and content descriptions for mathematics published by the Indiana Department of Education.**

About the Author

   Colleen Pintozzi has taught mathematics at the middle school, junior high, senior high, and adult level for 22 years. She holds a B.S. degree from Wright State University in Dayton, Ohio and has done graduate work at Wright State University, Duke University, and the University of North Carolina at Chapel Hill. She is the author of eleven mathematics books including such best-sellers as *Basics Made Easy: Mathematics Review, Passing the New Alabama Graduation Exam in Mathematics, Passing the Georgia High School Graduation Test in Mathematics, Writing, and English Language Arts, Passing the TCAP Competency Test in Mathematics, Passing the Louisiana LEAP 21 Graduation Exit Exam, Passing the Indiana ISTEP+ Graduation Qualifying Exam in Mathematics, Passing the Minnesota Basic Standards Test in Mathematics, Passing the South Carolina HSAP in Mathematics, California Math Review, Passing the FCAT in Mathematics* and *Passing the Nevada High School Proficiency Exam in Mathematics.*

TO THE TEACHER

   As you know, there are 4 kinds of problems on the ISTEP✦ GQE test: multiple choice, griddables, open-ended questions with a calculator, and open-ended questions without a calculator. To simulate the test taking experience we have included all 4 kinds of problems on the Diagnostic test and Practice tests 1 and 2. To practice griddable items we have added a griddable answer sheet to the back of the answer key so you can just put it on a copier and make the answer sheets you need. For more practice, we suggest using that answer sheet for various assignments throughout the instructional chapters. For your convenience, we grouped the multiple choice problems together, so you could use a scantron card to grade that much of the test; we also grouped the griddable items together for the griddable answer sheet. The open-ended items are grouped so students can use their own paper to answer them. Please note in each Session 1, students may NOT use a calculator; in each Session 2, they MAY use a calculator.

   Next to each problem in the Diagnostic test and Practice test 1 and 2, we have added in small print the standard covered by the problem. That way at any time during the school year, the teacher could pick and choose the problems for the standards already covered that year and design a test similar to the ISTEP✦ by putting a list of problems on the board for the students to do. We hope this is a help to you and if you have any more ideas to add to future printings, please call and share them with us toll free at 888-264-5877.

ISTEP+ Grades 9 and 10 Mathematics Review Sheet

Shape		Formulas for Area (A) and Circumference (C)
Triangle:		$A = \frac{1}{2}bh = \frac{1}{2} \times$ base $\times$ height
Rectangle:		$A = lw =$ length $\times$ width
Trapezoid:		$A = \frac{1}{2}(b_1 + b_2)h = \frac{1}{2} \times$ sum of bases $\times$ height
Parallelogram:		$A = bh =$ base $\times$ height
Square:		$A = s^2 =$ side $\times$ side
Circle:		$A = \pi r^2 = \pi \times$ square of radius $C = 2\pi r = 2 \times \pi \times$ radius $\pi =$ pi ≈ 3.14 or $\frac{22}{7}$

Figure		Formulas for Volume (V) and Surface Area (SA)	
Rectangular Prism:		$V = lwh =$ length $\times$ width $\times$ height $SA = 2lw + 2hw + 2lh$ $\quad = 2$(length $\times$ width) $+ 2$(width $\times$ height) $+ 2$(length $\times$ height)	
General Prisms:		$V = Bh =$ area of base $\times$ height $SA =$ sum of the areas of the faces	
Cylinder:		$V = \pi r^2 h = \pi \times$ square of radius $\times$ height $SA = 2\pi r(r + h) = 2 \times \pi \times$ radius (radius $+$ height) $\pi =$ pi ≈ 3.14 or $\frac{22}{7}$	$\pi \approx 3.14$ or $\pi \approx \frac{22}{7}$
Sphere:		$V = \frac{4}{3}\pi r^3 = \frac{4}{3} \times \pi \times$ cube of radius $SA = 4\pi r^2 = 4 \times \pi \times$ square of radius	
Right Cylinder Cone:		$V = \frac{1}{3}\pi r^2 h = \frac{1}{3} \times \pi \times$ square of radius $\times$ height	
Regular Pyramid:		$V = \frac{1}{3}Bh = \frac{1}{3} \times$ area of base $\times$ height	

Equation of a Line

Slope-Intercept Form:

$y = mx + b$

where m = slope and b = y-intercept

Point-Slope Form:

Where m = slope and (x_1, y_1)
and (x_2, y_2) are two points in the plane

Slope of a Line

Let (x_1, y_1) and (x_2, y_2) be two points
in the plane

$$\text{slope} = \frac{\text{change in } y}{\text{change in } x} = \frac{y_2 - y_1}{x_2 - x_1} \text{ where } x_2 \neq x_1$$

Pythagorean Theorem

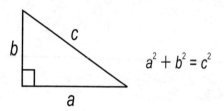

$$a^2 + b^2 = c^2$$

Distance Formula

$$d = rt$$

where d = distance, r = rate, and t = time

Temperature Formulas

$$°C = \frac{5}{9}(F - 32)$$
$$°\text{Celsius} = \frac{5}{9} \times (°\text{Fahrenheit} - 32)$$

$$°F = \frac{9}{5}C + 32$$
$$°\text{Fahrenheit} = \frac{9}{5} \times °\text{Celsius} + 32$$

Simple Interest Formula

$$i = prt$$

where i = interest, p = principle,
r = rate, and t = time

Quadratic Formula

$$x = \frac{-b \pm \sqrt{b^2 - 4ac}}{2}$$

where $ax^2 + bx + c = 0$, $a \neq 0$,
and $b^2 - 4ac \geq 0$

Conversions:

1 yard = 3 feet = 36 inches
1 mile = 1,760 yards = 5,280 feet
1 acre = 43,560 square feet
1 hour = 60 minutes
1 minute = 60 seconds

1 cup = 8 fluid ounces
1 pint = 2 cups
1 quart = 2 pints
1 gallon = 4 quarts

1 liter = 1,000 millimeters = 1,000 cubic centimeters
1 meter = 100 centimeters = 1,000 millimeters
1 kilometer = 1,000 meters
1 gram = 1,000 milligrams
1 kilogram = 1,000 grams

1 pound = 16 ounces
1 ton = 2,000 pounds

Diagnostic Test

1. Simplify:

$$\frac{\sqrt{20}}{\sqrt{35}}$$

   A. $\dfrac{2\sqrt{7}}{7}$ C. $\dfrac{2\sqrt{5}}{\sqrt{7}}$

   B. $\dfrac{2}{\sqrt{7}}$ D. $\dfrac{4}{7}$

   A1.1.2

2. Simplify the expression shown below:

$$3x^{-2}$$

   A. $(3x)^{-1}(3x)^{-1}$

   B. $\dfrac{9}{x^2}$

   C. $\dfrac{1}{3x^2}$

   D. $\dfrac{3}{x^2}$

   A1.1.4

3. Solve for a: $-2(-3-5) = 3 - a$

   A. -13
   B. 19
   C. 13
   D. -19

   A1.2.1

4. Jeff was making $6.25 per hour. His boss gave him a $.75 per hour raise. What percent raise did Jeff get?

   A. 12%   C. 40%
   B. 25%   D. 70%

   7.2.2

5. Find c: $\dfrac{c}{-2} > -6$

   A. $c > -12$
   B. $c < 12$
   C. $c > 12$
   D. $c < 3$

   A1.2.4

6. The regular price of a stereo (r) is $560. The stereo is on sale for 25% off. Which equation will help you find the sale price (s) of the stereo?

   A. $s = r - .25$
   B. $s = r - .25s$
   C. $s = r - .25r$
   D. $s = r - s$

   A1.2.4

7. Tom's school was considering making uniforms mandatory starting with the next school year. Tom hated the idea and wanted to do his own survey to see if parents were really in favor of it. He considered 4 places to conduct his survey. Which would give the most valid results?

   A. He would stop people at random walking through the mall.
   B. He would survey parents in the car pool lanes picking up students after school.
   C. He would survey the teachers after school.
   D. He would survey the students in his biology class to ask what their parents thought.

   8.6.2

8. Simplify:

$$(3x^2 - 5x + 6) - (x^2 + 4x - 7)$$

    A. $4x^2 - x - 1$
    B. $4x^2 - x + 13$
    C. $2x^2 - 9x - 1$
    D. $2x^2 - 9x + 13$

A1.6.1

9. In the following equation, which are the variable, the terms and the coefficient?

$$9x - 3 = 78$$

    A. Variable = x
        Terms = $9x$, -3
        Coefficient = 9

    B. Variable = -3
        Terms = 78, -3
        Coefficient = -3

    C. Variable = 9
        Terms = 9, -3
        Coefficient = x

    D. Variable = 78
        Terms = 78, 9
        Coefficient = x

7.3.3

10. What is the value of the expression $5(x + 6)$ when $x = -3$?

    A. -9
    B. 15
    C. 9
    D. 45

7.3.7

11. What is the median of the following set of data?

33, 31, 35, 24, 38, 30

    A. 32
    B. 31
    C. 30
    D. 29

7.6.5

12. Find the equation of the line perpendicular to the line containing the points (-2, -3) and (1, 4) and passing through the point (0, 3).

    A. $y = \frac{3}{7}x - 3$

    B. $y = -\frac{3}{7}x + 3$

    C. $y = -\frac{3}{7}x - 3$

    D. $y = -\frac{7}{3}x + 3$

A1.4.4

13. An arrow is shot upward with an initial velocity of 128 feet per second. The height (h) of the arrow is a function of time (t) in seconds since the arrow left the ground and can be expressed by the equation $h = 128t - 16\,t^2$. When will the arrow be at a height of 240 feet?

    A. At 3 sec and at 5 sec
    B. Only at 3 sec
    C. Only at 5 sec
    D. Only at 8 seconds

A1.5.6

4

14. Which of the following is the prime factorization of 40?

    A. $2^3 \times 5$
    B. $3^2 \times 5$
    C. 4×10
    D. $2 \times 5 \times 4$

    7.1.5

15. Find:
    $(-a^3 + 2a^2 - 8) - (-4a^3 + 5a - 2)$

    A. $3a^3 + 2a^2 - 5a - 6$
    B. $-5a^3 + 7a^2 - 10$
    C. $3a^3 + 7a^2 - 6$
    D. $-5a^3 + 7a^2 - 6$

    A1.1.3

16. Solve $I = PRT$ for R.

    A. $R = IPT$
    B. $R = I + PT$
    C. $R = I - PT$
    D. $R = \frac{I}{PT}$

    A1.1.2
    7.3.5

17. Which order of operations should be used to simplify the following expression?

    $12 \div 2 + 4(7 - 5)$

    A. subtract, multiply, add, divide
    B. divide, add, subtract, multiply
    C. divide, add, multiply, subtract
    D. Subtract, divide, multiply, add

    7.3.4

18. Which of the following computations will result in an irrational number?

    A. 7π
    B. $3\frac{1}{2} + 7\frac{1}{4}$
    C. $6.8 - 3.9$
    D. $5 - \frac{1}{2}$

    8.1.3

19. Which table of values below represents a linear function?

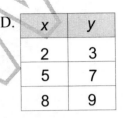

A.

x	y
0	−3
1	−1
2	1

C.

x	y
0	2
−1	0
−2	−1

B.

x	y
−1	5
−2	0
1	3

D.

x	y
2	3
5	7
8	9

7.3.9

20. Which equation matches the following graph?

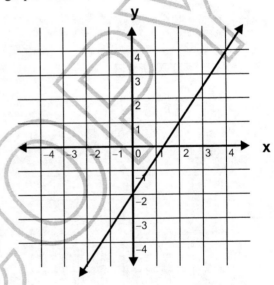

    A. $y = 3x - 2$
    B. $y + 2 = -3x$
    C. $3(y + 2) = 2x$
    D. $2(y + 2) = 3x$

    A1.4.1

21. Which graph shows a line that has a slope of −1 and *y*-intercept of (0, 1)?

A.

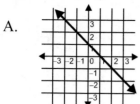

B.

C.

D.

A1.4.2
7.3.8

22. What is the relationship between figure 1 and 2?

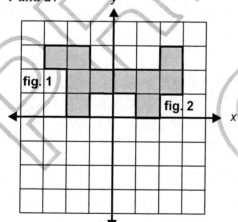

A. The figures are similar.
B. The figures are congruent and similar.
C. Figure 2 is larger than Figure 1.
D. Figure 2 is smaller than Figure 1.

7.4.2
8.4.4

23. Which graph represents a line containing the points (− 2, −3) and (1, 4)?

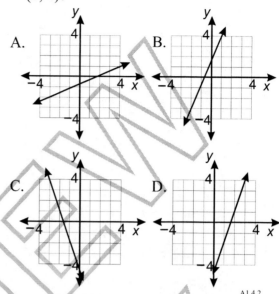

A.

B.

C.

D.

A1.4.2
7.3.7

24.

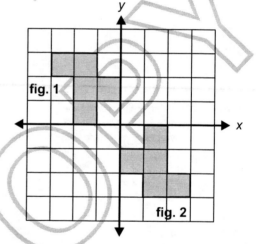

Figure 1 goes through a transformation to form Figure 2. Which of the following descriptions fits the transformation shown?

A. reflection across the *x*-axis
B. reflection across the *y*-axis
C. $\frac{1}{2}$ clockwise rotation around the origin
D. translation right 3 units and down 3 units

7.4.1

25. What is the distance from Point P to the origin?

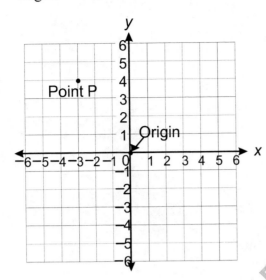

A. 5
B. $\sqrt{35}$
C. 7
D. 9

7.4.3
8.4.5

26. The area of a picture frame is 2 square feet. How many square inches is that?

A. 288 square inches
B. 24 square inches
C. 48 square inches
D. 96 square inches

7.5.1
8.5.1

27. A 25 foot ladder is leaning against a building. The base of the ladder is 7 feet from the base of the building. How high up the building does the ladder reach?

A. 21 feet
B. 22 feet
C. 23 feet
D. 24 feet

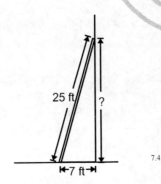

7.4.3

28. The scale drawing of an advertising sign is drawn with a scale of 1 inch = 4 feet.

What is the actual width of the advertising sign?

A. $5\frac{3}{8}$ feet
B. $13\frac{1}{2}$ feet
C. $14\frac{1}{2}$ feet
D. 58 feet

7.5.3

29. Emily needs to make a glass case with the following measurements:

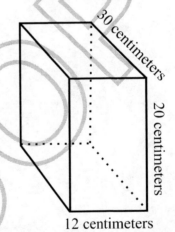

How many square centimeters of glass would it take to construct the case enclosed on all sides?

A.   60 square centimeters
B.   612 square centimeters
C.   2,400 square centimeters
D.   6,200 square centimeters

7.5.4

30. Cynda wants to buy a daycare center with the measurements below. How many square feet are in the building?

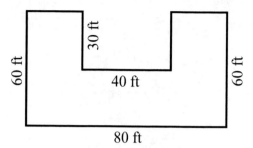

A. 340 square feet
B. 480 square feet
C. 3,600 square feet
D. 4,800 square feet

7.5.5

31. Find the volume of the figure below.

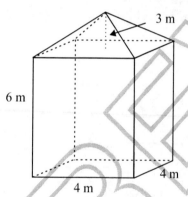

A. 96 m³
B. 99 m³
C. 112 m³
D. 288 m³

7.5.4

32. Write 3^{-2} as a fraction.

A. $\frac{1}{6}$

B. $\frac{1}{9}$

C. $\frac{2}{3}$

D. $\frac{3}{2}$

8.1.4

33. Which of the following nets represents a cube?

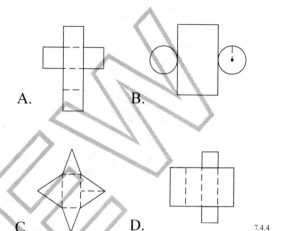

A.   B.

C.   D.

7.4.4

34. Transpiration is the process by which a plant loses water vapor through its leaves.

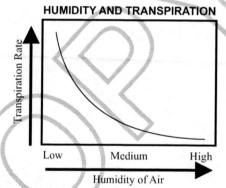

According to the graph, which of the following would be an accurate conclusion?

A. On days of low humidity, the transpiration rate is highest.
B. On days of medium humidity, the transpiration rate is above average.
C. On very humid days, the transpiration rate is highest.
D. On days of low humidity, the transpiration rate is lowest.

A1.3.2
7.6.1

8

35. Aaron wants to increase the number of pushups he does each day. The graph shows the number of pushups Aaron did Monday, Tuesday, and Wednesday.

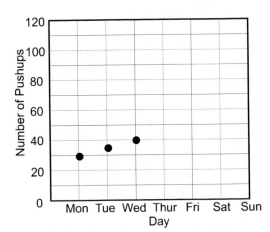

If the number of pushups continues to increase at the same rate, on which day of the week will the number of pushups reach 50 or more?

A. Thursday

B. Friday

C. Saturday

D. Sunday

7.6.2

36. Amanda has test scores of 44, 77, 44 and 68 in her history class. If she scores 100% on her next test, how will this affect the mean, median and mode of her scores?

A. The mean will go up.
The median will go up.
The mode will remain the same.

B. The mean will remain the same.
The median will remain the same.
The mode will remain the same.

C. The mean, median and mode will all go up.

D. The mean will go up.
The mode and median will remain
The same.

7.6.3

CURRENT WARMING TREND
Global average temperature
in Fahrenheit degrees

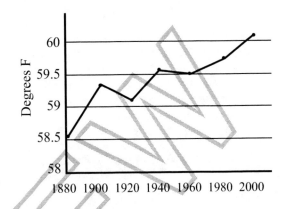

37. Why is the graph shown above misleading?

A. The bar graph does not take other factors such as industrialization into account.

B. The graph begins at 58°, not 0°, on the y-axis.

C. The lines on the graph do not connect the values.

D. The graph values are measured in °F instead of the more scientific measurement °C.

7.6.4

38. Solve the equation $(x-4)^2 = 25$

A. $x = -4, -5$
B. $x = 9, -1$
C. $x = 5, -5$
D. $x = 4, -4$

A1.8.3

39. Sally, Janet, and Nancy are helping with a reforestation project. Sally plants between 50 and 90 seedlings per hour. Janet and Nancy each plant between 40 and 75 trees per hour. Which of these inequalities represents the possible range of total number of trees the three workers can plant in one hour?

    A. $90 \leq x \leq 165$

    B. $130 \leq x \leq 240$

    C. $150 \leq x \leq 280$

    D. $180 \leq x \leq 330$

A1.2.6

40. Simplify $2^3 \times 2^5$

    A. 2

    B. 2^{-2}

    C. 2^{15}

    D. 2^8

A1.1.4
8.1.5

41. Simplify $(\sqrt{64})^2$

    A. 64

    B. 16

    C. $\frac{1}{64}$

    D. $\frac{1}{8}$

A1.1.2
8.1.6

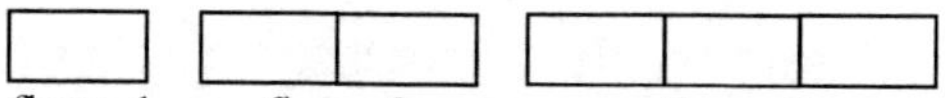

figure 1   figure 2   figure 3

42. The figures above are formed by placing rectangles side by side. The rectangle is 1 unit wide and 2 units long.

Look at the chart below.

Figure	Perimeter
1	6
2	10
3	14
4	
5	

Choose the correct perimeters for figures 4 and 5, and the mathematical sentence for the perimeter of the nth figure.

    A.   Figure 4 P = 18
          Figure 5 P = 22
          P = 4n + 1

    B.   Figure 4 P = 18
          Figure 5 P = 22
          P = 4n + 2

    C.   Figure 4 P = 17
          Figure 5 P = 21
          P = 2n + 1

    D.   Figure 4 P = 20
          Figure 5 P = 24
          P = n + 5

7.7.1

43. Which members of the set $\{-3, -2, -1, 0, 1, 2, 3\}$ are solutions for the inequality $-2x + 5 > 10$?

    A. $\{-3, -2\}$
    B. $\{0, 1, 2, 3\}$
    C. $\{-3, -2, -1\}$
    D. $\{-3\}$

A1.2.3

10

44. If Jenny puts $150.00 into her savings account which pays 5% simple interest per year on January 1st, how much interest will she earn by December 31st of the same year?

Use the formula I = PRT

    I = Amount of interest
    P = Principle (amount of money invested or borrowed)
    R = Rate of interest
    T = Time in years

    A. $ 5.00
    B. $ 7.50
    C. $ 17.50
    D. $157.50

8.2.1
8.2.2

45. Solve the equation $x^2 - 6x + 7 = 0$ by completing the square.

    A. $x = 7, -1$
    B. $x = \sqrt{3}, -\sqrt{3}$
    C. $x = 3 - \sqrt{2}, \sqrt{2} + 3$
    D. $x = 3 + 2i, 3 - 2i$

A1.8.4

TOY STORE
Sale
$\frac{1}{4}$ off all toys and games

46. According to the sale ad above, about how much could you save on a board game regularly priced at $12.47?

    A. $ 3.00
    B. $ 6.00
    C. $ 9.00
    D. $15.00

8.2.4

47. Solve the equation $x = \sqrt{4x - 3}$

    A. $x = 1, 3$
    B. $x = \pm\sqrt{3}$
    C. $x = -1, -3$
    D. $x = 2\sqrt{3}, -2\sqrt{3}$

A1.8.8

48. What is the intercept of the following linear equations?

$$y = 3x - 1$$
$$y = 4x + 2$$

    A. $(-3, 10)$
    B. $(-3, -10)$
    C. $(10, -3)$
    D. $(3, -10)$

A1.5.3

49. Evaluate: $4x^3 - 2x^2 - 3x + 1$ given $x = -2$

    A.   533
    B.   489
    C.   -33
    D.   22

8.3.4

50. Fido gets 2 doggy treats every time he sits and 4 doggy treats when he rolls over on command. Throughout the week, he has sat 6 times as often as he has rolled over. In total, he has earned 80 doggy treats. How many times has Fido sat?

    A.   4
    B.   5
    C.   24
    D.   30

8.7.3

51. Florence follows Great Aunt Emma's instructions for making coffee in various size coffee urns.

Capacity of the Coffee Urn	Number of Scoops of Coffee
4 quarts	18 scoops
6 quarts	26 scoops
10 quarts	42 scoops

Which of these graphs correctly plots the number of scoops as a function of the capacity of the urn?

A.
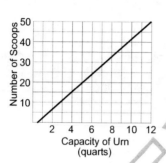

B.
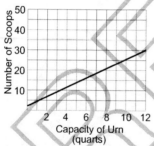

C.
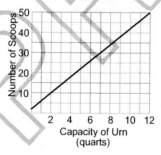

D.
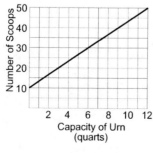

8.3.9

52. Which equation does the graph of the line below represent?

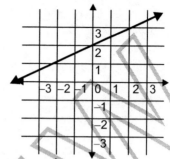

A. $y = \frac{1}{2}x$
B. $y = \frac{1}{2}x - 2$
C. $y = \frac{1}{2}x + 2$
D. $y = -2x$

8.3.5

53. Solve the following quadratic equation by factoring.

$$3x^2 = 4x + 7$$

A. $-\frac{1}{3}, \frac{7}{3}$

B. $-1, \frac{7}{3}$

C. $-1, -2$

D. $-\frac{7}{3}, \frac{3}{7}$

A1.8.2

54.
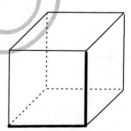

What is the relationship between the bold parts in the figure above?

A. parallel line segments
B. diagonal line segments
C. skew line segments
D. perpendicular line segments

8.4.1

55. The speed of sound in dry air at 32°F is 331.6 m/sec. How many cm/min is that? (Hint: use dimensional analysis)

    A.   663,200 cm/min
    B.   1,061,120 cm/min
    C.   1,989,600 cm/min
    D.   3,979,200 cm/min

    A1.1.5
    8.5.2

56. If you double the radius of a circle, how much does the area increase?

    A.   The area remains the same.
    B.   The area doubles.
    C.   The area is three times greater.
    D.   The area is four times greater.

    8.5.3

57. The following graph depicts the height of a projectile as a function of time.

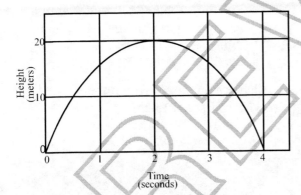

    What is the domain (D) of this function?

    A.   0 meters $\leq$ D $\leq$ 20 meters

    B.   4 seconds $\leq$ D $\leq$ 20 meters

    C.   20 meters $\leq$ D $\leq$ 4 seconds

    D.   0 seconds $\leq$ D $\leq$ 4 seconds

    A1.3.4

58.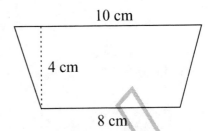

    Find the area of the trapezoid above.

    A.   22 square centimeters
    B.   36 square centimeters
    C.   72 square centimeters
    D.   320 square centimeters

    8.5.4

59. Multiply and simplify:
    $(3x + 2)(x - 4)$

    A. $3x^2 - 10x - 8$
    B. $3x^2 + 5x - 8$
    C. $3x^2 + 5x - 6$
    D. $8x^2 - 2$

    A1.6.4

60. Translate the following sentence into an algebraic equation:

    Two times the sum of a number, x, and 6 is 14.

    A. $2x + 6 = 14$
    B. $2x - 6 = 14$
    C. $2(x + 6) = 14$
    D. $2(x - 6) = 14$

    7.3.1

61. Examine the following two data sets:

    Set #1: 49, 55, 68, 72, 98
    Set #2: 20, 36, 47, 68, 75, 82, 89

    Which of the following statements is true?

    A.   They have the same mode.
    B.   They have the same median.
    C.   They have the same mean.
    D.   None of the above.

    8.6.3

13

62. Beth did a stem and leaf plot of her daily math grades for first semester.

Stem	Leaves
3	1
4	5
5	2
6	0,2,3,5,5,5,8,9,9
7	1,3,3,3,4,4,5,5,5,5,5,5,6,6,7,7,7,9,9,9
8	0,1,2,2,4,4,4,4,5,5,6,6,7,7,7,8,9
9	0,2,2,3,6,8,9

What grade was the mode of her data?

A. 31 C. 75
B. 65 D. 77

8.6.4

63. Sarah deposits 50¢ into Miss Clucky, a machine that makes chicken squawks and gives Sarah one plastic egg with a toy surprise. In the machine, 30 eggs contain a rubber frog, 43 eggs contain a plastic ring, 23 eggs contain a necklace, and 18 eggs contain a plastic car. What is the probability that Miss Clucky will give Sarah a necklace in her egg?

A. $\frac{1}{114}$

B. $\frac{23}{114}$

C. $\frac{23}{91}$

D. $\frac{1}{23}$

8.6.6

64. How many different 3 letter patterns can be formed using an A, a B, and a C?

A. 4
B. 6
C. 9
D. 27

8.6.7

65. Use the following statements to answer the following question.

Tammy got a higher grade on the math test than **Susanna**.
Susanna got a lower grade on the math test than **Nancy**.
Laurie got the same score on the math test as **Peter.**
Peter got a lower grade on the math test than **Susanna**.
Nancy got a higher grade on the math test than **Tammy**.

Which of the following is **not** true?

A. Laurie got a lower grade on the math test than Susanna but not lower than Peter.
B. Susanna got a lower grade on the math test than Tammy and Peter.
C. Nancy got a higher grade on the math test than Peter and Tammy.
D. Tammy got a higher grade than Susanna and Peter.

8.7.1

66. What are the next two numbers in the numerical pattern below?

1, 3, 9, 27, 81, □, □

A. 115, 162
B. 135, 189
C. 180, 372
D. 243, 729

8.7.2

67. If $2^{\square} = 5$, $3^{\square} = 10$ and $4^{\square} = 17$, what will $5^{\square} = ?$

A. 21
B. 19
C. 24
D. 26

8.7.4

14

68. Which of the following is a graph of the inequality $y \leq x - 3$?

A.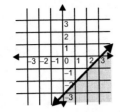

B.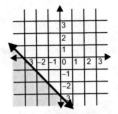

C.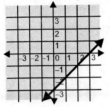

D.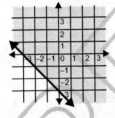

A.1.4.6
8.7.9

Use the grid sheets provided by your teacher to record your answers to the following questions.

69. What is the missing number in this sequence? Answer as a decimal.

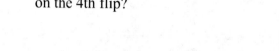
$$80, 20, 5, \underline{\quad ? \quad}, \frac{5}{16}$$

7.7.1
8.7.1

70. Marilyn flipped an honest coin 3 times and it came up heads each time. What is the probability that it will come up heads on the 4th flip?

7.6.6
8.6.6

71. Wanda's pet snail can crawl 4 feet in one hour. At this rate, how many days would it take the snail to crawl one mile?

8.5.2

72. The estimated number of fish in a lake decreased from 3.4 million to 2 million. What was the percent decrease in the fish population of the lake?

7.2.2

73. What is the value of the expression below?

$$4^3 + (-3)^2 - \sqrt{16}$$

7.3.4

74. Julia is paid $30 per hour on her new job. She made $20 per hour on her previous job. What percent increase is her new hourly pay over her previous job?

7.2.2

75. A pine tree casts a shadow 9 feet long. At the same time, a rod measuring 4 feet casts a shadow 1.5 feet long. How tall is the pine tree?

7.5.2

76. What is the length of the master bedroom?

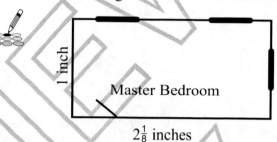

Scale: $\frac{1}{8}$ inch = 1 foot

7.5.3

77. Brandi spins the spinner twice. What is the probability that the sum of the two spins is an even number? Grid your answer as a fraction.

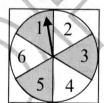

7.6.6

78.

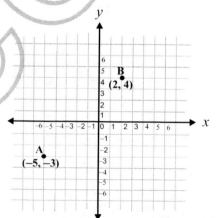

What is the distance between point A and point B? Answer to the nearest tenth of a unit.

7.4.3
8.4.5

16

79. The table shows values of x and y for the equation

$$y = x^3 - x^2 - 2x$$

What is the value of y when $x = 5$?

x	y
0	0
1	-2
2	0
3	12
4	40
5	?

8.3.8

80. Jason is a member of the Sunshine Health Spa. The monthly charge (C) for a member is $40 per month plus $5 for each visit ($V$). The bookkeeping department uses the formula $C = 40 + 5V$ to calculate each member's monthly charge. Jason's charge for last month was $85. How many times did Jason visit the Sunshine Health Spa last month?

7.3.2

81. If you were going to make a stem-and-leaf plot for the following set of data, how many stems would be needed?

21	51	64	29	33
42	57	72	15	39
51	28	36	30	52

8.6.4

82. $5^3 =$

7.1.4

83. Change .80 to a fraction.

7.1.7

Diagnostic Test - Session 1

Directions: Write out your response to each of the following problems on your own Paper. Show all of your work. You may NOT use a calculator on this portion of the diagnostic exam.

84. Four runners are lining up for the 100 meter dash. How many ways can the runners line up on the start line?

7.67
8.67

85. David earns $8.00 per hour plus 8% of sales over $2,000 per week. Last week, his Sales totaled $2,800. He earned $352. Using h to represent the number of hours David worked, write an equation that will show David his total earnings last week.

8.3.1

86. Divide $4x^2y^3 - 8xy^2 + 12xy$ by $4xy$.

87. When Brent started on his trip, the odometer read 47,682. At the end of his trip, it read 47,952. Brent's car averages 30 miles per gallon. He paid $1.20 per gallon for gas. Explain in words how Brent can find out how much he spent on gasoline for his trip. You do NOT have to solve this problem.

A1.6.5

7.2.1

88. Factor $16a^2 - 64$.

A1.6.7

89. On your own paper, draw a circle. Draw a chord in the circle. Draw a central angle in the circle.

8.4.1

90. Write an equation of a line with a slope of $\frac{1}{2}$ and a y-intercept of -3.

8.3.6
7.3.8

91. Factor $48xy^2 + 24xy^4 - 36xy^6$

92. Matt has a cube that measures 7 cm on each edge. What is the surface area of the cube?

7.3.5

93. Kristina measured a line to be 30 millimeters. How many centimeters is that?

8.5.4
7.5.4

94. Simplify: $\dfrac{4a^2 - 9}{10a^2 - 13a - 3}$

8.5.1

A1.7.1

18

Copyright © American Book Company

Diagnostic Test - Session 2

Directions: Write out your response to each of the following problems on your own paper. Show all of your work. You MAY use a calculator on this portion of the diagnostic exam.

95. Rob has a part-time job earning $10.00 per hour. He made a chart of his hours, earnings, and federal taxes taken out of his paycheck. If the pattern continues, how much will be taken from his check for federal taxes if he worked 32 hours.

Hours	Earned	Taxes
25	$250	$23
26	$260	$25
27	$270	$26
28	$280	$28
29	$290	$29

7.7.1
8.7.2

96. Terry built the box to the right for his sister. He wants to build a box just like it, only doubling each measurement. How many times larger will the volume be of the new box?

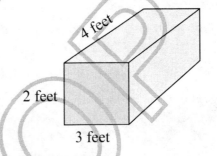

4 feet
2 feet
3 feet

8.5.3
7.5.3

97. Solve the equation $3x^2 + 8x - 3 = 0$ using the quadratic equation. Show your work.

A1.8.6

98. The city built a swimming pool. City officials then noticed that the builders left a big oak tree too close to the pool. It dropped leaves, twigs, and bugs into the water constantly. Keeping the pool clean was a problem. Finding out if there was enough room to cut down the tree was also a problem. City officials measured the shadow of the tree and found that its shadow was 36 feet long at the same time that an 8 foot pole standing straight up had a shadow of 3 feet. Write a proportion and then solve it to find out how tall the tree is.

7.5.3
8.5.3

99. A cubic centimeter of candle wax has a mass of 0.9 grams. Find how many grams of wax are needed to make the candle below. Round your answer to the nearest gram.

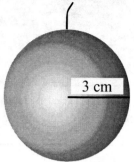

3 cm

Volume of a sphere: $V = \frac{4}{3}\pi r^3$ use $\pi = 3.14$

8.5.4

100. Find the circumference of the circle in the grid to the right.

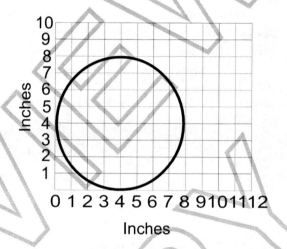

Inches

0 1 2 3 4 5 6 7 8 9 10 11 12

Inches

8.5.4

101. On a map drawn to scale, 2 centimeters represents 300 kilometers. How long would a line measure between two cities that are 500 kilometers apart?

8.5.3

102. Solve the equation $5x - 62 = -27$ and check your answer in the original equation.

103. Allie lives 190,080 inches from her friend Kelly. Would it be reasonable for Allie to 7.3.2 ride her bicycle to Kelly's house after school? Why or why not?

7.5.1

104. Marla found a dress she liked for $80. The next week it was on the sale rack for $44. What percent discount is that?

7.2.3

105. Graph the equations $y = 3x + 2$ and $x - 2y = 6$ on the same graph to find where the lines intersect. Use the graphs to estimate the solution of the pair of equations.

A1.5.1

106. Write the first twelve numbers in Base 2 notation. Use expanded notation to confirm Your list.

8.7.10

EVALUATION CHART
DIAGNOSTIC MATHEMATICS TEST

Directions: On the following chart, circle the question numbers that you answered incorrectly, and evaluate the results. Then turn to the appropriate topics (listed by chapters), read the explanations, and complete the exercises. Review the other chapters as needed. Finally, complete the Mathematics Progress Tests to check how much you have learned.

		QUESTIONS	PAGES
Chapter 1:	**Fractions & Decimals**	46	22-32
Chapter 2:	**Percents**	4,72,74,104	33-43
Chapter 3:	**Integers and Order of Operations**	17	44-52
Chapter 4:	**Number Sense**	1,2,14,18,32,40,41,82,83	53-67
Chapter 5:	**Introduction to Algebra**	6,9,10,49,60,73,79,80,85	68-78
Chapter 6:	**Solving One-Step Equations and Inequalities**	3,5	79-89
Chapter 7:	**Solving Multi-Step Equations and Inequalities**	43,102	90-106
Chapter 8:	**Algebra Word problems**	39,50	107-123
Chapter 9:	**Polynomials**	8,15,59,86	124-144
Chapter 10:	**Factoring Polynomials**	45,47,91	145-167
Chapter 11:	**Solving Quadratic Equations**	13,38,53,88,97	168-179
Chapter 12:	**Using Formulas**	16,44	180-185
Chapter 13:	**Graphing and Writing Equations**	12,20,21,23,51,52,90,105	186-212
Chapter 14:	**Graphing Inequalities**	68	213-216
Chapter 15:	**Systems of Equations and Systems of Inequalities**	48	217-225
Chapter 16:	**Relations and Functions**	19,57,79	226-232
Chapter 17:	**Plane Geometry**	25,27,29,30,56,58,78,89,100,101	233-247
Chapter 18:	**Solid Geometry**	31,33,54,92,96,99	248-273
Chapter 19:	**Measurement**	26,28,55,71,75,76,93,94,98,103	274-293
Chapter 20:	**Data Analysis and Probability**	11,34,35,36,37,61,62,63,70,77,81	294-320
Chapter 21:	**Permutations and Combinations**	64,84	321-326
Chapter 22:	**Problem Solving**	7,42,65,66,67,69,87,95,106	327-358
Chapter 23:	**Transformations and Plotted Shapes**	22,24	359-372

Chapter 1 | Fractions and Decimals

TWO-STEP WORD PROBLEMS

1. There are 25 miniature chocolate bars in a bag. There are 20 bags in a carton. Damon needs to order 10,000 miniature chocolate bars. How many cartons will he need to order?

2. LeAnn needs 2,400 boxes for her business. The boxes she needs come in bundles of 50 that weigh 45 pounds per bundle. What will be the total weight of the 2,400 boxes she needs?

3. Seth uses 20 nails to make a birdhouse. He wants to make 60 birdhouses to sell at the county fair. There are 30 nails in a box. How many boxes will he need?

4. There are 12 computer disks in a box. There are 10 boxes in a carton. John ordered 16 cartons. How many disks is he getting?

5. The Do-Nut Factory packs 13 doughnuts in each baker's dozen box. They also sell cartons of doughnuts which have 6 baker's dozen boxes. Duncan needs to feed 780 people. Assuming each person eats only 1 doughnut, how many cartons will he need to buy from the Do-Nut Factory?

6. Brittany has 2 dogs, a Saint Bernard and a Golden Retriever. The Saint Bernard eats twice as much as the Golden Retriever. The retriever eats 5 pounds of food in 6 days. How many pounds of food do the two dogs eat in 30 days?

7. Each of the 4 engines on a jet uses 500 gallons of fuel per hour. How many gallons of fuel are needed for a 5 hour flight with enough extra fuel for an additional 2 hours as a safety precaution?

8. The Farmer's Dairy has 1,620 pounds of butter to package. They are packaging the butter in five-pound tubs to distribute to restaurants. If they put 12 tubs in a case, how many cases of butter can they fill?

9. Tom has 155 head of cattle. Each animal eats 8 pounds of grain per day. How many pounds of grain does Tom need to feed his cattle for 10 days?

10. When you grind 3 cups of grain, you get 5 cups of flour. How many cups of grain must you grind to get 40 cups of flour?

11. Elaine has made 8 interest-free monthly payments of $115 on a new plasma TV. She still owes $865. How much did the TV cost originally?

8.2.1
7.2.1

ESTIMATED SOLUTIONS

In the real world, estimates can be very useful. The best approach to finding estimates is to round off all numbers in the problem. Then, solve the problem, and choose the closest answer. If money problems have both dollars and cents, round to the nearest dollar or ten dollars. $44.86 rounded to the nearest dollar is $45 and rounded to the nearest ten dollars is $40.

EXAMPLE: Which is a reasonable answer? $1580 \div 21$

      A. 80   B. 800   C. 880   D. 8,000

Step 1:   Round off the numbers in the problem.   1,580 rounds to 1,600   21 rounds to 20

Step 2:   Work the problem. $1,600 \div 20 = 80$ The closest answer is **A. 80**.

Choose the best answer below.

1. Which is a reasonable answer? 544×12 A. 54   B. 500   C. 540   D. 5,400

2. Jeff bought a pair of pants for $45.95, a belt for $12.97, and a dress shirt for $24.87. Estimate about how much he spent.   A. $60   B. $70   C. $80   D. $100

3. For lunch, Marcia ate a sandwich with 187 calories, a glass of skim milk with 121 calories, and 2 brownies with 102 calories each. About how many calories did she consume?
   A. 300   B. 350   C. 480   D. 510

4. Which is a reasonable answer? $89,990 \div 28$
   A. 300   B. 500   C. 1,000   D. 3,000

5. Which is a reasonable answer? $74,295 - 62,304$
   A. 12,000   B. 11,000   C. 10,000   D. 1,000

6. Delia bought 4 cans of soup at 99¢ each, a box of cereal for $4.78, and 2 frozen dinners at $3.89 each. About how much did she spend?
   A. $10.00   B. $11.00   C. $13.00   D. $17.00

7. Which is the best estimate? $22,480 + 5,516$
   A. 2,800   B. 17,000   C. 28,000   D. 32,000

School Store Price List

Pencils	Erasers	Folders	Binders	Compass	Protractor	Paper	Pens
2 for 78¢	59¢	21¢	$2.79	$1.59	89¢	$1.29	$1.10

8. Jake needs 2 pencils, 3 erasers, a binder, and a compass. About how much money will he need according to the chart?   A. $7.00   B. $8.00   C. $9.00   D. $10.00

9. Tracy needs a pack of paper, 2 folders, a protractor, and 6 pencils. About how much money does she need?   A. $3.00   B. $4.00   C. $5.00   D. $6.00

10. Which is the best estimate? $23,895 \div 599$ A. 4   B. 40   C. 400   D. 4,000

8.2.3

23

FRACTION REVIEW

Change to mixed numbers or whole numbers.

1. $\dfrac{11}{2} =$ 3. $\dfrac{9}{4} =$ 5. $\dfrac{16}{4} =$ 7. $\dfrac{13}{4} =$ 9. $\dfrac{30}{10} =$ 11. $\dfrac{10}{4} =$

2. $\dfrac{8}{3} =$ 4. $\dfrac{20}{9} =$ 6. $\dfrac{21}{5} =$ 8. $\dfrac{42}{7} =$ 10. $\dfrac{18}{7} =$ 12. $\dfrac{11}{3} =$

Change to an improper fraction.

13. $2\dfrac{1}{8} =$ 15. $4\dfrac{2}{3} =$ 17. $9\dfrac{2}{5} =$ 19. $6\dfrac{2}{7} =$ 21. $2\dfrac{5}{8} =$ 23. $8\dfrac{1}{9} =$

14. $8\dfrac{1}{4} =$ 16. $9\dfrac{1}{3} =$ 18. $6\dfrac{3}{4} =$ 20. $2\dfrac{3}{4} =$ 22. $9\dfrac{1}{2} =$ 24. $7\dfrac{2}{7} =$

Reduce to lowest terms.

25. $\dfrac{2}{8} =$ 27. $\dfrac{5}{15} =$ 29. $\dfrac{10}{12} =$ 31. $\dfrac{6}{18} =$ 33. $\dfrac{4}{8} =$ 35. $\dfrac{6}{14} =$

26. $\dfrac{10}{15} =$ 28. $\dfrac{6}{24} =$ 30. $\dfrac{16}{20} =$ 32. $\dfrac{8}{24} =$ 34. $\dfrac{8}{14} =$ 36. $\dfrac{7}{28} =$

Perform the following operations. Reduce each answer to lowest terms.

37. $\begin{array}{r} 5\frac{1}{2} \\ + \ 3\frac{5}{8} \\ \hline \end{array}$ 38. $\begin{array}{r} 7\frac{3}{8} \\ + \ 4\frac{1}{2} \\ \hline \end{array}$ 39. $\begin{array}{r} 9\frac{3}{4} \\ + \ 4\frac{1}{2} \\ \hline \end{array}$ 40. $\begin{array}{r} 2\frac{2}{3} \\ + \ 2\frac{1}{4} \\ \hline \end{array}$ 41. $\begin{array}{r} 4\frac{5}{6} \\ + \ 6\frac{1}{4} \\ \hline \end{array}$

42. $\begin{array}{r} 7 \\ - \ 3\frac{1}{4} \\ \hline \end{array}$ 43. $\begin{array}{r} 6\frac{1}{4} \\ - \ 2\frac{3}{5} \\ \hline \end{array}$ 44. $\begin{array}{r} 9\frac{1}{8} \\ - \ 7\frac{3}{4} \\ \hline \end{array}$ 45. $\begin{array}{r} 5\frac{1}{4} \\ - \ 2\frac{2}{3} \\ \hline \end{array}$ 46. $\begin{array}{r} 8\frac{1}{8} \\ - \ 2\frac{1}{10} \\ \hline \end{array}$

47. $\dfrac{3}{4} \times 100$ 48. $\dfrac{2}{5} \times 80$ 49. $\dfrac{3}{4} \times \dfrac{7}{9}$ 50. $\dfrac{1}{3} \times 16\dfrac{1}{2}$

51. $10\dfrac{1}{2} \times 4\dfrac{2}{3}$ 52. $4\dfrac{1}{6} \times 2\dfrac{4}{5}$ 53. $9 \times 7\dfrac{2}{3}$ 54. $2 \times 1\dfrac{7}{8}$

55. $28 \div 2\dfrac{2}{3}$ 56. $1\dfrac{1}{8} \div 2$ 57. $1\dfrac{1}{4} \div 2\dfrac{2}{9}$

58. $15 \div 1\dfrac{1}{3}$ 59. $8\dfrac{1}{4} \div \dfrac{1}{2}$ 60. $20 \div 2\dfrac{2}{5}$

8.2.1
7.2.1

DEDUCTIONS - FRACTION OFF

Sometimes sale prices are advertised as a fraction off, such as $\frac{1}{4}$ off or $\frac{1}{3}$ off. To find out how much you will save, just multiply the original price by the fraction off.

EXAMPLE: CD players are on sale for $\frac{1}{3}$ off. How much can you save on a $240 CD player?

$$\frac{1}{\overset{\displaystyle 3}{\underset{\displaystyle 1}{}}} \times \frac{\overset{\displaystyle 80}{240}}{1} = 80 \quad \text{You can save \$80.00.}$$

Find the amount of savings in the problems below.

J.P. Nichols is having a liquidation sale on all furniture. Sale prices are $\frac{1}{2}$ off the regular price. How much can you save on the following furniture items?

Liquidation Furniture Sale

$\frac{1}{2}$ **off all items in the store**

Item	Regular Price	Savings
1. Couch	$850	_____
2. Loveseat	$624	_____
3. Recliner	$457	_____
4. Dining Room Set	$1352	_____
5. Bedroom Set	$2648	_____

Buy Rite Computer Store is having a $\frac{1}{3}$ off sale on selected computer items in the store. How much can you save on the following items?

BUY RITE COMPUTER STORE

SALE: $\frac{1}{3}$ **off selected items in the store**

Item	Regular Price	Savings
6. Midline Computer	$1383	_____
7. Notebook Computer	$2280	_____
8. Tape Backup Drive	$210	_____
9. Laser Printer	$855	_____
10. Digital Camera	$690	_____

7.2.1

FINDING A FRACTION OF A TOTAL

Mathematicians use the word "of" in word problems to indicate that you need to multiply to find the answer.

EXAMPLE 1: Two-thirds of male high school seniors will be taller than their fathers by the time they graduate. In a sample of 400 male seniors, about how many will be taller than their fathers on graduation day?

Solution: Multiply the fraction by the total. $\frac{2}{3} \times \frac{400}{1} = \frac{800}{3} = 267$

About 267 out of 400 male seniors will be taller than their fathers.

EXAMPLE 2: **Favorite Cake Flavors**

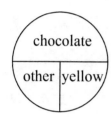

In a lunchroom survey of 360 students, about how many students preferred yellow cake?

Step 1: Estimate the fraction of the circle that shows the number of students that prefer yellow cake. It looks like about $\frac{1}{4}$ of the circle.

Step 2: Multiply $\frac{1}{4} \times 360$, the total number of students surveyed. About 90 students prefer yellow cake.

Solve the following problems.

1. This year $\frac{2}{3}$ of the seniors went to the prom. Out of a class of 438 seniors, how many went to the prom?

2. The North End Diner surveyed its customers one day to see which flavor of ice cream they preferred. Out of 512 customers, how many preferred chocolate chip?

3. It has rained 5 out of every 6 days for the last month! If there were 30 days in the last month, how many days did it rain?

4. Ryan worked 5 hours on his homework Tuesday. Two-thirds of the time was spent on algebra word problems. How much time did he spend on algebra?

5. Beth earned $20 babysitting. She spent $\frac{2}{5}$ of it on a paperback book. How much did she spend on the book?

6. Three-fourths of the graduating class at Lakewood High School plan on going to college. There are 680 graduating seniors. How many are planning on going to college?

7. At West End High School, $\frac{1}{5}$ of the students in the band are also in the choir. There are 205 band members. How many are also in the choir?

8. Allen bought a $3\frac{1}{4}$ pound roast to cook for dinner. How much did the roast cost at $2.00 per pound?

8.2.4
7.2.5

FRACTION WORD PROBLEMS

1. Sara Beth bought 30 hamburgers. Each hamburger weighed a quarter of a pound. How many pounds of meat did she buy?

2. When Jonathan took a trip, he wanted to check his mileage. At the start of his trip, his gas tank was full. He bought $10\frac{1}{4}$, $12\frac{1}{8}$, and $11\frac{3}{4}$, gallons. At the end of his trip he bought $5\frac{3}{8}$ gallons to fill up the tank again. How much gas did his trip take?

3. Jodene bought a 10 pound bag of flour. She used $3\frac{3}{4}$ pounds to make bread. How much flour did she have left?

4. Cindy needed a total of $5\frac{1}{4}$ yards of trim for cushions she was making. She already had $3\frac{3}{4}$ yards. How much more did she need to buy?

5. Mary bought a 10 ounce bottle of cough syrup. If each dose is $\frac{1}{4}$ ounce, how many doses are in the bottle?

6. It took $7\frac{1}{2}$ hours to go 390 miles on our vacation. What was our average speed in miles per hour?

7. Stella bought a 100 pound sack of flour to make bread for the bakery. If $\frac{7}{8}$ of a pound of flour is needed for each loaf, how many loaves can she make? What fraction of a pound of flour is left over?

8. Seth took a 30 inch piece of wire and cut it into $1\frac{1}{4}$ inch pieces. How many pieces did he get?

9. Keisha wanted to bake 10 cakes for a fund-raiser. She needed $1\frac{3}{4}$ cups of sugar for each cake. How many cups of sugar did she need in all?

10. Sal works for a movie theater and sells candy by the pound. Her first customer bought $1\frac{1}{3}$ pounds of candy, the second bought $\frac{3}{4}$ of a pound, and the third bought $\frac{1}{2}$ of a pound. How many pounds did she sell to the first three customers?

11. Beth has a bread machine that makes a loaf of bread that weighs $1\frac{1}{2}$ pounds. If she makes a loaf of bread for each of her three sisters, how many pounds of bread will she make?

12. A farmer hauled in 120 bales of hay. Each of his cows ate $1\frac{1}{4}$ bales. How many cows did the farmer feed?

13. John was competing in a 1000 meter race. He had to pull out of the race after running $\frac{3}{4}$ of it. How many meters did he run?

14. Tad needs to measure where the free-throw line should be in front of his basketball goal. He knows his shoes are $1\frac{1}{8}$ feet long, and the free-throw line should be 15 feet from the backboard. How many toe-to-toe steps does Tad need to take to mark off 15 feet?

15. A chemical plant takes in $5\frac{1}{2}$ million gallons of water from a local river and discharges $3\frac{2}{3}$ million directly back into the river. How much water does not go directly back into the river?

16. In January, Jeff filled his car with $11\frac{1}{2}$ gallons of gas the first week, $13\frac{1}{3}$ gallons the second week, $12\frac{1}{4}$ gallons the third week, and $10\frac{1}{6}$ gallons the fourth week of January. How many gallons of gas did he buy in January?

8.2.1
7.2.1

CHANGING DECIMALS TO FRACTIONS

EXAMPLE: 0.25

Step 1: Copy the decimal without the point.
This will be the top number of the fraction.

$$\frac{25}{\Box}$$

Step 2: The bottom number is a 1 with as many
0's after it as there are digits in the top
number.

$$\frac{25}{100} \leftarrow \text{Two digits} \atop \leftarrow \text{Two 0's}$$

$$\frac{25}{100} = \frac{1}{4}$$

Step 3: You then need to reduce the fraction.

EXAMPLES: $.2 = \frac{2}{10} = \frac{1}{5}$ $.65 = \frac{65}{100} = \frac{13}{20}$ $.125 = \frac{125}{1000} = \frac{1}{8}$

Change the following decimals to fractions.

1. .55	5. .75	9. .71	13. .35
2. .6	6. .82	10. .42	14. .96
3. .12	7. .3	11. .56	15. .125
4. .9	8. .42	12. .24	16. .375

CHANGING DECIMALS WITH WHOLE NUMBERS TO MIXED NUMBERS

EXAMPLE: Change 14.28 to a mixed number.

Step 1: Copy the portion of the number that is whole. 14

Step 2: Change .28 to a fraction $14\frac{28}{100}$

Step 3: Reduce the fraction $14\frac{28}{100} = 14\frac{7}{25}$

Change the following decimals to mixed numbers.

1. 7.125	5. 16.95	9. 6.7	13. 13.9
2. 99.5	6. 3.625	10. 45.425	14. 32.65
3. 2.13	7. 4.42	11. 15.8	15. 17.25
4. 5.1	8. 15.84	12. 8.16	16. 9.82

8.1.4

ESTIMATING DIVISION OF DECIMALS

EXAMPLE: The following division of decimals problem has the decimal point missing from the answer. Estimate the answer to determine where the decimal point should go.

$$2489 \div 5.8 = 42913793$$

Step 1: Round off the numbers in the problem to numbers that are divisible without a remainder.

5.8 rounds to 6
2489 rounds to 2400

$2400 \div 6 = 400$

Step 2: The answer should be close to 400, so put the decimal point after the third whole number.

Solution: $2489 \div 5.8 = 429.13793$

For each of the following problems, round off the numbers to determine where the decimal point belongs in the answer. Practice using a gridded answer sheet.

1. $15.63 \div 4.2 = 3\ 7\ 2\ 1\ 4$
2. $476.3 \div 5.81 = 8\ 1\ 9\ 7\ 9$
3. $7561.5 \div 10.6 = 7\ 1\ 3\ 3\ 4\ 9$
4. $6259 \div 8.1 = 7\ 7\ 2\ 7\ 1\ 6$
5. $11.78 \div .94 = 1\ 2\ 5\ 3\ 1\ 9$
6. $45.69 \div 4.67 = 9\ 7\ 8\ 3\ 7$
7. $768 \div 22.35 = 3\ 4\ 3\ 6\ 2\ 4$
8. $5.16 \div 1.78 = 2\ 8\ 9\ 8\ 8\ 7$
9. $87.32 \div 56.7 = 1\ 5\ 4\ 0\ 0\ 3\ 5$
10. $144.92 \div 12.4 = 1\ 1\ 6\ 8\ 7\ 0\ 9$
11. $456.98 \div 21.5 = 2\ 1\ 2\ 5\ 4\ 8\ 8$
12. $19 \div 8.6 = 2\ 2\ 0\ 9\ 3$

13. $79.19 \div 7.8 = 1\ 0\ 1\ 5\ 2\ 5\ 6$
14. $856.3 \div 8.2 = 1\ 0\ 4\ 4\ 2\ 6\ 8$
15. $11.235 \div .48 = 2\ 3\ 4\ 0\ 6$
16. $9.63 \div 4.1 = 2\ 3\ 4\ 8\ 7\ 8$
17. $96.68 \div 32.56 = 2\ 9\ 6\ 9\ 2\ 8\ 7$
18. $162.3 \div 87.5 = 1\ 8\ 5\ 4\ 8\ 5\ 7$
19. $45.98 \div 2.9 = 1\ 5\ 8\ 5\ 5$
20. $32.65 \div 1.689 = 1\ 9\ 3\ 3\ 0\ 9\ 7$
21. $26.5 \div 5.1 = 5\ 1\ 9\ 6$
22. $6.59 \div 2.147 = 3\ 0\ 6\ 9\ 3\ 9\ 9$
23. $75.26 \div 8.36 = 9\ 0\ 0\ 2\ 3\ 9$
24. $158.4 \div 3.09 = 5\ 1\ 2\ 6\ 2$

8.2.3

BEST BUY

When products come in different sizes, you need to figure out the cost per unit to see which is the best buy. Often the box marked "economy size" is not really the best buy.

SMITHFIELD'S
Instant Coffee

8 ounces $3.60
12 ounces $5.52
16 ounces $7.44

EXAMPLE:

Smithfield's Instant Coffee comes in three sizes. Which one has the lowest cost per unit? The coffee comes in 8, 12 and 16 ounce sizes. To figure the lowest cost per unit, you need to see how much each unit, in this case ounce, costs in each size. If 8 ounces of coffee costs $3.60, then 1 ounce costs $3.60 ÷ 8 or $.45. $.45 is the unit cost, the cost of 1 ounce. We need to figure the unit cost for each size:

$3.60 ÷ 8 = $.45
$5.52 ÷ 12 = $.46
$7.44 ÷ 16 = $.465

The 8 ounce size is the best one to buy because it has the lowest cost per unit.

Figure the unit cost of each item in each question below to find the best buy.

1. Which costs the most per ounce, 60 ounces of peanut butter for $5.40, 28 ounces for $2.24, or 16 ounces for $1.76?

2. Which is the least per pound, 5 pounds of chicken for $9.45, 3 pounds for $5.97, or 1 pound for $2.05?

3. Which costs the most per disk, a 10 pack of $3\frac{1}{2}$ inch floppy disks for $5.99, a 25 pack for $12.50, or a 50 pack for $18.75?

4. Which is the best buy, 6 ballpoint pens for $4.80 or 8 for $6.48?

5. Which costs the least per ounce, a 20 ounce soda for $0.60, 68 ounces for $2.38, or 100 ounces for $3.32?

6. Which costs more, oranges selling at 3 for $1.00 or oranges selling 4 for $1.36?

7. Which is the best buy, 1 roll of paper towels for $2.13, 3 rolls for $5.88, or 15 rolls for $29.55?

8. Which costs the most per tablet, 50 individually wrapped pain reliever tablets for $9.50, 100 tablets in a bottle for $6.32, or 500 tablets in a bottle for $9.50?

9. Which costs the least per can, a 24 pack of cola for $5.52, a 12 pack of cola for $2.64, or a 6 pack of cola for $1.35?

10. Which costs less per bag, 18 tea bags for $2.70 or 64 tea bags for $9.28?

11. Which is the best buy, a 3 pack of correction fluid for $2.97 or a 12 pack for $11.76?

12. Which is the least per roll, a roll of masking tape for $2.45, a 3-roll pack for $7.38, or a 12-roll pack for $29.16?

13. White rice sells for $5.64 for a 20 pound bag. A three pound bag costs $1.59. Which is the better buy?

14. Daisy's Discount Mart sells 400 sheet packs of notebook paper for $40.00, 125 sheet packs for $1.19, and 200 sheet packs for $1.40. Which is the best buy?

15. A new set of golf clubs at Ted's Sports Club sells for $149.50. A catalog offers the same set for $129.50 plus $12.25 for shipping and handling. How much can be saved by purchasing the golf clubs through the catalog?

8.2.1

DECIMAL WORD PROBLEMS

1. Micah can have his oil changed in his car for $19.99, or he can buy the oil and filter and change it himself for $8.79. How much would he save by changing the oil himself?

2. Megan bought 5 boxes of cookies for $3.75 each. How much did she spend?

3. Will subscribes to a monthly auto magazine. His one year subscription cost $29.97. If he pays for the subscription in 3 equal installments, how much is each payment?

4. Pat purchases 2.5 pounds of hamburger at $0.98 per pound. What is the total cost of the hamburger?

5. The White family took $650 cash with them on vacation. At the end of their vacation, they had $4.67 left. How much cash did they spend on vacation?

6. Acer Middle School spent $1,443.20 on 55 math books. If each book costs the same price, how much did each book cost?

7. The Junior Beta Club needs to raise $1513.75 to go to a national convention. If they decide to sell candy bars at $1.25 each, how many will they need to sell to meet their goal?

8. Fleta owns a candy store. On Monday, she sold 6.5 pounds of chocolate, 8.34 pounds of jelly beans, 4.9 pounds of sour snaps, and 5.64 pounds of yogurt-covered raisins. How many pounds of candy did she sell total?

9. Randal purchased a rare coin collection for $1803.95. He sold it at auction for $2700. How much money did he make on the coins?

10. A leather jacket that normally sells for $259.99 is on sale now for $197.88. How much can you save if you buy it now?

11. At the movies, Gigi buys 0.6 pounds of candy priced at $2.10 per pound. How much did she spend on candy?

12. George has $6.00 to buy candy. If each candy bar costs $.60, how many bars can he buy?

13. Monica's monthly salary is $2,100. Her deductions are $157.50 for FICA, $302 for federal income tax, and $57.80 for state income tax. What is her take-home pay?

14. Richard buys a camcorder for $229.95. Later, he sees the same camcorder on sale for $207.99. How much could he have saved if he waited to buy the camcorder on sale?

15. Sprayberry High School decided to sell boxes of oranges to earn money for new football uniforms. They ordered a truckload of 500 boxes of oranges from a Florida grower for $16.00 per box. They sold 450 boxes for $19.00 per box. On the last day of the sale, they sold the oranges they had left for $17.00 per box. How much profit did they make?

16. Super-X sells tires for $24.56 each. All-Wheel sells the identical tire for $21.97. How much can you save by purchasing a tire from All-Wheel?

8.2.1
7.2.1

31

CHAPTER 1 REVIEW

Solve the following word problems.

1. The animal keeper feeds Mischief, the monkey, 5 pounds of bananas per day. The gorilla eats 4 times as many bananas as the monkey. How many pounds of bananas does the animal keeper need to feed both animals for a week?

2. You and 4 friends are going to split a restaurant bill evenly. The total bill is $46.80. How much is each of you going to pay?

3. The Bing family's odometer read 65453 before driving to Disney World for vacation. After their vacation, the odometer read 66245. How many miles did they drive during their vacation?

4. Jonathan can assemble 47 widgets per hour. How many can he assemble in an 8 hour day?

5. Jacob drove 252 miles, and his average speed was 42 miles per hour. How many hours did he drive?

6. The Jones family traveled 300 miles in 5 hours. What was their average speed?

7. Alisha climbed a mountain that was 4,760 feet high in 14 hours. What was her average speed per hour?

8. Xandra bought a mechanical pencil for $2.38, 3 pens for $0.89 each, and a pack of graph paper for $3.42. She paid $0.42 tax. What was her change from a ten dollar bill?

9. Charlie makes $13.45 per hour repairing lawn mowers part-time. If he worked 15 hours, how much was his gross pay?

10. Gene works for his father sanding wooden rocking chairs. He earns $6.35 per chair. How many chairs does he need to sand in order to buy a portable radio/CD player for $146.05?

11. Margo's Mint Shop has a machine that produces 4.35 pounds of mints per hour. How many pounds of mints are produced in each 8 hour shift if the machine is constantly running?

12. The Carter Junior High track team ran the first leg of a 400 meter relay race in 10.23 seconds, the second leg in 11.4 seconds, the third leg in 10.77 seconds, and the last leg in 9.9 seconds. How long did it take for them to complete the race?

13. Brent drives to and from school Monday through Friday. He lives 10.8 miles from school. His car gets 22 miles per gallon, and he is paying $1.42 per gallon at the gas pump for gas right now. How much does gasoline cost to drive to and from school for 2 weeks?

14. Doug is a plumber. He charges $63 per hour for labor plus $50 per service call. He went to Mary's house to fix a pipe. The job took him 2.5 hours to complete. How much did he charge Mary to fix the pipe? The materials cost $12.00.

8.2.1
7.2.1

CHANGING PERCENTS TO DECIMALS AND DECIMALS TO PERCENTS

Change the following percents to **decimal** numbers.

Directions: Move the **decimal** point two places to the left and drop the **percent** sign. If there is no decimal point written, it is after the number and before the percent sign. Sometimes you will need to add a "0". (See 5% below.)

EXAMPLES: $14\% = 0.14$ $5\% = 0.05$ $100\% = 1$ $103\% = 1.03$
(decimal point)

Change the following percents to decimal numbers.

1. 18% = _____	8. 119% = _____	15. 5% = _____
2. 23% = _____	9. 2% = _____	16. 25% = _____
3. 9% = _____	10. 55% = _____	17. 410% = _____
4. 63% = _____	11. 80% = _____	18. 1% = _____
5. 4% = _____	12. 17% = _____	19. 50% = _____
6. 45% = _____	13. 66% = _____	20. 99% = _____
7. 2% = _____	14. 13% = _____	21. 107% = _____

Change the following decimal numbers to percents.

Directions: To change a **decimal** number to a **percent**, move the **decimal** point two places to the right, and add a **percent** sign. You may need to add a "0". (See 0.8 below.)

EXAMPLES: $0.62 = 62\%$ $0.07 = 7\%$ $0.8 = 80\%$ $0.166 = 16.6\%$ $1.54 = 154\%$

Change the following decimal numbers to percents.

22. 0.15 = _____	30. 0.58 = _____	38. 5.09 = _____
23. 0.62 = _____	31 0.86 = _____	39. 0.75 = _____
24. 1.53 = _____	32. 0.29 = _____	40. 0.3 = _____
25. 0.22 = _____	33. 0.06 = _____	41. 2.9 = _____
26. 0.35 = _____	34. 0.48 = _____	42. 0.06 = _____
27. 0.375 = _____	35. 3.089 = _____	43. 0.122 = _____
28. 0.648 = _____	36. 0.042 = _____	44. 0.575 = _____
29. 0.044 = _____	37. 0.375 = _____	45. 0.478 = _____

7.2.2

FRACTIONS, DECIMALS, AND PERCENTS

We have learned to change fractions to percents, percents to fractions, fractions to decimals, and decimals to fractions. In this lesson, we examine the relationships of common fractions, percents, and decimals. Percent means per one hundred and the percent sign represents hundredths. When relating fractions to decimals, and decimals to percents, it is best to change them to hundredths.

Example 1: Write $\frac{3}{10}$ as a decimal and as a percent.

$$\frac{3 \times 10}{10 \times 10} = \frac{30}{100} = .30 = 30\%$$

Example 2: Write 50% as a decimal and a reduced fraction.

$$50\% = .50 = \frac{50 \div 50}{100 \div 50} = \frac{1}{2}$$

Example 3: Write .25 as a percent and a common fraction.

$$.25 = 25\% = \frac{25}{100}, \quad \frac{25 \div 25}{100 \div 25} = \frac{1}{4}$$

Complete the following.

	fraction	decimal	percent
1.	$\frac{1}{5}$	= _____	= _____
2.	_____	= .8	= _____
3.	_____	= _____	= 75%
4.	$\frac{1}{10}$	= _____	= _____
5.	_____	= $.33\frac{1}{3}$	= _____
6.	_____	= _____	= 60%
7.	$\frac{5}{8}$	= _____	= _____
8.	_____	= .2	= _____
9.	_____	= _____	= 25%
10.	$\frac{2}{5}$	= _____	= _____
11.	_____	= .125	= _____

	fraction	decimal	percent
12.	_____	= _____	= 30%
13.	$\frac{3}{8}$	= _____	= _____
14.	_____	= .7	= _____
15.	_____	= _____	= 90%
16.	$\frac{2}{3}$	= _____	= _____
17.	_____	= .625	= _____
18.	_____	= _____	= 37.5%
19.	$\frac{7}{8}$	= _____	= _____
20.	_____	= .3	= _____
21.	_____	= _____	= 40%
22.	$\frac{4}{5}$	= _____	= _____

MIXED NUMBERS AND PERCENTS GREATER THAN 100%

In a previous lesson, we looked at some of the most frequently used fractions, decimals, and percent equivalents. It would benefit you to memorize them. In this lesson, we will expand this by using them as part of a reduced mixed number.

Example 1: Write $2\frac{3}{10}$ as a decimal fraction and as a percent.

$$2\frac{3 \times 10}{10 \times 10} = 2\frac{30}{100} = 2.30 = 230\%$$

Example 2: Write 150% as a decimal and a reduced mixed number.

$$150\% = 1.50 = 1\frac{50}{100} \div \frac{50}{50} = 1\frac{1}{2}$$

Example 3: Write 3.25 as a percent and a reduced mixed number.

$$3.25 = 325\% = 3\frac{25}{100} = 3\frac{25}{100} \div \frac{25}{25} = 3\frac{1}{4}$$

Complete the following

	fraction	decimal	percent
1.	$4\frac{3}{8}$ =	_____ =	_____
2.	_____ =	1.3 =	_____
3.	_____ =	_____ =	637.5%
4.	$2\frac{2}{3}$ =	_____ =	_____
5.	_____ =	5.625 =	_____
6.	_____ =	_____ =	190%
7.	$3\frac{2}{5}$ =	_____ =	_____
8.	_____ =	1.2 =	_____
9.	_____ =	_____ =	425%
10.	$7\frac{4}{5}$ =	_____ =	_____
11.	_____ =	2.125 =	_____

	fraction	decimal	percent
12.	_____ =	_____ =	830%
13.	$9\frac{1}{5}$ =	_____ =	_____
14.	_____ =	2.7 =	_____
15.	_____ =	_____ =	260%
16.	$6\frac{7}{8}$ =	_____ =	_____
17.	_____ =	$3.33\frac{1}{3}$ =	_____
18.	_____ =	_____ =	1075%
19.	$4\frac{1}{10}$ =	_____ =	_____
20.	_____ =	6.8 =	_____
21.	_____ =	_____ =	440%
22.	$3\frac{5}{8}$ =	_____ =	_____

CHANGING RATIOS TO PERCENTS

EXAMPLE: The ratio of girls to boys in the class is 9 to 11. What percent of the class is girls?

Step 1: We need to change the ratio to a fraction. Copy down the part of the ratio that are girls. Girls are mentioned first, so the **top** number of the fraction will be 9.

Step 2: The bottom number of the fraction is the sum of both numbers of the ratio: $9 + 11 = 20$. The bottom number of the fraction is 20.

Step 3: Change the fraction $\frac{9}{20}$ to a percent. $9 \div 20 = .45$ or 45%

Change the following ratios to percents. Round to the nearest percent or whole number.

1. The ratio of students who walk to school to students who take the bus is 3 to 5. What percent of the students take the bus?

2. The survey showed that pizza is preferred to peanut butter sandwiches 7 to 3. What percent of those surveyed preferrred pizza?

3. The ratio of black rabbits to white rabbits in a pet store is 7 to 3. What percent are black?

4. Nicholas collected 11 empty orange drink cans and 9 empty cola cans for his school's recycling program. What percent of the cans are cola cans?

5. The principal of Northside High School presented perfect attendance awards to 25 boys and 50 girls. What percent of those who received awards were girls?

6. Volunteers at the Notre Dame Homeless Shelter are planning a Thanksgiving dinner for the residents. The residents include 12 adults and 28 children. What percent of the residents are children?

FINDING THE PERCENT OF THE TOTAL

EXAMPLE: There were 75 customers at Bill's gas station this morning. Thirty-two percent used a credit card to make their purchase. How many customers used credit cards this morning at Bill's?

Step 1: Change 32% to a decimal. .32

Step 2: Multiply by the total number mentioned.

$$\begin{array}{r} .32 \\ \times\ 75 \\ \hline 160 \\ 224 \\ \hline 24.00 \end{array}$$

24 customers used credit cards.

---

7. Eighty-five percent of Mrs. Coomer's math class passed her final exam. There were 40 students in her class. How many passed?

8. Fifteen percent of a bag of chocolate candies have a red coating on them. How many red pieces are in a bag of 60 candies?

9. Sixty-eight percent of Valley Creek School students attended this year's homecoming dance. There are 675 students. How many attended the dance?

10. Out of the 4,500 people who attended the rock concert, forty-six percent purchased a T-shirt. How many people bought T-shirts?

11. Nina sold ninety-five percent of her 500 cookies at the bake sale. How many cookies did she sell?

12. Twelve percent of yesterday's customers purchased premium grade gasoline from GasCo. If GasCo had 200 customers, how many purchased premium grade gasoline?

FINDING THE PERCENT

EXAMPLE: 15 is what **percent** of 60?

Step 1: To solve these problems, simply divide the smaller amount by the larger amount. In this example, you will need to add a decimal point and two 0's to the fifteen when dividing.

$$
\begin{array}{r}
.25 \\
60{\overline{\smash{\big)}\,15.00}} \\
\underline{-12\,0} \\
3\,00 \\
\underline{-3\,00} \\
0
\end{array}
$$

Step 2: Change the answer, .25, to a percent by moving the decimal point two places to the right.
.25 = 25%   15 is 25% of 60.

Remember: **To change a decimal to a percent, you will sometimes have to add a zero when moving the decimal point two places to the right.**

Find the following percents.

1. What percent of 50 is 16?

2. 20 is what percent of 80?

3. 9 is what percent of 100?

4. 19 is what percent of 95?

5. Ruth made 200 cookies for the picnic. Only 25 were left at the end of the day. What percent of the cookies was left?

6. Asad made 116 bird houses to sell at the county fair. The first day he sold 29. What percent of the bird houses did he sell?

7. Eileen planted 90 sweet corn seeds, but only 18 plants came up. What percent of the seeds germinated?

8. Tomika invests $36 of her $240 paycheck in a retirement account. What percent of her pay is she investing?

9. Ray sold a house for $115,000, and his commission was $9,200. What percent commission did he make?

10. Julio was making $16.00 per hour. After one year, he received a $2.00 per hour raise. What percent of a raise did he get?

11. Calvin budgets $235 per month for food. If his salary is $940 per month, what percent of his salary does he budget for food?

12. Katie earned $45 on commission for her sales totaling $225. What percent was her commission?

13. Among the students taking band this year, 2 out of 5 are freshmen. What percent of the band are freshmen?

14. Of the donuts we have left to sell, the ratio of chocolate donuts to non-chocolate is 3 to 5. What percent of the donuts left to sell are chocolate?

15. The school bought 340 new history books for 400 students. What percent of the students got new history books?

16. Of the 48 dogs enrolled in obedience school, 36 successfully completed training. What percent of the dogs completed training?

FINDING THE PERCENT INCREASE AND DECREASE

EXAMPLE 1: Office Supply Co. purchased paper wholesale for $18.00 per case. They sold the paper for $20.00 per case. By what percent did the store increase the price of the paper (or what is the percent markup)?

$$\text{Percent change} = \frac{\text{Amount of change}}{\text{Original Amount}}$$

Step 1: Find the amount of change. In this problem, the price was marked up $2.00. The amount of change is 2.

Step 2: Divide the amount of change, 2, by the wholesale cost, 18. $\frac{2}{18} = .111$

Step 3: Change the decimal, .111, to a percent. .111 = 11.1% The price increase was 11.1%.

EXAMPLE 2: The price of gas went from $2.40 per gallon to $1.30. What is the percent of decrease in the price of gas?

$$\text{Percent change} = \frac{\text{Amount of change}}{\text{Original Amount}}$$

Step 1: Find the amount of change. In this problem, the price decreased $1.10 The amount of change is 1.10.

Step 2: Divide the amount of change, 1.10, by the original cost, $2.40. $\frac{1.10}{2.40} = .46$

Step 3: Change the decimal, .46, to a percent. .46 = 46% The price of gas has decreased 46%.

Find the percent increase or decrease for each of the problems below.

1. Mary was making $25,000 per year. Her boss gave her a $3,000 raise. What percent increase is that?

2. Last week Matt's total sales were $12,000. This week his total sales were only $2,000. By what percent did his sales for this week decrease?

3. Emil was making $16.00 per hour. After one year, he received a $2.00 per hour raise. What percent raise did he get?

4. Sara owned an office supply store. She marked down pens from $1.50 to $1.20. What percent discount is that?

5. Rosa bought a clock marked $18.00. After sales tax, the total came to $19.08. What percent sales tax did she pay?

6. Cowboys bought boots wholesale for $103.35. They sold the boots in their store for $159. What percent was the markup on the boots?

7. Blakeville has a population of 1,600. According to the last census, Blakeville had a population of 1,850. What has been the percent decrease in population?

8. Last year, Roswell High School had 680 graduates. This year they graduated 812. What has been the percent increase in graduates?

9. Michi got a new job that pays $52,000 per year. That is $16,000 more than his last job. What percent pay increase is that?

COMMISSIONS

Commission: In many businesses, sales people are paid on **commission,** which is a percent of the total sales they make.

EXAMPLE: Derek made a 4% commission on an $8,000 pickup truck he sold. What was the dollar amount of his commission?

Step 1 Change the percent commission to a decimal. $4\% = .04$

Step 2 Multiply the percent commission by the total sale.

TOTAL SALE	**$ 8,000**
× **RATE OF COMMISSION**	× **.04**
COMMISSION	**$ 320.00**

Solve each of the following problems.

1. Mabel is a real estate agent who gets a 6% commission when she sells a house. How much will make on the sale of a $225,000 house?

_____

2. Dan makes a 12% commission on the men's clothes that he sells. Last week his sales totaled $1,860. How much did he earn on commission?

_____

3. Micah earns 2% commission on the life insurance policies he sells. How much will he earn on a $80,000 policy?

_____

4. Lane sells skin care products for a 35% commission. Last month, she sold $560.00 worth. How much was her commission?

_____

5. Bailey earned 19% commission on her yearly sales of $158,500. How much commission did she earn for the year?

_____

6. Carter sells vacuum cleaners for 8% commission. How much will he make on each $690 vacuum he sells?

_____

7. Kent sells encyclopedias for $1,540 per set. He earns 15% commission on each set he sells. How much does he make per set?

_____

8. Leslie earns 10% commission on airline tickets she sells as a travel agent. Last week she sold $9,540 worth of tickets. How much was her commission?

_____

9. Pam earns 25% commission as a salesperson for her company. If she sells $1,570 worth of a product, how much is her commission?

_____

10. Caleb earns 5% commission as a car salesman. How much will he make if he sells a car for $15,590?

_____

7.2.3

FINDING THE DISCOUNTED SALE PRICE

Read the **EXAMPLE** below to learn how to figure **discount** prices.

EXAMPLE: A $74.00 chair is on sale for 25% off. How much will the chair cost after the discount?

Step 1   Change 25% to a decimal.   25% = .25

Step 2   Multiply the original price by the discount.

	ORIGINAL PRICE	$74.00	
×	**% DISCOUNT**	× .25	
	SAVINGS	$18.50	

Step 3   Subtract the savings amount from the original price to find the sale price.

	ORIGINAL PRICE	$74.00
−	**SAVINGS**	− 18.50
	SALE PRICE	$55.50

Figure the sale price of the items below. The first one is done for you.

ITEM	PRICE	% OFF	SAVINGS	SUBTRACT	SALE PRICE
1. pen	$1.50	20%	1.50 × .2 = $0.30	1.50 − 0.30 = 1.20	$1.20
2. recliner	$325	25%			
3. juicer	$55	15%			
4. blanket	$14	10%			
5. earrings	$2.40	20%			
6. figurine	$8	15%			
7. boots	$159	35%			
8. calculator	$80	30%			
9. candle	$6.20	50%			
10. camera	$445	20%			
11. VCR	$235	25%			
12. video game	$25	10%			

7.2.3

UNDERSTANDING SIMPLE INTEREST *

I = PRT is a formula to figure out the **cost of borrowing money** or the **amount you earn** when you **put money in a savings account**. When you want to buy a used truck or car, you go to the bank to borrow the $7,000 you need. The bank will charge you interest on the $7,000. If the simple interest rate is 9% for four years, you can figure the cost of the interest with this formula.

First, you need to understand these terms:

I = Interest = The amount charged by the bank or other lender
P = Principal = The amount you borrow
R = Rate = The interest rate the bank is charging you
T = Time = How many years you will take to pay off the loan

EXAMPLE:

In the problem above: **I = PRT** This means the **interest** equals the **principal** times the **rate** times the **time** in **years.**

$$I = \$7,000 \times 9\% \times 4 \text{ years}$$
$$I = \$7,000 \times .09 \times 4$$
$$I = \$2,520$$

Use the formula I = PRT to work the following problems:

1. Craig borrowed $1,800 from his parents to buy a stereo. His parents charged him 3% simple interest for 2 years. How much interest did he pay his parents? _____

2. Raul invested $5,000 in a savings account that earned 2% simple interest. If he kept the money in the account for 5 years, how much interest did he earn? _____

3. Bridgette borrowed $11,000 to buy a car. The bank charged 12% simple interest for 7 years. How much interest did she pay the bank? _____

4. A tax accountant invested $25,000 in a money market account for 3 years. The account earned 5% simple interest. How much interest did the accountant make on his investment? _____

5. Linda Kay started a savings account for her nephew with $2,000. The account earned 6% simple interest. How much interest did the account accumulate in 3 years? _____

6. Renada bought a living room set on credit. The set sold for $2,300 and the store charged her 9% simple interest for one year. How much interest did she pay? _____

7. Duane took out a $3,500 loan at 8% simple interest for 3 years. How much interest did he pay for borrowing the $3,500? _____

* Simple interest is not commonly used by banks and other lending institutions. Compound interest is more commonly used, but its calculations are more complicated and beyond the scope of the material presented in this text.

7.2.2

CHAPTER 2 REVIEW

Change the following percents to decimals.

1. 45% _____
2. 219% _____
3. 22% _____
4. 1.25% _____

Change the following decimals to percents.

5. 0.52 _____
6. 0.64 _____
7. 1.09 _____
8. 0.625 _____

Change the following percents to fractions.

9. 25% _____
10. 3% _____
11. 68% _____
12. 102% _____

Change the following fractions to percents.

13. $\frac{9}{10}$ _____
14. $\frac{5}{16}$ _____
15. $\frac{1}{8}$ _____
16. $\frac{1}{4}$ _____

17. What is 1.65 written as a percent?

_____

18. What is $2\frac{1}{4}$ written as a percent?

_____

19. Change 5.65 to a percent.

_____

Fill in the equivalent numbers represented by the shaded area.

20. fraction _____

21. decimal _____

22. percent _____

Fill in the equivalent numbers represented by the shaded area.

23. fraction _____

24. decimal _____

25. percent _____

26. Uncle Howard left his only niece 56% of his assets according to his will. If his assets totaled $564,000 when he died, how much did his niece inherit?

_____

27. Celeste makes 6% commission on her sales. If her sales for a week total $4,580, what is her commission?

_____

28. Peeler's Jewelry is offering a 30% off sale on all bracelets. How much will you save if you buy a $45.00 bracelet during the sale?

_____

29. How much would an employee pay for a $724.00 stereo if the employee got a 15% discount?

_____

30. Misha bought a CD for $14.95. If sales tax was 7%, how much did she pay total?

_____

31. The Pep band made $640 during a fund-raiser. The band spent $400 of the money on new uniforms. What percent of the total did the band members spend on uniforms?

_____

32. Linda took out a simple interest loan for $7,000 at 11% interest for 5 years. How much interest did she have to pay back?

_____

33. McMartin's is offering a deal on fitness club memberships. You can pay $999 up front for a 3 year membership, or pay $200 down and $30 per month for 36 months. How much would you save by paying up front?

_____

34. Patton, Patton, and Clark, a law firm, won a malpractice law suit for $4,500,000. Sixty-eight percent went to the law firm. How much did the law firm make?

_____

35. Jeneane earned $340.20 commission by selling $5,670 worth of products. What percent commission did she earn?

_____

36. Tara put $500 in a savings account that earned 3% simple interest. How much interest did she make after 5 years?

_____

37. Ms. Clark put $3,000 in an account that compounded 5% interest twice a year. If she left the money in for 1 year, how much would she have?

_____

38. Hank got 10 miles to the gallon in his vintage 1966 Mustang. After getting a new carburetor installed in his car, he got 16 miles to the gallon. What percent increase in mileage did he get?

_____

39. A department store is selling all swimsuits for 40% off in August. How much would you pay for a swimsuit that is normally priced at $35.80?

_____

40. High school students voted on where they would go on a field trip. For every 3 students who wanted to see Calaveras Big Trees State Park, 8 students wanted to see Columbia State Historic Park. What percent of the students wanted to go to Columbia State Historic Park?

_____

41. An increase from 20 to 36 is what percent of increase?

_____

42. An auto parts store buys air filters for $12.00 each. The store sells air filters for $18.60 each. What is the percent markup on the air filters?

_____

<table>
<tr><td>Chapter 3</td><td># Integers and Order of Operations</td></tr>
</table>

INTEGERS

In elementary school, you learned to use whole numbers.

Whole numbers = { 0, 1, 2, 3, 4, 5 . . . }

For most things in life, whole numbers are all we need to use. However, when a checking account falls below zero or the temperature falls below zero, we need a way to express that. Mathematicians have decided that a negative sign, which looks exactly like a subtraction sign, would be used in front of a number to show that the number is below zero. All the negative whole numbers and positive whole numbers plus zero make up the set of integers.

Integers = { . . . −4, −3, −2, −1, 0, 1, 2, 3, 4 . . . }

ABSOLUTE VALUE

The absolute value of a number is the distance the number is from zero on the number line.

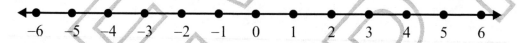

$$-6 \quad -5 \quad -4 \quad -3 \quad -2 \quad -1 \quad 0 \quad 1 \quad 2 \quad 3 \quad 4 \quad 5 \quad 6$$

The absolute value of 6 is written $|6|$. $|6| = 6$
The absolute value of −6 is written $|-6|$. $|-6| = 6$

Both 6 and −6 are the same distance, 6 spaces, from zero, so their absolute value is the same, 6.

EXAMPLES:

$\lvert -4 \rvert = 4$	$-\lvert -4 \rvert = -4$	$\lvert -9 \rvert + 5 = 9 + 5 = 14$
$\lvert 9 \rvert - \lvert 8 \rvert = 9 - 8 = 1$	$\lvert 6 \rvert - \lvert -6 \rvert = 6 - 6 = 0$	$\lvert -5 \rvert + \lvert -2 \rvert = 5 + 2 = 7$

Simplify the following absolute value problems.

1. $\lvert 9 \rvert = $ _____

2. $-\lvert 5 \rvert = $ _____

3. $\lvert -25 \rvert = $ _____

4. $-\lvert -12 \rvert = $ _____

5. $-\lvert 64 \rvert = $ _____

6. $\lvert -2 \rvert = $ _____

7. $-\lvert -3 \rvert = $ _____

8. $\lvert -4 \rvert - \lvert 3 \rvert = $ _____

9. $\lvert -8 \rvert - \lvert -4 \rvert = $ _____

10. $\lvert 5 \rvert + \lvert -4 \rvert = $ _____

11. $\lvert -2 \rvert + \lvert 6 \rvert = $ _____

12. $\lvert 10 \rvert + \lvert 8 \rvert = $ _____

13. $\lvert -2 \rvert + \lvert 4 \rvert = $ _____

14. $\lvert -3 \rvert + \lvert -4 \rvert = $ _____

15. $\lvert 7 \rvert - \lvert -5 \rvert = $ _____

7.2.1
8.2.1

ADDING INTEGERS

First, we will see how to add integers on the number line; then, we will learn rules for working the problems without using a number line.

EXAMPLE 1:   Add: $(-3) + 7$

Step 1:   The first integer in the problem tells us where to start.
Find the first integer, -3, on the number line.

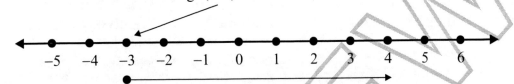

Step 2: $(-3) + 7$ The second integer in the problem, $+7$, tells us the direction to go, positive (toward positive numbers), and how far, 7 places.
$(-3) + 7 = 4$

EXAMPLE 2:   Add: $(-2) + (-3)$

Step 1:   Find the first integer, (-2), on the number line.

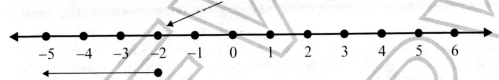

Step 2: $(-2) + (-3)$ The second integer in the problem, (-3), tells us the direction to go, negative (toward the negative numbers), and how far, 3 places.
$(-2) + (-3) = (-5)$

Solve the problems below using this number line.

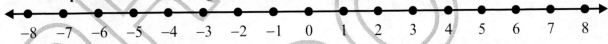

1. $2 + (-3) =$ _____

2. $4 + (-2) =$ _____

3. $(-3) + 7 =$ _____

4. $(-4) + 4 =$ _____

5. $(-1) + 5 =$ _____

6. $(-1) + (-4) =$ _____

7. $3 + 2 =$ _____

8. $(-5) + 8 =$ _____

9. $3 + (-7) =$ _____

10. $(-2) + (-2) =$ _____

11. $6 + (-7) =$ _____

12. $2 + (-5) =$ _____

13. $(-5) + 3 =$ _____

14. $(-6) + 7 =$ _____

15. $(-3) + (-3) =$ _____

16. $(-8) + 6 =$ _____

17. $(-2) + 6 =$ _____

18. $(-4) + 8 =$ _____

19. $(-7) + 4 =$ _____

20. $(-5) + 8 =$ _____

21. $-2 + (-2) =$ _____

22. $8 + (-6) =$ _____

23. $5 + (-3) =$ _____

24. $1 + (-8) =$ _____

7.2.1
8.2.1

RULES FOR ADDING INTEGERS WITH THE SAME SIGNS

To add integers of the same sign without using the number line, use these simple rules:

> 1. **Add the numbers together.**
> 2. **Give the answer the same sign.**

EXAMPLE 1: $(-2) + (-5) =$ _____   Both integers are negative. To find the answer, add the numbers together $(2 + 5)$, and give the answer a negative sign.

$(-2) + (-5) = (-7)$

EXAMPLE 2: $3 + 4 =$ _____   Both integers are positive, so the answer is positive.

$3 + 4 = 7$ **NOTE:** Sometimes positive signs are placed in front of positive numbers. For example $3 + 4 = 7$ may be written $(+3) + (+4) = +7$. Positive signs are optional. If a number has no sign, it is considered positive.

Solve the problems below using the rules for adding integers with the same signs.

1. $(-18) + (-4) =$ _____
2. $(-12) + (-3) =$ _____
3. $(-2) + (-7) =$ _____
4. $(22) + (11) =$ _____
5. $(-7) + (-6) =$ _____
6. $13 + 12 =$ _____
7. $16 + 11 =$ _____
8. $(-9) + (-8) =$ _____
9. $8 + 4 =$ _____
10. $(-4) + (-7) =$ _____

11. $(-15) + (-5) =$ _____
12. $(+7) + (+4) =$ _____
13. $(-4) + (-2) =$ _____
14. $(-15) + (-1) =$ _____
15. $(-8) + (-12) =$ _____
16. $6 + 9 =$ _____
17. $9 + 7 =$ _____
18. $(-9) + (-7) =$ _____
19. $(-14) + (-6) =$ _____
20. $(6) + (+19) =$ _____

21. $(-11) + (-7) =$ _____
22. $(+8) + (+6) =$ _____
23. $(+5) + 7 =$ _____
24. $(-4) + (-9) =$ _____
25. $(2) + (8) =$ _____
26. $(+18) + 5 =$ _____
27. $14 + (+7) =$ _____
28. $(-11) + (-19) =$ _____
29. $13 + (+11) =$ _____
30. $(-8) + (-21) =$ _____

7.2.1
8.2.1

RULES FOR ADDING INTEGERS WITH OPPOSITE SIGNS

> 1. Ignore the signs and find the difference.
> 2. Give the answer the sign of the larger number.

EXAMPLE 1: $(-4) + 6 = $ _____   To find the difference, take the larger number minus the smaller number: $6 - 4 = 2$. Looking back at the original problem, the larger

$(-4) + 6 = 2$ number, 6, is positive, so the answer is positive.

EXAMPLE 2: $3 + (-7) = $ _____   Find the difference. $7 - 3 = 4$ Looking at the problem, the larger number, 7, is a negative

$3 + (-7) = (-4)$ number, so the answer is negative.

Solve the problems below using the rules of adding integers with opposite signs.

1. $(-4) + 8 = $ _____	11. $(-11) + 1 = $ _____	21. $-14 + 8 = $ _____
2. $-10 + 12 = $ _____	12. $(-12) + 8 = $ _____	22. $-11 + 15 = $ _____
3. $9 + (-3) = $ _____	13. $-14 + 9 = $ _____	23. $(-8) + 16 = $ _____
4. $(+3) + (-3) = $ _____	14. $14 + (-11) = $ _____	24. $2 + (-15) = $ _____
5. $+8 + (-7) = $ _____	15. $(-20) + 12 = $ _____	25. $-2 + 8 = $ _____
6. $(-5) + (+12) = $ _____	16. $-19 + 21 = $ _____	26. $(-5) + 15 = $ _____
7. $-14 + (+7) = $ _____	17. $-4 + 18 = $ _____	27. $2 + (-11) = $ _____
8. $15 + (-3) = $ _____	18. $3 + (-6) = $ _____	28. $-3 + 7 = $ _____
9. $7 + (-8) = $ _____	19. $4 + (-10) = $ _____	29. $4 + (-12) = $ _____
10. $6 + (-12) = $ _____	20. $(-2) + 8 = $ _____	30. $-12 + 5 = $ _____

Solve the mixed addition problems below using the rules for adding integers.

31. $-7 + 8 = $ _____	37. $8 + (-5) = $ _____	43. $(-7) + (+10) = $ _____
32. $5 + 6 = $ _____	38. $(-6) + 13 = $ _____	44. $(+4) + 11 = $ _____
33. $(-2) + (-6) = $ _____	39. $(-9) + (-12) = $ _____	45. $11 + 6 = $ _____
34. $3 + (-5) = $ _____	40. $(-7) + (+12) = $ _____	46. $-4 + (-10) = $ _____
35. $(-7) + (-9) = $ _____	41. $+8 + (-9) = $ _____	47. $(+6) + (+2) = $ _____
36. $14 + 9 = $ _____	42. $(-13) + (-18) = $ _____	48. $1 + (-17) = $ _____

7.2.1
8.2.1

SUBTRACTING INTEGERS

The easiest way to subtract integers is to change the problem to an addition problem and follow the rules you already know.

RULES FOR SUBTRACTING INTEGERS

> 1. Change the subtraction sign to addition.
> 2. Change the sign of the second number to the opposite sign.

EXAMPLE 1: $-6 - (-2) = $ _____   Change the subtraction sign to addition and -2 to 2. $-6 - (-2) = (-6) + 2$

$(-6) + 2 = (-4)$

EXAMPLE 2: $5 - 6 = $ _____   Change the subtraction sign to addition and 6 to -6. $5 - 6 = 5 + (-6)$

$5 + (-6) = (-1)$

Solve the problems using the rules above.

1. $(-3) - 8 = $ _____
2. $5 - (-9) = $ _____
3. $8 - (-5) = $ _____
4. $(-2) - (-6) = $ _____
5. $8 - (-9) = $ _____
6. $(-4) - (-1) = $ _____

7. $(-5) - (-13) = $ _____
8. $6 - (-7) = $ _____
9. $8 - (-6) = $ _____
10. $(-2) - (-2) = $ _____
11. $(-3) - 7 = $ _____
12. $(-4) - 8 = $ _____

13. $(-7) - 4 = $ _____
14. $1 - (-9) = $ _____
15. $(-5) - 12 = $ _____
16. $(-1) - 9 = $ _____
17. $6 - (-7) = $ _____
18. $(-8) - (-12) = $ _____

Solve the addition and subtraction problems below.

19. $4 - (-2) = $ _____
20. $(-3) + 7 = $ _____
21. $(-4) + 14 = $ _____
22. $(-1) - 5 = $ _____
23. $(-1) + (-4) = $ _____
24. $(-12) + (-2) = $ _____
25. $0 - (-6) = $ _____
26. $2 - (-5) = $ _____

27. $(-5) + 3 = $ _____
28. $(-6) + 7 = $ _____
29. $(-4) + 8 = $ _____
30. $(-4) - 11 = $ _____
31. $(-5) + 8 = $ _____
32. $2 - (-2) = $ _____
33. $(-8) + 9 = $ _____
34. $0 + (-10) = $ _____

35. $30 + (-15) = $ _____
36. $-40 - (-5) = $ _____
37. $25 - 50 = $ _____
38. $-13 + 12 = $ _____
39. $(-21) - (-1) = $ _____
40. $62 - (-3) = $ _____
41. $(-16) + (-2) = $ _____
42. $(-25) + 5 = $ _____

7.2.1
8.2.1

MULTIPLYING INTEGERS

You are probably used to seeing multiplication written with a "×" sign, but multiplication can be written two other ways. A "·" between numbers means the same as "×", and parentheses ( ) around a number without a "×" or a "·" also means to multiply.

EXAMPLES: $2 \times 3 = 6$ or $2 \cdot 3 = 6$ or $(2)(3) = 6$

All of these mean the same thing, multiply.

DIVIDING INTEGERS

Division is commonly indicated two ways: with a "÷" or in the form of a fraction.

EXAMPLE: $6 \div 3 = 2$ means the same thing as $\frac{6}{3} = 2$

RULES FOR MULTIPLYING AND DIVIDING INTEGERS

> 1. **If the numbers have the same sign, the answer is positive.**
> 2. **If the numbers have different signs, the answer is negative.**

EXAMPLES: $6 \times 8 = 48$ $(-6) \times 8 = (-48)$ $(-6) \times (-8) = 48$

$48 \div 6 = 8$ $(-48) \div 6 = (-8)$ $(-48) \div (-6) = 8$

Solve the problems below using the rules of adding integers with opposite signs.

1. $(-4) \div 2 =$ _____

2. $12 \div (-3) =$ _____

3. $\frac{(-14)}{(-2)} =$ _____

4. $-15 \div 3 =$ _____

5. $(-3) \times (-7) =$ _____

6. $(-1) \cdot (5) =$ _____

7. $-1 \times (-4) =$ _____

8. $2 (-5) =$ _____

9. $3 \times (-7) =$ _____

10. $(-12) \cdot (-2) =$ _____

11. $\frac{(-18)}{(-6)} =$ _____

12. $21 \div (-7) =$ _____

13. $-5 \times 3 =$ _____

14. $(-6) (7) =$ _____

15. $(-5) \times 8 =$ _____

16. $\frac{-12}{6} =$ _____

17. $8 (-4) =$ _____

18. $1 \cdot (-8) =$ _____

19. $(-7) \cdot (-4) =$ _____

20. $(-2) \div (-2) =$ _____

21. $\frac{18}{(-6)} =$ _____

7.2.1
8.2.1

MIXED INTEGER PRACTICE

1. $(-6) + 13 =$ _____
2. $(-3) + (-9) =$ _____
3. $(-4) \times 4 =$ _____
4. $(-18) \div 3 =$ _____
5. $(-1) - 5 =$ _____
6. $(-1) \times (-4) =$ _____
7. $3 + (-5) =$ _____
8. $6 + (-5) =$ _____
9. $(-9) - (-12) =$ _____

10. $2 + (-5) =$ _____
11. $\dfrac{(-24)}{(-6)} =$ _____
12. $(-5) + 3 =$ _____
13. $(-6) - 7 =$ _____
14. $(-33) \div (-11) =$ _____
15. $(-21)(-3) =$ _____
16. $(-7) + (-14) =$ _____
17. $(-5) - 8 =$ _____
18. $1(-8) =$ _____

19. $(-2) \cdot (-2) =$ _____
20. $8 + (-6) =$ _____
21. $\dfrac{-14}{7} =$ _____
22. $(+7) \cdot (-2) =$ _____
23. $(10)(4) =$ _____
24. $24 \div (-4) =$ _____
25. $6(-5) =$ _____
26. $\dfrac{12}{(-3)} =$ _____
27. $36 \div 12 =$ _____

INTEGER WORD PROBLEMS

Solve the following word problems.

1. If it is 2° outside, and the temperature will drop 15° tonight, how cold will it get?

2. What is the difference in elevation between Mount McKinley (20,320 ft) and Death Valley (-282 ft)?

3. It is -24° tonight, but the weather reporter predicted it would be 16° warmer tomorrow. What will the temperature be tomorrow?

4. The average temperature of the earth's stratosphere is -70°F. The average temperature on the earth's surface is 57°F. How much warmer is the average temperature on the surface of the earth than in the stratosphere?

5. The elevation of the Dead Sea is $-1,286$ feet. (The Dead Sea is below sea level.) Mt. McKinley has an elevation of 20,320 feet. What is the difference in the elevation between the Dead Sea and Mt. McKinley?

6. A submarine dives 462 feet beneath the surface of the ocean. It then climbs up 257 feet. What depth is the submarine now?

7. Eratosthenes was born about 274 BC. Sir Isaac Newton was born in 1642 AD. About how many years apart were they born?

8. The average daily low temperature in International Falls, Minnesota, during the month of January is -9°F. The average high is 14°F. What is the temperature difference between the average low and the average high?

7.2.1
8.2.1

50

ORDER OF OPERATIONS

In long math problems with $+$, $-$, $\times$, $\div$, ( ), and exponents in them, you have to know what to do first. Without following the same rules, you could get different answers. If you will memorize the silly sentence "Please Excuse My Dear Aunt Sally," you can determine what order you must follow.

Please   "P" stands for parentheses. You must get rid of parentheses first.
        Examples: $3(1+4) = 3 \times 5 = 15$
              $6(10-6) = 6 \times 4 = 24$

Excuse   "E" stands for exponents. You must eliminate exponents next.
        Example: $4^2 = 4 \times 4 = 16$

My Dear   "M" stands for multiply. "D" stands for divide. Start on the left of the equation, and perform all multiplications and divisions in the order they appear.

Aunt Sally "A" stands for add. "S" stands for subtract. Start on the left, and perform all additions and subtractions in the order they appear.

EXAMPLE: $12 \div 2(6-3) + 3^2 - 1$

Please	Eliminate **parentheses**. $6-3 = 3$, so now we have	$12 \div 2 \times 3 + 3^2 - 1$
Excuse	Eliminate **exponents**. $3^2 = 9$, so now we have	$12 \div 2 \times 3 + 9 - 1$
My Dear	**Multiply** and **divide** next in order from left to right.	$12 \div 2 = 6$ then $6 \times 3 = 18$
Aunt Sally	Last, we **add** and **subtract** in order from left to right.	$18 + 9 - 1 = 26$

Simplify the following problems.

1. $6 + 9 \times 2 - 4 =$ _____

2. $3(4+2) - 6^2 =$ _____

3. $3(6-3) - 2^3 =$ _____

4. $49 \div 7 - 3 \times 3 =$ _____

5. $10 \times 4 - (7-2) =$ _____

6. $2 \times 3 \div 6 \times 4 =$ _____

7. $50 - 8(4+2) =$ _____

8. $7 + 8(14-6) \div 4 =$ _____

9. $(2 + 8 - 12) \times 4 =$ _____

10. $4(8-13) \times 4 =$ _____

11. $8 + 4^2 \times 2 - 6 =$ _____

12. $3^2(4+6) + 3 =$ _____

13. $(12-6) + 27 \div 3^2 =$ _____

14. $82^0 - 1 + 4 \div 2^2 =$ _____

15. $1 - (2-3) + 8 =$ _____

16. $12 - 4(7-2) =$ _____

17. $18 \div (6+3) - 12 =$ _____

18. $10^2 + 3^3 - 2 \times 3 =$ _____

19. $4^2 + (7+2) \div 3 =$ _____

20. $7 \times 4 - 9 \div 3 =$ _____

CHAPTER 3 REVIEW

Simplify the following problems, using the correct order of operations.

1. $(-9) \times (-10) =$ _____

2. $12 + (-22) =$ _____

3. $10 - (-13) =$ _____

4. $12 \div (-3) =$ _____

5. $(-5) + (4) =$ _____

6. $-6 - 5 =$ _____

7. $(-7) \cdot (6) =$ _____

8. $(-9) - (-2) =$ _____

9. $4 - 9 =$ _____

10. $\dfrac{(-25)}{(-5)} =$ _____

11. $(-13) + (-4) =$ _____

12. $(-10)(-6) =$ _____

13. $-10 + (-2) =$ _____

14. $(-4)(-22) =$ _____

15. $3 + (-9) =$ _____

16. $\dfrac{(-16)}{4} =$ _____

17. $(6)(-11) =$ _____

18. $-4 + (-10) =$ _____

19. $(-24) \div (-6) =$ _____

20. $13 - 18 =$ _____

21. $14 + (-20) =$ _____

22. $\dfrac{45}{(-9)} =$ _____

23. $(-7) + (-5) =$ _____

24. $-3 - (-3) =$ _____

25. $(-1) \times 12 =$ _____

26. $-7 + (-27) =$ _____

27. $9 - (-4) =$ _____

28. $(12)(-3) =$ _____

29. $\dfrac{(-60)}{(-12)} =$ _____

30. $13 - 27 =$ _____

31. $(-15) \times (-2) =$ _____

32. $-4 - (-8) =$ _____

33. $(-19) + 8 =$ _____

34. $-13 - (-13) =$ _____

35. $(-7) \cdot (-8) =$ _____

36. $36 \div (-6) =$ _____

Simplify the following problems using the correct order of operations.

37. $2^3 + (2^2)(8 - 12) =$ _____

38. $20 \div (-2 - 8) + 2 =$ _____

39. $14 + (7)(3 - 5) \div 2 =$ _____

40. $8 - 7^2 + (3 - 12) =$ _____

41. $(18 - 5) \times (2 - 3) - 10 =$ _____

42. $24 - (6^2 - 6) \div 5 =$ _____

43. $3^3 + (7)(9 - 5) =$ _____

44. $-10(1 - 9) \div (-20) + 1 =$ _____

45. $42 \div (12 - 5) - 2 =$ _____

46. $2 + 5^2 \div (15 - 20) =$ _____

Solve the following word problems.

47. Aristotle, an ancient Greek philosopher, was born in 384 BC. Roger Bacon, a famous English philosopher, was born about 1214 AD. About how many years apart were they born?

48. Echo River flows through a deep cavern in Mammoth Cave in southwest Kentucky, 360 feet below the surface. The highest peak of the Appalachian Mountains is Mount Mitchell with a height of 6684 feet. What is the difference in elevation between Mammoth Cave's Echo River and the top of Mount Mitchell?

Chapter 4 | Number Sense

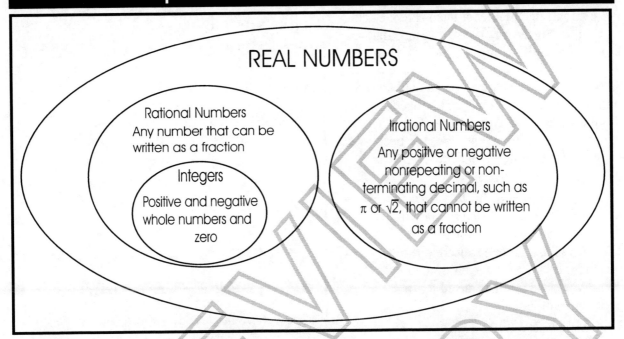

REAL NUMBERS

Rational Numbers
Any number that can be written as a fraction

Integers
Positive and negative whole numbers and zero

Irrational Numbers
Any positive or negative nonrepeating or non-terminating decimal, such as π or $\sqrt{2}$, that cannot be written as a fraction

Real numbers include all positive and negative numbers and zero. Included in the set of real numbers are positive and negative fractions, decimals, and rational and irrational numbers.

Use the diagram above and your calculator to answer the following questions.

1. Using your calculator, find the square root of 7. Does it repeat? Does it end? Is it a rational or an irrational number?
2. Find $\sqrt{25}$. Is it rational or irrational? Is it an integer?
3. Is an integer an irrational number?
4. Is an integer a real number?
5. Is $\frac{1}{8}$ a real number? Is it rational or irrational?

Identify the following numbers as rational (R) or irrational (I).

6. 5π
7. $\sqrt{8}$
8. $\frac{1}{3}$
9. -7.2
10. $-\frac{3}{4}$

11. $\frac{\sqrt{2}}{2}$
12. $9 + \pi$
13. 1.0004
14. $-\frac{4}{5}$
15. $1.1\overline{8}$

16. $\sqrt{81}$
17. $\frac{\pi}{4}$
18. $-\sqrt{36}$
19. $17\frac{1}{2}$
20. $-\frac{5}{3}$

8.1.2
8.1.3
7.1.3

UNDERSTANDING EXPONENTS

Sometimes it is necessary to multiply a number by itself one or more times. For example, in a math problem, you may need to multiply 3×3 or $5 \times 5 \times 5 \times 5$. In these situations, mathematicians have come up with a shorter way of writing out this kind of multiplication. Instead of writing 3×3, you can write 3^2; also, instead of $5 \times 5 \times 5 \times 5$, 5^4 means the same thing. The first number is the **base**. The small, raised number is called the **exponent**. The exponent tells how many times the base should be multiplied by itself.

EXAMPLE 1: 6^3

This means multiply 6 three times: $6 \times 6 \times 6$. You also need to know two special properties of exponents:
1.   Any base number raised to the exponent of 1 equals the base number.
2.   Any base number raised to the exponent of 0 equals 1.

EXAMPLE 2: $4^1 = 4$ $10^1 = 10$ $25^1 = 25$
$\qquad\qquad\quad 4^0 = 1$ $10^0 = 1$ $25^0 = 1$

Rewrite the following problems using exponents.

Example: $2 \times 2 \times 2 = 2^3$

1. $4 \times 4 \times 4$ 4. 8×8
2. $5 \times 5 \times 5 \times 5$ 5. $17 \times 17 \times 17$
3. 11×11 6. 12×12

Use your calculator to figure what number each number with an exponent represents.

Example: $2^3 = 2 \times 2 \times 2 = 8$

7. 2^5 10. 3^4
8. 10^2 11. 25^1
9. 8^3 12. 10^5

Express each of the following numbers as a number with an exponent.

Example: $4 = 2 \times 2 = 2^2$

13. $32 = $____ or ___   15. $1000 = $_____
14. $64 = $____ or ____   16. $27 = $_____

MULTIPLYING EXPONENTS

To multiply two expressions with the same base, add the exponents together and keep the base the same.

EXAMPLE 1: $2^3 \times 2^5 = 2^{3+5} = 2^8$
EXAMPLE 2: $3a^2 \times 2a^3 = 6a^{2+3} = 6a^5$
Notice that only the a's are raised to a power and not the 3 or the 2.

Simplify each of the expressions below.

1. $2^3 \times 2^5$ 7. $10^5 \times 10^4$
2. $x^5 \times x^3$ 8. $5^2 \times 5^4$
3. $2a^3 \times 3a^3$ 9. $3^3 \times 3^2$
4. $4^5 \times 4^3$ 10. $4x \times x^2$
5. $2x^3 \times x^5$ 11. $a^2 \times 3a^4$
6. $4b^3 \times 2b^4$ 12. $2^3 \times 2^4$

If a product in parentheses is raised to a power, then each factor is raised to the power when parentheses are eliminated.

EXAMPLE 1: $(2 \times 4)^2 = 2^2 \times 4^2$

$\qquad\qquad\qquad = 4 \times 16 = 64$

EXAMPLE 2: $(3a)^3 = 3^3 \times a^3 = 27a^3$

EXAMPLE 3: $(7b^5)^2 = 7^2 b^{10} = 49b^{10}$

Simplify each of the expressions below.

1. $(2^3)^2$ 11. $(2a)^4$
2. $(7a^5)^2$ 12. $(2^2)^3$
3. $(6b^2)^2$ 13. $(3 \times 2)^3$
4. $(3^2)^2$ 14. $(5x^3)^2$
5. $(3 \times 5)^2$ 15. $(4r^7)^3$
6. $(3x^4)^2$ 16. $(2m^3)^2$
7. $(6y^7)^2$ 17. $(6 \times 4)^2$
8. $(11w^3)^2$ 18. $(9a^5)^2$
9. $(3^3)^2$ 19. $(7b^5)^2$
10. $(3 \times 3)^2$ 20. $(9^2)^2$

7.1.4
8.1.5

NEGATIVE EXPONENTS

Expressions can also have negative exponents. Negative exponents do not indicate negative numbers. They indicate **reciprocals**. The **reciprocal** of a number is one divided by that number. For example, the reciprocal of 2 is $\frac{1}{2}$. (A number multiplied by its reciprocal is equal to 1.) If the negative exponent is in the bottom of a fraction, the reciprocal will put the expression on the top of the fraction without the negative sign.

EXAMPLE 1: $2^{-3} = \frac{1}{2^3} = \frac{1}{8}$

EXAMPLE 2: $3a^{-5} = 3 \times \frac{1}{a^5} = \frac{3}{a^5}$ Notice that the 3 is not raised to the −5 power, only the a.

EXAMPLE 3: $\frac{6}{5x^{-2}} = \frac{6x^2}{5}$ Notice that the 5 is not raised to the −2 power, only the x.

Rewrite using only positive exponents.

1. $5m^{-6}$
2. $\frac{5x^{-4}}{7}$
3. $14z^{-8}$

4. $\frac{1}{5s^{-4}}$
5. $14h^{-5}$
6. $\frac{h^{-3}}{5}$

7. $\frac{2y^{-3}}{4}$
8. x^{-4}
9. $-2y^{-2}$

10. $5y^{-5}$
11. $\frac{x^{-3}}{5}$
12. $10z^{-7}$

13. $7x^{-3}$
14. r^{-2}
15. $\frac{m^{-4}}{6}$

MULTIPLYING WITH NEGATIVE EXPONENTS

Multiplying with negative exponents follows the same rules as multiplying with positive exponents.

EXAMPLE 1: $6^2 \times 6^{-3} = 6^{2+(-3)} = 6^{-1} = \frac{1}{6}$

EXAMPLE 2: $(5a \times 2)^{-3} = (10a)^{-3} = \frac{1}{(10a)^3} = \frac{1}{1000a^3}$

EXAMPLE 3: $(7a^2)^{-3} = 7^{-3}a^{-6} = \frac{1}{7^3 a^6}$

Simplify the following. Answers should <u>not</u> have any negative exponents.

1. $5^{-2} \times 5^5$
2. $(6^3 \times 6^{-2})^{-2}$
3. $10^{-4} \times 10^2$
4. $11^{-5} \times 11^7$

5. $4^7 \times 4^{-10}$
6. $20^8 \times 20^{-6}$
7. $5^{-8} \times 5^4$
8. $(2^{-2} \times 2^3)^{-4}$

9. $7^{-2} \times 7^{-1}$
10. $(3x^4)^{-3}$
11. $12^{-10} \times 12^8$
12. $(10^8 \times 10^{-10})^2$

13. $3^{-2} \times 2^{-2}$
14. $(8x^5)^{-4}$
15. $(6b^3)^{-6}$
16. $(9y)^{-2}$

8.1.4

SQUARE ROOT

Just as working with exponents is related to multiplication, so finding square roots is related to division. In fact, the sign for finding the square root of a number looks similar to a division sign. The best way to learn about square roots is to look at examples.

EXAMPLES: **This is a square root problem:** $\sqrt{64}$

It is asking, "What is the square root of 64?"

It means, "What number multiplied by itself equals 64?"

The answer is 8.   8 × 8 = 64.

Look at the square root of the following numbers.

$\sqrt{36}$ 6 × 6 = 36 so $\sqrt{36}$ = 6 $\sqrt{144}$ 12 × 12 = 144 so $\sqrt{144}$ = 12

Find the square roots of the following numbers.

1.	$\sqrt{49}$ _____	5.	$\sqrt{121}$ _____	9.	$\sqrt{900}$ _____
2.	$\sqrt{81}$ _____	6.	$\sqrt{100}$ _____	10.	$\sqrt{64}$ _____
3.	$\sqrt{25}$ _____	7.	$\sqrt{36}$ _____	11.	$\sqrt{9}$ _____
4.	$\sqrt{16}$ _____	8.	$\sqrt{4}$ _____	12.	$\sqrt{144}$ _____

SIMPLIFYING SQUARE ROOTS

Square roots can sometimes be simplified even if the number under the square root is not a perfect square. One of the rules of roots is that if a and b are two positive real numbers, then it is always true that $\sqrt{a \cdot b} = \sqrt{a} \cdot \sqrt{b}$. You can use this rule to simplify square roots.

EXAMPLE 1: $\sqrt{100} = \sqrt{4 \cdot 25} = \sqrt{4} \cdot \sqrt{25} = 2 \cdot 5 = 10$

EXAMPLE 2: $\sqrt{200} = \sqrt{100 \cdot 2} = 10\sqrt{2}$ ⟵ Means 10 multiplied by the square root of 2

EXAMPLE 3: $\sqrt{160} = \sqrt{10 \cdot 16} = 4\sqrt{10}$

Simplify.

1.	$\sqrt{98}$ _____	5.	$\sqrt{8}$ _____	9.	$\sqrt{54}$ _____	13.	$\sqrt{90}$ _____
2.	$\sqrt{600}$ _____	6.	$\sqrt{63}$ _____	10.	$\sqrt{40}$ _____	14.	$\sqrt{175}$ _____
3.	$\sqrt{50}$ _____	7.	$\sqrt{48}$ _____	11.	$\sqrt{72}$ _____	15.	$\sqrt{18}$ _____
4.	$\sqrt{27}$ _____	8.	$\sqrt{75}$ _____	12.	$\sqrt{80}$ _____	16.	$\sqrt{20}$ _____

ADDING AND SUBTRACTING ROOTS

You can add and subtract terms with square roots only if the number under the square root sign is the same.

EXAMPLE 1: $2\sqrt{2} + 3\sqrt{2} = 5\sqrt{2}$
EXAMPLE 2: $12\sqrt{7} - 3\sqrt{7} = 9\sqrt{7}$

Or, look at the following examples where you can simplify the square roots and then add or subtract.

EXAMPLE 3: $2\sqrt{25} + \sqrt{36}$
Step 1:   Simplify. You know that $\sqrt{25} = 5$, and $\sqrt{36} = 6$ so the problem simplifies to $2(5) + 6$

Step 2:   Solve: $2(5) + 6 = 10 + 6 = 16$

EXAMPLE 4: $2\sqrt{72} - 3\sqrt{2}$

Step 1:   Simplify what you know. $\sqrt{72} = \sqrt{36 \cdot 2} = 6\sqrt{2}$
Step 2:   Substitute $6\sqrt{2}$ for $\sqrt{72}$ and simplify
                   $2(6)\sqrt{2} - 3\sqrt{2} = 12\sqrt{2} - 3\sqrt{2} = 9\sqrt{2}$

Simplify the following addition and subtraction problems.

1. $3\sqrt{5} + 9\sqrt{5}$

2. $3\sqrt{25} + 4\sqrt{16}$

3. $4\sqrt{8} + 2\sqrt{2}$

4. $3\sqrt{32} - 2\sqrt{2}$

5. $\sqrt{25} - \sqrt{49}$

6. $2\sqrt{5} + 4\sqrt{20}$

7. $5\sqrt{8} - 3\sqrt{72}$

8. $\sqrt{27} + 3\sqrt{27}$

9. $3\sqrt{20} - 4\sqrt{45}$

10. $4\sqrt{45} - \sqrt{125}$

11. $2\sqrt{28} + 2\sqrt{7}$

12. $\sqrt{64} + \sqrt{81}$

13. $5\sqrt{54} - 2\sqrt{24}$

14. $\sqrt{32} + 2\sqrt{50}$

15. $2\sqrt{7} + 4\sqrt{63}$

16. $8\sqrt{2} + \sqrt{8}$

17. $2\sqrt{8} - 4\sqrt{32}$

18. $\sqrt{36} + \sqrt{100}$

19. $\sqrt{9} + \sqrt{25}$

20. $\sqrt{64} - \sqrt{36}$

21. $\sqrt{75} + \sqrt{108}$

22. $\sqrt{81} + \sqrt{100}$

23. $\sqrt{192} - \sqrt{75}$

24. $3\sqrt{5} + \sqrt{245}$

MULTIPLYING ROOTS

You can also multiply square roots. To multiply square roots, you just multiply the numbers under the square root sign and then simplify. Look at the examples below.

EXAMPLE 1: $\sqrt{2} \times \sqrt{6}$

Step 1: $\sqrt{2} \times \sqrt{6} = \sqrt{2 \times 6} = \sqrt{12}$ Multiply the numbers under the square root sign.

Step 2: $\sqrt{12} = \sqrt{4 \times 3} = 2\sqrt{3}$ Simplify.

EXAMPLE 2: $3\sqrt{3} \times 5\sqrt{6}$

Step 1: $(3 \times 5)\sqrt{3 \times 6} = 15\sqrt{18}$ Multiply the numbers in front of the square root, and multiply the numbers under the square root sign.

Step 2: $15\sqrt{18} = 15\sqrt{3 \times 9}$ Simplify.
$15 \times 3\sqrt{3} = 45\sqrt{3}$

EXAMPLE 3: $\sqrt{14} \times \sqrt{42}$ For this more complicated multiplication problem, use the rule of roots that you learned on page 97, $\sqrt{a \cdot b} = \sqrt{a} \cdot \sqrt{b}$.

Step 1: $\sqrt{14} = \sqrt{7} \times \sqrt{2}$ and   Instead of multiplying 14 by 42, divide these
$\sqrt{42} = \sqrt{2} \times \sqrt{3} \times \sqrt{7}$ numbers into their roots.

$\sqrt{14} \times \sqrt{42} = \sqrt{7} \times \sqrt{2} \times \sqrt{2} \times \sqrt{3} \times \sqrt{7}$

Step 2:   Since you know that $\sqrt{7} \times \sqrt{7} = 7$, and $\sqrt{2} \times \sqrt{2} = 2$, the problem simplifies to
$(7 \times 2)\sqrt{3} = 14\sqrt{3}$

Simplify the following multiplication problems.

1. $\sqrt{5} \times \sqrt{7}$

2. $\sqrt{32} \times \sqrt{2}$

3. $\sqrt{10} \times \sqrt{14}$

4. $2\sqrt{3} \times 3\sqrt{6}$

5. $4\sqrt{2} \times 2\sqrt{10}$

6. $\sqrt{5} \times 3\sqrt{15}$

7. $\sqrt{45} \times \sqrt{27}$

8. $5\sqrt{21} \times \sqrt{7}$

9. $\sqrt{42} \times \sqrt{21}$

10. $4\sqrt{3} \times 2\sqrt{12}$

11. $\sqrt{56} \times \sqrt{24}$

12. $\sqrt{11} \times 2\sqrt{33}$

13. $\sqrt{13} \times \sqrt{26}$

14. $2\sqrt{2} \times 5\sqrt{5}$

15. $\sqrt{6} \times \sqrt{12}$

DIVIDING ROOTS

When dividing a number or a square root by another square root, you cannot leave the square root sign in the denominator (the bottom number) of a fraction. You must simplify the problem so that the square root is not in the denominator. Look at the examples below.

EXAMPLE 1: $\dfrac{\sqrt{2}}{\sqrt{5}}$

Step 1: $\dfrac{\sqrt{2}}{\sqrt{5}} \times \dfrac{\sqrt{5}}{\sqrt{5}}$ ⟵ The fraction $\dfrac{\sqrt{5}}{\sqrt{5}}$ is equal to 1, and multiplying by 1 does not change the value of a number

Step 2: $\dfrac{\sqrt{2 \times 5}}{5} = \dfrac{\sqrt{10}}{5}$ Multiply and simplify. Since $\sqrt{5} \times \sqrt{5}$ equals 5, you no longer have a square root in the denominator.

EXAMPLE 2: $\dfrac{6\sqrt{2}}{2\sqrt{10}}$ In this problem, the numbers outside of the square root will also simplify.

Step 1: $\dfrac{6}{2} = 3$ so you have $\dfrac{3\sqrt{2}}{\sqrt{10}}$

Step 2: $\dfrac{3\sqrt{2}}{\sqrt{10}} \times \dfrac{\sqrt{10}}{\sqrt{10}} = \dfrac{3\sqrt{2 \times 10}}{10} = \dfrac{3\sqrt{20}}{10}$

Step 3: $\dfrac{3\sqrt{20}}{10}$ will further simplify because $\sqrt{20} = 2\sqrt{5}$ so you then have $\dfrac{3 \times 2\sqrt{5}}{10}$ which reduces to $\dfrac{3 \times \overset{1}{2}\sqrt{5}}{\underset{5}{10}}$ or $\dfrac{3\sqrt{5}}{5}$

Simplify the following division problems.

1. $\dfrac{9\sqrt{3}}{\sqrt{5}}$

2. $\dfrac{16}{\sqrt{8}}$

3. $\dfrac{24\sqrt{10}}{12\sqrt{3}}$

4. $\dfrac{\sqrt{121}}{\sqrt{6}}$

5. $\dfrac{\sqrt{40}}{\sqrt{90}}$

6. $\dfrac{33\sqrt{15}}{11\sqrt{2}}$

7. $\dfrac{\sqrt{32}}{\sqrt{12}}$

8. $\dfrac{\sqrt{11}}{\sqrt{5}}$

9. $\dfrac{\sqrt{2}}{\sqrt{6}}$

10. $\dfrac{2\sqrt{7}}{\sqrt{14}}$

11. $\dfrac{5\sqrt{2}}{4\sqrt{8}}$

12. $\dfrac{4\sqrt{21}}{7\sqrt{7}}$

13. $\dfrac{9\sqrt{22}}{2\sqrt{2}}$

14. $\dfrac{\sqrt{35}}{2\sqrt{14}}$

15. $\dfrac{\sqrt{40}}{\sqrt{15}}$

EXPRESSING ROOTS AS EXPONENTS

We have seen that a square root indicates that the number under the radical ($\sqrt{}$) is expressed as some number multiplied by itself. This idea can be extended to expressing a number multiplied by itself three times, called a cube root ($\sqrt[3]{}$).

EXAMPLE 1: $\sqrt{144} = 12$ because $12 \times 12 = 144$

$\sqrt{225} = 15$ because $15 \times 15 = 225$

$\sqrt[3]{27} = 3$ because $3 \times 3 \times 3 = 27$

$\sqrt[3]{125} = 5$ because $5 \times 5 \times 5 = 125$.

This same idea can be expressed for 4th roots, 5th roots, or any number of roots greater than 0.

EXAMPLE 2: $\sqrt[5]{32} = 2$ because $2 \times 2 \times 2 \times 2 \times 2 = 32$

$\sqrt[4]{256} = 4$ because $4 \times 4 \times 4 \times 4 = 256$

Sometimes mathematicians express these roots as **fractional exponents**. All you do is take the radical sign off the number, and put a 1 over the number of the root you are taking. Make this fraction your exponent for the number instead of having the radical over it. For square roots, the exponent is always $\frac{1}{2}$. For cube roots, it is always $\frac{1}{3}$, and so on.

EXAMPLE 3: $\sqrt{144} = 144^{\frac{1}{2}} = 12$

$\sqrt[3]{27} = 27^{\frac{1}{3}} = 3$

$\sqrt[4]{256} = 256^{\frac{1}{4}} = 4$

$\sqrt[5]{32} = 32^{\frac{1}{5}} = 2$

These roots and exponents work exactly the same with variables as they do with numbers. Often, it is easier to simplify expressions using exponents than it is using radicals.

EXAMPLE 4: $\sqrt{3x^2y^2} = (3x^2y^2)^{\frac{1}{2}}$

$= (3)^{\frac{1}{2}}(x^2)^{\frac{1}{2}}(y^2)^{\frac{1}{2}} = 3^{\frac{1}{2}}xy$

$= \sqrt{3}\ xy$

EXAMPLE 5: $216^{\frac{1}{3}} = \sqrt[3]{216} = \sqrt[3]{6 \times 6 \times 6} = 6$

$$64^{\frac{1}{6}} = \sqrt[6]{64} = \sqrt[6]{2 \times 2 \times 2 \times 2 \times 2 \times 2} = 2$$

What if the number in the numerator is something other than a one? It means the number under the radical is raised to the power indicated in the numerator. If the fractional exponent is $\frac{2}{3}$, you still take the cube root, but you square the number under the radical.

EXAMPLE 6: $8^{\frac{2}{3}} = \sqrt[3]{8^2} = \sqrt[3]{8 \times 8} = \sqrt[3]{64} = \sqrt[3]{4 \times 4 \times 4}$

$$= 4$$

$$25^{\frac{3}{2}} = \sqrt{25^3} = \sqrt{25 \times 25 \times 25} = \sqrt{15{,}625} = 125$$

Express the following roots as fractional exponents. The first one is done for you.

1. $\sqrt{7} = 7^{\frac{1}{2}}$

2. $\sqrt{3} =$

3. $\sqrt{10} =$

4. $\sqrt{17} =$

5. $\sqrt[3]{13} =$

6. $\sqrt[3]{17} -$

7. $\sqrt[4]{19} =$

8. $\sqrt[4]{21} =$

9. $\sqrt[4]{5} =$

10. $\sqrt[5]{5} =$

11. $\sqrt[4]{y} =$

12. $\sqrt{3^3} =$

13. $\sqrt[3]{5^5} =$

14. $\sqrt[3]{6^2} =$

15. $\sqrt[5]{7^2} =$

16. $\sqrt{2^5} -$

17. $\sqrt{x^3} =$

18. $\sqrt[3]{y^2} =$

19. $\sqrt[4]{w^3} =$

20. $\sqrt{z^5} =$

Express the following fractional exponents as roots. The first one is done for you.

1. $3^{\frac{1}{2}} = \sqrt{3}$

2. $5^{\frac{1}{3}} =$

3. $7^{\frac{1}{2}} =$

4. $2^{\frac{2}{3}} =$

5. $4^{\frac{1}{6}} =$

6. $11^{\frac{1}{4}} =$

7. $14^{\frac{3}{2}} =$

8. $19^{\frac{4}{5}} =$

9. $4^{\frac{1}{7}} =$

10. $x^{\frac{1}{2}} =$

11. $y^{\frac{1}{3}} =$

12. $3^{\frac{4}{3}} =$

13. $w^{\frac{3}{5}} =$

14. $x^{\frac{1}{3}} y^{\frac{1}{3}} =$

15. $21^{\frac{5}{3}} =$

16. $13^{\frac{1}{9}} =$

17. $6^{\frac{3}{4}} =$

18. $17^{\frac{2}{3}} =$

19. $(xyz)^{\frac{4}{3}} =$

20. $(\frac{1}{x})^{\frac{1}{2}} =$

PRIME FACTORIZATION

Prime factorization is the process of factoring a number into **prime numbers**. A prime number, also called a prime, is a number that can only be divided by itself and 1. There are two main ways of finding the primes of a number: dividing and splitting.

EXAMPLE 1:   Find the primes of 66 by division.

Step 1:   To find the primes by division, you must only divide 66 by prime numbers until you can only divide by one.

$66 \div \mathbf{2} = 33$
$33 \div \mathbf{3} = 11$
$11 \div \mathbf{11} = 1$
$11 \div 1 = 11$ (Remember, 1 is not a prime number.)

Step 2:   All the prime numbers used as divisors make up the prime factorization of 66.
$\mathbf{66 = 2 \times 3 \times 11}$

Check: To check, multiply the prime numbers together, and you should get the original value.

EXAMPLE 2:   Find the primes of 66 using the splitting method.

Step 1:   In this method, the number must be split by any two factors until all of the factors are prime.

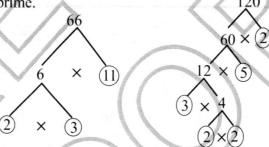

$\mathbf{66 = 2 \times 3 \times 11}$ $\mathbf{120 = 2 \times 2 \times 2 \times 5 \times 3 = 2^3 \times 3 \times 5}$

Step 2:   All the prime numbers used in splitting 66 make up the prime factorization of 66.
All the prime numbers used in splitting 120 make up the prime factorization of 120.

Hint: The factors found during prime factorization should always be prime, should always multiply together to get the correct answer, and should always be listed from least to greatest.

7.1.3
7.1.5

Find the prime factorization of each number using the division method. The first one has been done for you.

1. $10 \div 2 = 5 \div 5 = 1$
    $10 = 2 \times 5$

2. 14 =

3. 55 =

4. 110 =

5. 126 =

6. 142 =

7. 8 =

8. 21 =

9. 32 =

10. 36 =

11. 51 =

12. 84 =

13. 125 =

14. 48 =

15. 77 =

16. 65 =

17. 200 =

18. 413 =

Find the prime factorization of each number using the splitting method. The first one has been done for you.

1.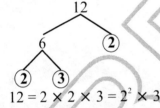
    $12 = 2 \times 2 \times 3 = 2^2 \times 3$

2. 24 =

3. 45 =

4. 120 =

5. 52 =

6. 91 =

7. 18 =

8. 67 =

9. 20 =

10. 15 =

11. 35 =

12. 122 =

SCIENTIFIC NOTATION

Mathematicians use **scientific notation** to express very large and very small numbers.
Scientific notation expresses a number in the following form:

$$x.xx \times 10^x$$

only one digit before the decimal →

multiplied by a multiple of ten

remaining digits not ending in zeros after the decimal →

USING SCIENTIFIC NOTATION FOR LARGE NUMBERS

Scientific notation simplifies very large numbers that have many zeros. For example, Pluto averages a distance of 5,900,000,000 kilometers from the sun. In scientific notation, a decimal is inserted after the first digit (5.); the rest of the digits are copied except for the zeros at the end (5.9), and the result is multiplied by 10^9. The exponent = the total number of digits in the original number minus 1 or the number of spaces the decimal point moved.

$5,900,000,000 = 5.9 \times 10^9$ The following are more examples:

EXAMPLES: $32,560,000,000 = 3.256 \times 10^{10}$ $5,060,000 = 5.06 \times 10^6$

decimal moves 10 spaces to the left ⏌   ⎿decimal moves 6 spaces to the left

Convert the following numbers to scientific notation.

1.   4,230,000,000 =   7.   450,000,000,000 =

2.   64,300,000 =   8.   6,200 =

3. 951,000,000,000 =   9.   87,000,000 =

4.   12,300 =   10.   105,000,000 =

5.   20,350,000,000 =   11. 1,083,000,000,000 =

6.   9,000 =   12.   304,000 =

To convert a number written in scientific notation back to conventional form, reverse the steps.
EXAMPLE: $4.02 \times 10^5 = 4.02000 = 402,000$ Move the decimal 5 spaces to the right and add zeros.

Convert the following numbers from scientific notation to conventional numbers.

13. $6.85 \times 10^8 =$ 19. $5.87 \times 10^7 =$

14. $1.3 \times 10^{10} =$ 20. $8.047 \times 10^8 =$

15. $4.908 \times 10^4 =$ 21. $3.81 \times 10^5 =$

16. $7.102 \times 10^6 =$ 22. $9.5 \times 10^{12} =$

17. $2.5 \times 10^3 =$ 23. $1.504 \times 10^6 =$

18. $9.114 \times 10^5 =$ 24. $7.3 \times 10^9 =$

7.1.1
8.1.1

USING SCIENTIFIC NOTATION FOR SMALL NUMBERS

Scientific notation also simplifies very small numbers that have many zeros. For example, the diameter of a helium atom is 0.000000000244 meters. It can be written in scientific notation as 2.44×10^{-10}. The first number is always greater than 0, and the first number is always followed by a decimal point. The negative exponent indicates how many digits the decimal point moved to the right. The exponent is negative when the original number is less than 1. To convert small numbers to scientific notation, follow the **EXAMPLES** below.

EXAMPLES: $0.00058 = 5.8 \times 10^{-4}$ $0.00003059 = 3.059 \times 10^{-5}$

decimal point moves   negative exponent   decimal moves 5 spaces to the right
4 spaces to the right   indicates the original
                            number is less than 1.

Convert the following numbers to scientific notation.

1. 0.00000254 = _____

2. 0.00000000508 = _____

3. 0.000008004 = _____

4. 0.00047 = _____

5. 0.000000005478 = _____

6. 0.00000059 = _____

7. 0.00000004712 = _____

8. 0.00025 = _____

9. 0.0000000501 = _____

10. 0.0000006 = _____

11. 0.0000000000875 = _____

12. 0.00004 = _____

Now convert small numbers written in scientific notation back to conventional form.

EXAMPLE: $3.08 \times 10^{-5} = .00003.08 = 0.0000308$ Move the decimal 5 spaces to the left, and add zeros.

Convert the following numbers from scientific notation to conventional numbers.

13. 1.18×10^{-7} = _____

14. 2.3×10^{-5} = _____

15. 6.205×10^{-9} = _____

16. 4.1×10^{-6} = _____

17. 7.632×10^{-4} = _____

18. 5.48×10^{-10} = _____

19. 2.75×10^{-8} = _____

20. 4.07×10^{-7} = _____

21. 5.2×10^{-3} = _____

22. 7.01×10^{-6} = _____

23. 4.4×10^{-5} = _____

24. 3.43×10^{-2} = _____

7.1.1
8.1.1

ORDERING NUMBERS IN SCIENTIFIC NOTATION

When you compare numbers such as 500 and 500,000 it is easy to see that the more 0's there are, the larger the number. Converting these two numbers to scientific notation, $500 = 5 \times 10^2$ and $500,000 = 5 \times 10^5$, you can see that the larger the exponent, the larger the number expressed in scientific notation. With negative exponents, the farther the exponent is from 0 on the number line, the smaller the number it represents. Thus 3.45×10^{-5} is smaller than 3.45×10^{-4}. You may be asked to order numbers in scientific notation from greatest to least or least to greatest.

EXAMPLE 1: Consider the median distance from the sun of the following 3 planets: Planet A is 7.783×10^8 km from the sun, planet B is 1.082×10^8 km from the sun, and planet C is 5.899×10^9 km from the sun. Put the planets in order of distance from the sun starting with the planet closest to the sun.

Step 1: Look at the exponents. Planets A and B have the smallest exponents, so they are closer to the sun than Planet C.

Step 2: Compare the decimal numbers. 1.082 is smaller than 7.783, so Planet B is the closest to the sun, followed by Planet A and then Planet C.

EXAMPLE 2: The radii of 3 atoms are as follows: Atom 1 measures 1.8×10^{-8}, Atom 2 measures 1.97×10^{-8}, and Atom 3 measures 5.3×10^{-7}. Put the atoms in order of the measure of their radii from greatest to smallest.

Step 1: Look at the negative exponents. The closer the exponent is to 0, the larger the number. 10^{-7} is larger than 10^{-8}, so Atom 3 is the largest.

Step 2: Compare the decimal numbers. 1.97 is larger than 1.8, so Atom 2 is the next largest and Atom 1 is the smallest.

For problems 1 - 6 below, put each number in scientific notation in order from smallest to largest.

1.	2.54×10^{-7}	1.082×10^8	1.97×10^{-8}		1.504×10^6
2.	4.4×10^{-5}	4.04×10^{-6}	4.14×10^{-5}		4.4×10^{-6}
3.	5.2×10^{-3}	5.632×10^{-4}	5.01×10^{-6}		5.4×10^{-5}
4.	10^5	10^{-3}	10^6	10^{-8}	10^7	10^{-5}
5.	1.3×10^{10}	1.102×10^6	1.81×10^5		1.047×10^8
6.	0.206×10^{-4}	2.5×10^{-3}	2.102×10^{-6}		2.114×10^{-5}

CHAPTER 4 REVIEW

Rewrite the following problems using exponents.

1. $3 \times 3 \times 3 \times 3$ _____

2. $5 \times 5 \times 5$ _____

3. $10 \times 10 \times 10 \times 10 \times 10$ _____

4. 25×25 _____

Use a calculator to figure the solution to the following.

5. 2^2 = _____

6. 5^3 = _____

7. 12^1 = _____

8. 15^0 = _____

9. 10^4 = _____

10. 7^2 = _____

Answer the following square root questions.

11. Is $\sqrt{6}$ closer to 2 or 3?

12. Is $\sqrt{22}$ closer to 4 or 5?

13. Is $\sqrt{12}$ closer to 3 or 4?

14. Is $\sqrt{28}$ closer to 5 or 6?

15. $2\sqrt{3} + 3\sqrt{3}$ is closest to

    a) 5   b) 7   c) 9   d) 11

16. $5\sqrt{10} + 4\sqrt{10}$ is closest to

    a) 26   b) 28   c) 24   d) 22

17. Add the following square roots.

    a. $2\sqrt{5} + 6\sqrt{5}$
    b. $5\sqrt{3} + \sqrt{12}$

18. Subtract the following square roots.

    a. $4\sqrt{7} - \sqrt{28}$
    b. $9\sqrt{5} - 5\sqrt{5}$

19. Multiply the following square roots.

    a. $4\sqrt{5} \times 2\sqrt{3}$
    b. $\sqrt{18} \times 2\sqrt{3}$

20. Divide the following square roots.

    a. $\dfrac{\sqrt{10}}{\sqrt{5}}$ b. $\dfrac{4\sqrt{7}}{7\sqrt{14}}$

Convert the following numbers to scientific notation.

21. $22{,}300{,}000 =$ _____

22. $5{,}340{,}000 =$ _____

23. $0.00000005874 =$ _____

24. $1{,}451 =$ _____

25. $0.0000041 =$ _____

26. $0.0004178 =$ _____

27. $105{,}000 =$ _____

28. $705{,}000{,}000 =$ _____

29. $0.0000747 =$ _____

30. $0.08 =$ _____

31. $105 =$ _____

32. $0.0048754 =$ _____

33. $62{,}400 =$ _____

Convert the following numbers from scientific notation to conventional numbers.

34. $5.204 \times 10^{-5} =$ _____

35. $1.02 \times 10^{7} =$ _____

36. $8.1 \times 10^{5} =$ _____

37. $2.0078 \times 10^{-4} =$ _____

38. $4.7 \times 10^{-3} =$ _____

39. $7.75 \times 10^{-8} =$ _____

40. $9.795 \times 10^{9} =$ _____

41. $3.51 \times 10^{2} =$ _____

42. $6.32514 \times 10^{3} =$ _____

43. $1.584 \times 10^{-6} =$ _____

44. $7.041 \times 10^{4} =$ _____

45. $4.09 \times 10^{-7} =$ _____

Chapter 5 | Introduction to Algebra

ALGEBRA VOCABULARY

Vocabulary Word	Example	Definition
variable	$4x$ (x is the variable)	a letter that can be replaced by a number
coefficient	$4x$ (4 is the coefficient)	a number multiplied by a variable or variables
term	$5x^2 + x - 2$ terms	numbers or variables separated by $+$ or $-$ signs
constant	$5x + 2y + 4$ constant	a term that does not have a variable
numerical expression	$2^3 + 6 - 5$	two or more terms using only constants (numbers)
algebraic expression	$2x + 5^2 - 7$	one or more terms that include one or more variables
sentence	$2x = 7$ or $5 \leq x$	two algebraic expressions connected by $=, \neq, <, >, \leq, \geq$, or $\approx$
equation	$4x = 8$	a sentence with an equal sign
inequality	$7x < 30$ or $x \neq 6$	a sentence with one of the following signs: $\neq, <, >, \leq,$ or $\geq$
base	$6^3 \longleftarrow$ base	the number used as a factor
exponent	$6^3 \longleftarrow$ exponent	the number of times the base is multiplied by itself

SUBSTITUTING NUMBERS FOR VARIABLES

There may be some problems in which you are told what numbers to substitute in place of variables (letters). Then you simply solve the problem. These problems may look difficult at first glance, but they are very easy.

EXAMPLE 1: In the following problems, substitute 10 for *a*.

PROBLEM	SUBSTITUTION	SOLUTION
1. $a + 1$	Simply replace the *a* with 10. $10 + 1$	11
2. $17 - a$	$17 - 10$	7
3. $9a$	This means multiply. 9×10	90
4. $\dfrac{30}{a}$	This means divide. $30 \div 10$	3
5. a^3	$10 \times 10 \times 10$	1000
* 6. $5a + 6$	$(5 \times 10) + 6$	56

* Note: Be sure to do all multiplying and dividing before adding and subtracting.

EXAMPLE 2: In the following problems, let *x* = 2, *y* = 4, and *z* = 5.

PROBLEM	SUBSTITUTION	SOLUTION
1. $5xy + z$	$5 \times 2 \times 4 + 5$	45
2. $xz^2 + 5$	$2 \times 5^2 + 5 = 2 \times 25 + 5$	55
3. $\dfrac{yz}{x}$	$(4 \times 5) \div 2 = 20 \div 2$	10

In the following problems, $t = 7$. Solve the problems.

1. $t + 3 =$ _____	4. $3t - 5 =$ _____	7. $9t \div 3 =$ _____
2. $18 - t =$ _____	5. $t^2 + 1 =$ _____	8. $\dfrac{t^2}{7} =$ _____
3. $\dfrac{21}{t} =$ _____	6. $2t - 4 =$ _____	9. $5t + 6 =$ _____

In the following problems $a = 4$, $b = -2$, $c = 5$, and $d = 10$.   Solve the problems.

10. $4a + 2c =$ _____	14. $a^2 - b =$ _____	18. $\dfrac{6b}{a} =$ _____
11. $3bc - d =$ _____	15. $abd =$ _____	19. $9a + b =$ _____
12. $\dfrac{ac}{d} =$ _____	16. $5c - ad =$ _____	20. $5 + 3bc =$ _____
13. $d - 2a =$ _____	17. $cd + bc =$ _____	21. $d^2 + d + 1 =$ _____

7.3.4

69

UNDERSTANDING ALGEBRA WORD PROBLEMS

The biggest challenge to solving word problems is figuring out whether to add, subtract, multiply, or divide. Below is a list of key words and their meanings. This list does not include every situation you might see, but it includes the most common examples.

Words Indicating Addition	Example	Add
and	6 **and** 8	$6 + 8$
increased	The original price of $15 **increased** by $5.	$15 + 5$
more	3 coins and 8 **more**	$3 + 8$
more than	Josh has 10 points. Will has 5 **more than** Josh.	$10 + 5$
plus	8 baseballs **plus** 4 baseballs	$8 + 4$
sum	the **sum** of 3 and 5	$3 + 5$
total	the **total** of 10, 14, and 15	$10 + 14 + 15$

Words Indicating Subtraction	Example	Subtract
decreased	$16 **decreased** by $5	$16 - 5$
difference	the **difference** between 18 and 6	$18 - 6$
less	14 days **less** 5	$14 - 5$
less than	Jose completed 2 laps **less than** Mike's 9.	* $9 - 2$
left	Ray sold 15 out of 35 tickets. How many did he have **left**?	* $35 - 15$
lower than	This month's rainfall is 2 inches **lower than** last month's rainfall of 8 inches.	* $8 - 2$
minus	15 **minus** 6	$15 - 6$

* In subtraction word problems, you cannot always subtract the numbers in the order that they appear in the problem. Sometimes the first number should be subtracted from the last. You must read each problem carefully.

Words Indicating Multiplication	Example	Multiply
double	Her $1000 profit **doubled** in a month.	1000×2
half	**Half** of the $600 collected went to charity.	$\frac{1}{2} \times 600$
product	the **product** of 4 and 8	4×8
times	Li scored 3 **times** as many points as Ted who only scored 4.	3×4
triple	The bacteria **tripled** its original colony of 10,000 in just one day.	$3 \times 10,000$
twice	Ron has 6 Cd's. Tom has **twice** as many.	2×6

Words Indicating Division	Example	Divide
divide into, by, or among	The group of 70 **divided into** 10 teams.	$70 \div 10$ or $\frac{70}{10}$
quotient	the **quotient** of 30 and 6	$30 \div 6$ or $\frac{30}{6}$

Match the phrase on the left with the correct algebraic expression on the right. The answers on the right will be used more than once.

1. _____ 2 more than y

2. _____ 2 divided into y

3. _____ 2 less than y

4. _____ twice y

5. _____ the quotient of y and 2

6. _____ y increased by 2

7. _____ 2 less y

8. _____ the product of 2 and y

9. _____ y decreased by 2

10. _____ y doubled

11. _____ 2 minus y

12. _____ the total of 2 and y

A. $y - 2$

B. $2y$

C. $y + 2$

D. $\dfrac{y}{2}$

E. $2 - y$

Now practice writing parts of algebraic expressions from the following word problems.

EXAMPLE: the product of 3 and a number, t Answer: $3t$

13. 3 less than x _____

14. y divided among 10 _____

15. the sum of t and 5 _____

16. n minus 14 _____

17. 5 times k _____

18. the total of z and 12 _____

19. double the number b _____

20. x increased by 1 _____

21. the quotient of t and 4 _____

22. half of a number y _____

23. bacteria culture, b, doubled _____

24. triple John's age, y _____

25. a number, n, plus 4 _____

26. quantity, t, less 6 _____

27. 18 divided by a number, x _____

28. n feet lower than 10 _____

29. 3 more than p _____

30. the product of 4 and m _____

31. a number, y, decreased by 20 _____

32. 5 times as much as x _____

7.3.1

If a word problem contains the word "sum" or "difference," put the numbers that "sum" or "difference" refer to in parentheses to be added or subtracted first. Do not separate them. Look at the examples below.

EXAMPLES:

	RIGHT	WRONG
sum of 2 and 4, times 5	$5(2 + 4) = 30$	$2 + 4 \times 5 = 22$
the sum of 4 and 6, divided by 2	$\dfrac{(4 + 6)}{2} = 5$	$4 + \dfrac{6}{2} = 7$
4 times the difference between 10 and 5	$4(10 - 5) = 20$	$4 \times 10 - 5 = 35$
20 divided by the difference between 4 and 2	$\dfrac{20}{(4 - 2)} = 10$	$20 \div 4 - 2 = 3$
the sum of x and 4, multiplied by 2	$2(x + 4) = 2x + 8$	$x + 4 \times 2 = x + 8$

Change the following phrases into algebraic expressions.

1. 5 times the sum of x and 6

   _____

2. the difference between 5 and 3, divided by 4

   _____

3. 30 divided by the sum of 2 and 3

   _____

4. twice the sum of 10 and x

   _____

5. the difference between x and 9, divided by 10

   _____

6. 7 times the difference between x and 4

   _____

7. 9 multiplied by the sum of 3 and 4

   _____

8. the difference between x and 5, divided by 6

   _____

9. x divided by the sum of 4 and 9

   _____

10. x minus 5, times 10

   _____

11. 100 multiplied by the sum of x and 6

   _____

12. twice the difference between 3 and x

   _____

13. 4 times the sum of 5 and 1

   _____

14. 5 times the difference between 4 and 2

   _____

15. 12 divided by the sum of 2 and 4

   _____

16. four minus x, multiplied by 2

   _____

7.3.1

Look at the examples below for more phrases that may be used in algebra word problems.

EXAMPLES:

one-half of the sum of x and 4	$\frac{1}{2}(x+4)$ or $\frac{x+4}{2}$
six more than four times a number, x	$6+4x$
100 decreased by the product of a number, x, and 5	$100-5x$
ten less than the product of 3 and x	$3x-10$

Change the following phrases into algebraic expressions.

1. one-third of the sum of x and 5

   _____

2. three more than the product of a number, x, and 7

   _____

3. ten less than the sum of t and 4

   _____

4. the product of 4 and n, minus 3

   _____

5. 15 less the sum of 3 and x

   _____

6. the difference of 10, and 3 times a number, n

   _____

7. one-fifth of t

   _____

8. the product of 3 and x, minus 14

   _____

9. x times the difference between 4 and x

   _____

10. five more than the quotient of x and 6

    _____

11. the sum of 5 and k, divided by 2

    _____

12. one less than the product of 3 and x

    _____

13. 5 increased by one-half of a number, n

    _____

14. 10 more than twice x

    _____

15. six subtracted from four times m

    _____

16. 8 times x, subtracted from 20

    _____

7.3.1

SETTING UP ALGEBRA WORD PROBLEMS

So far, you have seen only the first part of algebra word problems. To complete an algebra problem, an equal sign must be added. The words "**is**" or "**are**" as well as "**equal(s)**" signal that you should add an equal sign.

EXAMPLE: Double Jake's age, x, minus 4 is 22.

$$2x - 4 = 22$$

Translate the following word problems into algebra problems. DO NOT find the solutions to the problems yet.

1. Triple the original number, n, is 2,700.

2. The product of a number, y, and 5 is equal to 15.

3. Four times the difference of a number, x, and 2 is 20.

4. The total, t, divided into 5 groups is 45.

5. The number of parts in inventory, p, minus 54 parts sold today is 320.

6. One-half an amount, x, added to $50 is $262.

7. One hundred seeds divided by 5 rows equals n number of seeds per row.

8. A number, y, less than 50 is 82.

9. His base pay of $200 increased by his commission, x, is $500.

10. Seventeen more than half a number, h, is 35.

11. This month's sales of $2,300 are double January's sales, x.

12. The quotient of a number, w, and 4 is 32.

13. Six less a number, d, is 12.

14. Four times the sum of a number, y, and 10 is 48.

15. We started with x number of students. When 5 moved away, we had 42 left.

16. A number, b, divided into 36 parts is 12.

7.3.1

MATCHING ALGEBRAIC EXPRESSIONS

Match each set of algebraic expressions with the correct phrase underneath them.

1. _____ $2x + 5$

2. _____ $2(x + 5)$

3. _____ $2x - 5$

4. _____ $2(x - 5)$

A.   twice the sum of x and 5

B.   five less than the product of 2 and x

C.   five more than the product of 2 and x

D.   two times the difference of x and 5

---

5. _____ $4(y - 2)$

6. _____ $\dfrac{y - 2}{4}$

7. _____ $4y - 2$

8. _____ $\dfrac{y}{4} - 2$

A.   two less than the product of y and 4

B.   the difference of y and 2 divided by 4

C.   two less than one-fourth of y

D.   four times the difference of y and 2

---

9. _____ $5y + 8$

10. _____ $5(y + 8)$

11. _____ $8y + 5$

12. _____ $8(y + 5)$

A.   eight times the sum of y and 5

B.   eight more than the product of 5 and y

C.   five more than eight times y

D.   five multiplied by the sum of y and 8

---

13. _____ $9 - x = 7$

14. _____ $x - 9 = 7$

15. _____ $9 - 7 = x$

A.   nine less than x is 7

B.   nine less x is 7

C.   the difference between 9 and 7 is x

---

16. _____ $\dfrac{n + 5}{2} = 10$

17. _____ $\dfrac{n}{2} + 5 = 10$

18. _____ $\tfrac{1}{2}n - 5 = 10$

A.   one-half the sum of n and 5 is 10

B.   five less than half of n is 10

C.   five added to half of n is 10

---

19. _____ $x + \dfrac{4}{5} = 8$

20. _____ $\dfrac{x}{5} + 4 = 8$

21. _____ $\dfrac{x + 4}{5} = 8$

A.   the sum of x and 4, divided by 5 is 8

B. x added to the quotient of 4 and 5 is 8

C.   four more than x divided by 5 is 8

---

22. _____ $7t + 1 = 5$

23. _____ $7(t + 1) = 5$

24. _____ $7t = 5$

A.   one more than seven times t is 5

B.   seven times the sum of t and 1 is 5

C.   the product of seven and t is 5

7.3.1

75

CHANGING ALGEBRA WORD PROBLEMS TO ALGEBRAIC EQUATIONS

EXAMPLE: There are 3 people who have a total weight of 595 pounds. Sally weighs 20 pounds less than Jessie. Rafael weighs 15 pounds more than Jessie. How much does Jessie weigh?

Step 1: Notice everyone's weight is given in terms of Jessie. Sally weighs 20 pounds less than Jessie. Rafael weighs 15 pounds more than Jessie. First, we write everyone's weight in terms of Jessie, j.

$$\text{Jessie} = j$$
$$\text{Sally} = j - 20$$
$$\text{Rafael} = j + 15$$

Step 2: We know that all three together weigh 595 pounds. We write the sum of everyone's weight equal to 595.

$$j + j - 20 + j + 15 = 595$$

We will learn to solve these problems in the next chapter.

Change the following word problems to algebraic equations.

1. Fluffy, Spot, and Shampy have a combined age in dog years of 91. Spot is 14 years younger than Fluffy. Shampy is 6 years older than Fluffy. What is Fluffy's age, f, in dog years?

2. Jerry Marcosi puts 5% of the amount he makes per week into a retirement account, r. He is paid $11.00 per hour and works 40 hours per week for a certain number of weeks, w. Write an equation to help him find out how much he puts into his retirement account.

3. A furniture store advertises a 40% off liquidation sale on all items. What would the sale price (p) be on a $2530 dining room set?

4. Kyle Thornton buys an item which normally sells for a certain price, x. Today the item is selling for 25% off the regular price. A sales tax of 6% is added to the equation to find the final price, f.

5. Tamika Francois runs a floral shop. On Tuesday, Tamika sold a total of $600 worth of flowers. The flowers cost her $100, and she paid an employee to work 8 hours for a given wage, w. Write an equation to help Tamika find her profit, p, on Tuesday.

6. Sharice is a waitress at a local restaurant. She makes an hourly wage, $3.50, plus she receives tips. On Monday, she worked 6 hours and received tip money, t. Write an equation showing what Sharice made on Monday, y.

7. Jenelle buys x shares of stock in a company at $34.50 per share. She later sells the shares at $40.50 per share. Write an equation to show how much money, m, Jenelle has made.

7.3.1

CHAPTER 5 REVIEW

Solve the following problems using $x = 2$.

1. $3x + 4 =$ _____

2. $\dfrac{6x}{4} =$ _____

3. $x^2 - 5 =$ _____

4. $\dfrac{x^3 + 8}{2} =$ _____

5. $12 - 3x =$ _____

6. $x - 5 =$ _____

7. $-5x + 4 =$ _____

8. $9 - x =$ _____

9. $2x + 2 =$ _____

Solve the following problems. Let $w = -1$, $y = 3$, $z = 5$.

10. $5w - y =$ _____

11. $wyz + 2 =$ _____

12. $z - 2w =$ _____

13. $\dfrac{3z + 5}{wz} =$ _____

14. $\dfrac{6w}{y} + \dfrac{z}{w} =$ _____

15. $25 - 2yz =$ _____

16. $-2y + 3 =$ _____

17. $4w - (yw) =$ _____

18. $7y - 5z =$ _____

Write out the algebraic expression given in each word problem.

19. three less the sum of x and 5 _____

20. double Amy's age, a _____

21. the number of bacteria, b, tripled _____

22. five less than the product of 5 and y _____

23. half of a number, n, less 15 _____

24. the quotient of a number, x, and 6 _____

For questions 25-27, write an equation to match each problem.

25. Calista earns \$450 per week for a 40 hour work week plus \$16.83 per hour for each hour of overtime after 40 hours. Write an equation that would be used to determine her weekly wages where w is her wages, and v is the number of overtime hours worked.

26. Daniel purchased a 1 year CD, c, from a bank. He bought it at an annual interest rate of 6%. After 1 year, Daniel cashes in the CD. What is the total amount it is worth?

27. Omar is a salesman. He earns an hourly wage of \$8.00 per hour plus he receives a commission of 7% on the sales he makes. Write an equation which would be used to determine his weekly salary, w, where x is the number of hours worked, and y is the amount of sales for the week.

7.3.1

28. Tom earns $500 per week before taxes are taken out. His employer takes out a total of 33% for state, federal, and Social Security taxes. Which expression below will help Tom figure his net pay?

   A. $500 - .33$
   B. $500 \div .33$
   C. $500 + .33(500)$
   D. $500 - .33(500)$

29. Rosa has to pay $100 of her medical expenses in a year before she qualifies for her insurance company to begin paying. After paying the $100 "deductible," her insurance company will pay 80% of her medical expenses. This year, her total medical expenses came to $960.00. Which expression below shows how much her insurance company will pay?

   A. $.80(960 - 100)$
   B. $100 + (960 \div .80)$
   C. $960(100 - .80)$
   D. $.80(960 + 100)$

30. A plumber charges $45.00 per hour plus a $25.00 service charge. If a represents his total charges in dollars, and b represents the number of hours worked, which formula below could the plumber use to calculate his total charges?

   A. $a = 45 + 25b$
   B. $a = 45 + 25 + b$
   C. $a = 45b + 25$
   D. $a = (45)(25) + b$
   E. $a = 70b$

31. In 2004, Bell Computers announced to its sales force to expect a 2.6% price increase on all computer equipment in the year 2005. A certain sales representative wanted to see how much the increase would be on a computer, c, that sold for $2,200 in 2004. Which expression below will help him find the cost of the computer in the year 2005?

   A. $.26(2200)$
   B. $2200 - .026(2200)$
   C. $2200 + .026(2200)$
   D. $.026(2200) - 2200$

32. Juan sold a boat that he bought 5 years ago. He sold it for 60% less than he originally paid for it. If the original cost is b, write an expression that shows how much he sold the boat for.

33. Toshi is going to get a 7% raise after he works at his job for 1 year. If s represents his starting salary, write an expression that shows how much he will make after his raise.

7.3.1

16. Keith drove west to go see his sister in California. He drove at an average of 50 miles per hour. His mom and dad left 4 hours later and drove an average of 70 miles per hour to catch up with him. How long did it take for them to catch up to Keith?

17. Priscilla bought 20 pounds of chocolate-covered peanuts for $3.50 per pound. How many pounds of chocolate-covered walnuts would she have to buy at $7.00 per pound to make boxes of nut mixtures that would cost her $4.50 per pound?

18. One solution is 15% ammonia. A second solution is 40% ammonia. How many ounces of each should be used to make 100 ounces of a 20% ammonia solution?

19. Joe, Craig, and Dylan have a combined weight of 429 pounds. Craig weighs 34 pounds more than Joe. Dylan weighs 13 pounds more than Craig. How many pounds does Craig weigh?

20. Tracie and Marcia drove to northern Florida to see Marcia's sister in Tallahassee. Tracie drove one hour more than three times as much as Marcia. The trip took a total of 17 driving hours. How many hours did Tracie drive?

21. Jesse and Larry entered a pie eating contest. Jesse ate 2 less than twice as many pies as Larry. They ate a total of 28 pies. How many pies did Larry eat?

22. Lena and Jodie are sisters and together they have 68 bottles of nail polish. Lena bought 5 more than half the bottles. How many did Jodie buy?

23. Janet and Artie wanted to play tug of war. Artie pulls with 150 pounds of force while Janet pulls with 40 pounds of force. In order to make this a fair contest, Janet enlists the help of her friends Trudi, Sherri, and Bridget who pull with 30, 25, and 40 pounds respectively. Write an inequality describing the minimum amount Janet's fourth friend, Tommy, must pull to beat Artie.

24. Jim takes great pride in decorating his float for the homecoming parade for his high school. With the $5,000 he has to spend, Jim bought 5,000 carnations at $.25 each, 4,000 tulips at $.50 each, and 300 irises at $.90 each. Write an inequality which describes how many roses, r, Jim can buy if roses cost $.80 each.

25. Mr. Chan wants to sell some or all of his shares of stock in a company. He purchased the 80 shares for $.50 last month, and the shares are now worth $4.50 each. Write an inequality which describes how much profit, p, Mr. Chan can make by selling his shares.

26. Kyle can deliver all of the newspapers on a given paper route in three hours. Jessica can deliver all the newspapers on the same route in five hours. How long would it take them to deliver newspapers if they both work together?

27. Chris can construct a robot on an erector set in 7 hours. His friend, Cristobal, can construct a robot using the same materials in 8 hours. How long would it take them to construct a robot if they both work together?

28. Anita and Reginald both enjoy rollerblading as an afternoon activity. Anita can run a particular course called the Silver Comet Trail in 7 hours. Reginald is able to rollerblade the same course in 5 hours. If Reginald and Anita start at opposite ends of the trail and rollerblade towards each other, how much time will have passed when they meet? (Hint: Think about this problem as a "working together" problem.)

Chapter 9

Polynomials

Polynomials are algebraic expressions that include **monomials**, which only have one term, **binomials**, which contain two terms, and **trinomials**, which contain three terms. Expressions with more than three terms are all called **polynomials**. **Terms** are separated by plus and minus signs.

EXAMPLES

Monomials:	Binomials:	Trinomials:	Polynomials:
$4f$	$4t + 9$	$x^2 + 2x + 3$	$x^3 - 3x^2 + 3x - 9$
$3x^3$	$9 - 7g$	$5x^2 - 6x - 1$	$p^4 + 2p^3 + p^2 - 5p + 9$
$4g^2$	$5x^2 + 7x$	$y^4 + 15y^2 + 100$	
2	$6x^3 - 8x$		

ADDING AND SUBTRACTING MONOMIALS

Two **monomials** can be added or subtracted as long as the **variable and its exponent** are the **same**. This is called combining like terms. Use the same rules you used for adding and subtracting integers.

EXAMPLES $4x + 5x = 9x$ $2x^2 - 9x^2 = -7x^2$ $6y^3 - 5y^3 = y^3$

$$\begin{array}{r} 5y \\ + 2y \\ \hline 7y \end{array} \qquad \begin{array}{r} 3x^4 \\ - 8x^4 \\ \hline -5x^4 \end{array}$$

> **Remember:** When the integer in front of the variable is "1", it is usually not written. $1x^2$ is the same as x^2, *and* $-1x$ is the same as $-x$.

Add or subtract the following monomials:

1. $2x^2 + 5x^2 = $ _____
2. $5t + 8t = $ _____
3. $9y^3 - 2y^3 = $ _____
4. $6g - 8g = $ _____
5. $7y^2 + 8y^2 = $ _____

6. $s^5 + s^5 = $ _____
7. $-2x - 4x = $ _____
8. $4w^2 - w^2 = $ _____
9. $z^4 + 9z^4 = $ _____
10. $-k + 2k = $ _____

11. $3x^2 - 5x^2 = $ _____
12. $9t + 2t = $ _____
13. $-7v^3 + 10v^3 = $ _____
14. $-2x^3 + x^3 = $ _____
15. $10y^4 - 5y^4 = $ _____

16. $\begin{array}{r} y^4 \\ + 2y^4 \end{array}$

17. $\begin{array}{r} 4x^3 \\ - 9x^3 \end{array}$

18. $\begin{array}{r} 8t^2 \\ + 7t^2 \end{array}$

19. $\begin{array}{r} -2y \\ - 4y \end{array}$

20. $\begin{array}{r} 5w^2 \\ + 8w^2 \end{array}$

21. $\begin{array}{r} 11t^3 \\ - 4t^3 \end{array}$

22. $\begin{array}{r} -5z \\ + 9z \end{array}$

23. $\begin{array}{r} 4w^5 \\ + w^5 \end{array}$

24. $\begin{array}{r} 7t^3 \\ - 6t^3 \end{array}$

25. $\begin{array}{r} 3x \\ + 8x \end{array}$

ADDING POLYNOMIALS

When adding **polynomials**, make sure the exponents and variables are the same on the terms you are combining. The easiest way is to put the terms in columns with **like exponents** underneath each other. Each column is added as a separate problem. Fill in the blank spots with zeros if it helps you keep the columns straight. You never carry to the next column when adding polynomials.

EXAMPLE 1: Add $3x^2 + 14$ and $5x^2 + 2x$

$$\begin{array}{r} 3x^2 + 0x + 14 \\ (+)\ \underline{5x^2 + 2x + \ \ 0} \\ 8x^2 + 2x + 14 \end{array}$$

EXAMPLE 2: $(4x^3 - 2x) + (-x^3 - 4)$

$$\begin{array}{r} 4x^3 - 2x + 0 \\ (+)\ \underline{-x^3 + 0x - 4} \\ 3x^3 - 2x - 4 \end{array}$$

Add the following polynomials.

1. $y^2 + 3y + 2$ and $2y^2 + 4$

2. $(5y^2 + 4y - 6) + (2y^2 - 5y + 8)$

3. $5x^3 - 2x^2 + 4x - 1$ and $3x^2 - x + 2$

4. $-p + 4$ and $5p^2 - 2p + 2$

5. $(w - 2) + (w^2 + 2)$

6. $4t^2 - 5t - 7$ and $8t + 2$

7. $t^4 + t + 8$ and $2t^3 + 4t - 4$

8. $(3s^3 + s^2 - 2) + (-2s^3 + 4)$

9. $(-v^2 + 7v - 8) + (4v^3 - 6v + 4)$

10. $6m^2 - 2m + 10$ and $m^2 - m - 8$

11. $-x + 4$ and $3x^2 + x - 2$

12. $(8t^2 + 3t) + (-7t^2 - t + 4)$

13. $(3p^4 + 2p^2 - 1) + (-5p^2 - p + 8)$

14. $12s^3 + 9s^2 + 2s$ and $s^3 + s^2 + s$

15. $(-9b^2 + 7b + 2) + (-b^2 + 6b + 9)$

16. $15c^2 - 11c + 5$ and $-7c^2 + 3c - 9$

17. $5c^3 + 2c^2 + 3$ and $2c^3 + 4c^2 + 1$

18. $-14x^3 + 3x^2 + 15$ and $7x^3 - 12$

19. $(-x^2 + 2x - 4) + (3x^2 - 3)$

20. $(y^2 - 11y + 10) + (-13y^2 + 5y - 4)$

21. $3d^5 - 4d^3 + 7$ and $2d^4 - 2d^3 - 2$

22. $(6t^5 - t^3 + 17) + (4t^5 + 7t^3)$

23. $4p^2 - 8p + 9$ and $-p^2 - 3p - 5$

24. $20b^3 + 15b$ and $-4b^2 - 5b + 14$

25. $(-2w + 11) + (w^3 + w - 4)$

26. $(25z^2 + 13z + 8) + (z^2 - 2z - 10)$

SUBTRACTING POLYNOMIALS

When you subtract polynomials, it is important to remember to change all the signs in the subtracted polynomial (the subtrahend), and then add.

EXAMPLE: $(4y^2 + 8y + 9) - (2y^2 + 6y - 4)$

Step 1: Copy the subtraction problem into vertical form. Make sure you line up the terms with like exponents under each other just like you did for adding polynomials.

$$\begin{array}{r} 4y^2 + 8y + 9 \\ (-)\ \underline{2y^2 + 6y - 4} \end{array}$$

Step 2: Change the subtraction sign to addition and all the signs of the subtracted polynomial to the opposite sign. The bottom polynomial in the problem becomes $-2y^2 - 6y + 4$.

Step 3: Add:
$$\begin{array}{r} 4y^2 + 8y + 9 \\ (+)\ \underline{-2y^2 - 6y + 4} \\ 2y^2 + 2y + 13 \end{array}$$

Subtract the following polynomials.

1. $(2x^2 + 5x + 2) - (x^2 + 3x + 1)$

2. $(8y - 4) - (4y + 3)$

3. $(11t^3 - 4t^2 + 3) - (-t^3 + 4t^2 - 5)$

4. $(-3w^2 + 9w - 5) - (-5w^2 - 5)$

5. $(6a^5 - a^3 + a) - (7a^5 + a^2 - 3a)$

6. $(14c^4 + 20c^2 + 10) - (7c^4 + 5c^2 + 12)$

7. $(5x^2 - 9x) - (-7x^2 + 4x + 8)$

8. $(12y^3 - 8y^2 - 10) - (3y^3 + y + 9)$

9. $(-3h^2 - 7h + 7) - (5h^2 + 4h + 10)$ 10. $(10k^3 - 8) - (-4k^3 + k^2 + 5)$

11. $(x^2 - 5x + 9) - (6x^2 - 5x + 7)$ 12. $(12p^2 + 4p) - (9p - 2)$

13. $(-2m - 8) - (6m + 2)$ 14. $(13y^3 + 2y^2 - 8y) - (2y^3 + 4y^2 - 7y)$

15. $(7g + 3) - (g^2 + 4g - 5)$ 16. $(-8w^3 + 4w) - (-10w^3 - 4w^2 - w)$

17. $(12x^3 + x^2 - 10) - (3x^3 + 2x^2 + 1)$ 18. $(2a^2 + 2a + 2) - (-a^2 + 3a + 3)$

19. $(c + 19) - (3c^2 - 7c + 2)$ 20. $(-6v^2 + 12v) - (3v^2 + 2v + 6)$

21. $(4b^3 + 3b^2 + 5) - (7b^3 - 8)$ 22. $(15x^3 + 5x^2 - 4) - (4x^3 - 4x^2)$

23. $(8y^2 - 2) - (11y^2 - 2y - 3)$ 24. $(-z^2 - 5z - 8) - (3z^2 - 5z + 5)$

A subtraction of polynomials problem may be stated in sentence form. Study the examples below.

EXAMPLE 1:   Subtract $-5x^3 + 4x - 3$ from $3x^3 + 4x^2 - 6x$.

   Step 1:   Copy the problem in columns with terms with the same exponent and variable under each other. Notice the second polynomial in the sentence will be the top polynomial of the problem.

$$\begin{array}{r} 3x^3 + 4x^2 - 6x \\ (-)\ \underline{-5x^3 \qquad + 4x - 3} \end{array}$$

   Step 2:   Since this is a subtraction problem, change all the signs of the terms in the bottom polynomial, then add.

$$\begin{array}{r} 3x^3 + 4x^2 - 6x \\ (+)\ \underline{5x^3 \qquad - 4x + 3} \\ 8x^3 + 4x^2 - 10x + 3 \end{array}$$

EXAMPLE 2:   From $6y^2 + 2$ subtract $4y^2 - 3y + 8$

In a problem phrased like this one, the first polynomial will be on top, and the second will be on bottom. Change the signs on the bottom polynomial, then add.

$$\begin{array}{r} 6y^2 \qquad + 2 \\ (-)\ \underline{4y^2 - 3y + 8} \end{array} \longrightarrow \begin{array}{r} 6y^2 \qquad + 2 \\ (+)\ \underline{-4y^2 + 3y - 8} \\ 2y^2 + 3y - 6 \end{array}$$

Solve the following subtraction problems.

1. Subtract $3x^2 + 2x - 5$ from $5x^2 + 2$

2. From $5y^3 - 6y + 9$ subtract $8y^3 - 10$

3. From $4m^2 - 4m + 7$ subtract $2m - 3$

4. Subtract $8z^2 + 3z + 2$ from $4z^2 - 7z + 8$

5. Subtract $10t^3 + t^2 - 5$ from $-2t^3 - t^2 - 5$

6. Subtract $-7b^3 - 2b + 4$ from $-b^2 + b + 6$

7. From $10y^3 + 20$ subtract $5y^3 - 5$

8. From $14t^2 - 6t - 8$ subtract $4t^2 - 3t + 2$

9. Subtract $3p^2 + p - 2$ from $-7p^2 - 5p + 2$

10. Subtract $x^3 + 8$ from $3x^3 - 2x^2 + 9$

11. Subtract $12a^2 + 10$ from $a^3 - a^2 - 1$

12. From $6m^2 + 3m + 1$ subtract $-6m^2 - 3m$

13. From $-13z^3 - 3z^2 - 2$ subtract $-20z^3 + 20$

14. Subtract $9c^2 + 10$ from $8c^2 - 5c + 3$

15. Subtract $b^2 + b - 5$ from $5b^2 - 4b + 5$

16. Subtract $-3x - 4$ from $3x^2 + x + 9$

17. From $15y^2 + 2$ subtract $4y^2 + 3y + 7$

18. Subtract $3g^2 - 5g + 5$ from $9g^2 - 3g - 4$

19. From $-7m^2 - 8m$ subtract $3m^2 + 7$

20. Subtract $x + 1$ from $5x + 5$

21. Subtract $c^2 + c + 2$ from $-c^2 - c - 2$

22. From $8t^3 + 6t^2 - 4t + 2$ subtract $t^3 + 3t$

ADDING AND SUBTRACTING POLYNOMIALS REVIEW

Practice adding and subtracting the polynomials below.

1. Add $-3x^2 + x$ and $4x^2 - 2$

2. Subtract $(-2y^3 + 9y)$ from $(6y^3 - y + 4)$

3. $(8t^3 - 3t^2 - 9) + (-7t^3 + t - 4)$

4. $(7p^2 + 3p + 1) - (5p^2 + 4p + 6)$

5. From $4w^3 + 5w - 2$ subtract $6w^2 - 4$

6. Add $-8a^3 - 7a^2 + 10$ and $6a^3 + 4a^2$

7. $(-14b^2 + b + 2) + (6b^2 - b + 3)$

8. Subtract $(g^3 - 7g^2 - 5)$ from $(5g^3 - 10)$

9. $(4c - 6) - (2c^2 - 3c + 9)$

10. From $-m^3 + 2m^2 + m$ subtract $9m^3 + 2m$

11. $(-3v^2 + 9v - 6) + (3v^2 - 4v + 6)$

12. Add $10s^2 + 4$ and $5s - 6$

13. Subtract $-x^3 - 9x^2 - x$ from $3x^3 + 2x + 4$

14. $(-5y^2 - 4y - 1) - (5y^2 - 2y - 8)$

MULTIPLYING MONOMIALS

When two monomials have the **same variable**, they can be multiplied. The **exponents** are **added together**. If the variable has no exponent, it is understood that the exponent is 1.

Add

EXAMPLE: $4x^4 \times 3x^2 = 12x^6$

Multiply

EXAMPLE: $2y \times 5y^2 = 10y^3$

Multiply the following monomials

1. $6a \times 9a^5$ = _____

2. $2x^6 \times 5x^3$ = _____

3. $4y^3 \times 3y^2$ = _____

4. $10t^2 \times 2t^2$ = _____

5. $2p^5 \times 4p^2$ = _____

6. $9b^2 \times 8b$ = _____

7. $3c^3 \times 3c^3$ = _____

8. $2d^8 \times 9d^2$ = _____

9. $6k^3 \times 5k^2$ = _____

10. $7m^5 \times m$ = _____

11. $11z \times 2z^7$ = _____

12. $3w^4 \times 6w^5$ = _____

13. $4x^4 \times 5x^3$ = _____

14. $5n^2 \times 3n^3$ = _____

15. $8w^7 \times w$ = _____

16. $10s^6 \times 5s^3$ = _____

17. $4d^5 \times 4d^5$ = _____

18. $5y^2 \times 8y^6$ = _____

19. $7t^{10} \times 3t^5$ = _____

20. $6p^8 \times 2p^3$ = _____

21. $x^3 \times 2x^3$ = _____

When problems include negative signs, follow the rules for multiplying integers.

22. $-7s^4 \times 5s^3$ = $-35s^7$

23. $-6a \times -9a^5$ = _____

24. $4x \times -x$ = _____

25. $-3y^2 \times -y^3$ = _____

26. $-5b^2 \times 3b^5$ = _____

27. $9c^4 \times -2c$ = _____

28. $-4t^3 \times 8t^3$ = _____

29. $10d \times -8d^7$ = _____

30. $-3g^6 \times -2g^3$ = _____

31. $-7s^4 \times 7s^3$ = _____

32. $-d^3 \times -2d$ = _____

33. $11p \times -2p^5$ = _____

34. $-5x^7 \times -3x^3$ = _____

35. $8z^4 \times 7z^4$ = _____

36. $-4w \times -5w^8$ = _____

37. $-5y^4 \times 6y^2$ = _____

38. $9x^3 \times -7x^5$ = _____

39. $-a^4 \times -a$ = _____

40. $-7k^2 \times 3k$ = _____

41. $-15t^2 \times -t^4$ = _____

42. $3x^8 \times 9x^2$ = _____

MULTIPLYING MONOMIALS BY POLYNOMIALS

In Chapter 7, you learned to remove parentheses by multiplying the number outside parentheses by each term inside parentheses, $2(4x - 7) = 8x - 14$. Multiplying monomials by polynomials works the same way.

EXAMPLE: $-5t(2t^2 - 7t + 9)$

Step 1: Multiply $-5t \times 2t^2 = \mathbf{-10t^3}$
Step 2: Multiply $-5t \times -7t = \mathbf{35t^2}$
Step 3: Multiply $-5t \times 9 = \mathbf{-45t}$
Step 4: Arrange the answers horizontally in order: $\mathbf{-10t^3 + 35t^2 - 45t}$

Remove parentheses in the following problems.

1. $3x(3x^2 + 4x - 1)$

2. $4y(y^3 - 7)$

3. $7a^2(2a^2 + 3a + 2)$

4. $-5d^3(d^2 - 5d)$

5. $2w(-4w^2 + 3w - 8)$

6. $8p(p^3 - 6p + 5)$

7. $-9b^2(-2b + 5)$

8. $2t(t^2 - 4t - 10)$

9. $10c(4c^2 + 3c - 7)$

10. $6z(2z^4 - 5z^2 - 4)$

11. $-9t^2(3t^2 + 5t + 6)$

12. $c(-3c - 5)$

13. $3p(p^3 - p^2 - 9)$

14. $-k^2(2k + 4)$

15. $-3(4m^2 - 5m + 8)$

16. $6x(-7x^3 + 10)$

17. $-w(w^2 - 4w + 7)$

18. $2y(5y^2 - y)$

19. $3d(d^5 - 7d^3 + 4)$

20. $-5t(-4t^2 - 8t + 1)$

21. $7(2w^2 - 9w + 4)$

22. $3y^2(y^2 - 11)$

23. $v^2(v^2 + 3v + 3)$

24. $8x(2x^3 + 3x + 1)$

25. $-5d(4d^2 + 7d - 2)$

26. $-k^2(-3k + 6)$

27. $3x(-x^2 - 5x + 5)$

28. $4z(4z^4 - z - 7)$

29. $-5y(9y^3 - 3)$

30. $2b^2(7b^2 + 4b + 4)$

DIVIDING POLYNOMIALS BY MONOMIALS

EXAMPLE: $\dfrac{-8wx + 6x^2 - 16wx^2}{2wx}$

Step 1: Rewrite the problem. Divide each term from the top by the denominator, $2wx$.

$$\frac{-8wx}{2wx} + \frac{6w^2}{2wx} + \frac{-16wx}{2wx}$$

Step 2: Simplify each term in the problem. Then, combine like terms.

$$-4 + \frac{3w}{x} - 8 = -12 + \frac{3w}{x}$$

Simplify each of the following:

1. $\dfrac{bc^2 - 8bc - 2b^2c^2}{2bc}$

2. $\dfrac{3jk^2 + 12k + 9j^2k}{3jk}$

3. $\dfrac{5x^2y - 8xy^2 + 2y^3}{2xy}$

4. $\dfrac{16st^2 + st - 12s}{4st}$

5. $\dfrac{4wx^2 + 6wx - 12w^3}{2wx}$

6. $\dfrac{cd^2 + 10cd^3 + 16c^2}{2cd}$

7. $\dfrac{y^2z^3 - 2yz - 8z^2}{-2yz^2}$

8. $\dfrac{a^2b + 2ab^2 - 14ab^3}{2a^2}$

9. $\dfrac{pr^2 + 6pr + 8p^2r^2}{2pr^2}$

10. $\dfrac{6xy^2 - 3xy + 18x^2}{-3xy}$

11. $\dfrac{6x^2y + 12xy - 24y^2}{6xy}$

12. $\dfrac{5m^2n - 10mn - 25n^2}{5mn}$

13. $\dfrac{st^2 - 10st - 16s^2t^2}{2st}$

14. $\dfrac{7jk^2 - 14jk - 63j^2}{7jk}$

---

15. Divide $12x^2y - 6xy^2 + 12xy$ by $6xy$

16. Divide $5ab^2 - 15a^2b^3 + 25ab^2$ by $5ab^2$

17. Divide $8m^3n^3 + 6m^2n^2 - 10mn^3$ by $2mn^2$

18. Divide $36c^5d^4 - 18c^3d^3 - 24c^4d^7$ by $6c^2d^3$

19. Divide $16v^2w - 8v^3w^2 + 20v^2w$ by $4vw$

20. Divide $14y^3z^5 + 28y^2z^4 - 35yz^3$ by $7yz^3$

REMOVING PARENTHESES AND SIMPLIFYING

In the following problem, you must multiply each set of parentheses by the numbers and variables outside the parentheses, and then add the polynomials to simplify the expressions.

EXAMPLE: $8x (2x^2 - 5x + 7) - 3x (4x^2 + 3x - 8)$

Step 1:   Multiply to remove the first set of parentheses.
$8x (2x^2 - 5x + 7) = 16x^3 - 40x^2 + 56x$

Step 2:   Multiply to remove the second set of parentheses.
$-3x (4x^2 + 3x - 8) = -12x^3 - 9x^2 + 24x$

Step 3:   Copy each polynomial in columns, making sure the terms with the same variable and exponent are under each other. Add to simplify.

$$16x^3 - 40x^2 + 56x$$
$$(+) \underline{-12x^3 - 9x^2 + 24x}$$
$$4x^3 - 49x^2 + 80x$$

Remove the parentheses, and simplify the following problems.

1. $4t (t + 7) + 5t (2t^2 - 4t + 1)$

2. $-5y (3y^2 - 5y + 3) - 6y (y^2 - 4y - 4)$

3. $-3 (3x^3 + 4x) + 5x (x^2 + 3x + 2)$

4. $2b (5b^2 - 8b - 1) - 3b (4b + 3)$

5. $8d^2 (3d + 4) - 7d (3d^2 + 4d + 5)$

6. $5a (3a^2 + 3a + 1) - (-2a^2 + 5a - 4)$

7. $3m (m + 7) + 8(4m^2 + m + 4)$

8. $4c^2 (-6c^2 - 3c + 2) - 7c (5c^3 + 2c)$

9. $-8w (-w + 1) - 4w (3w - 5)$

10. $6p (2p^2 - 4p - 6) + 3p (p^2 + 6p + 9)$

MULTIPLYING TWO BINOMIALS
USING AREA REPRESENTATION

When you multiply two binomials such as $(2x + 1)(x + 3)$, every term in the first binomial must be multiplied by every term in the second binomial. One way to find the product of two binomials is area representation.

EXAMPLE: $(2x + 1)(x + 3)$

Step 1:   Set up the grid. $(2x + 1)$ will be used to determine the number of columns you will have. Since $2x + 1 = x + x + 1$, you can conclude that there will be three columns, two for the two x's and one for the 1. Do the same for the rows. Use $(x + 3)$ to determine the number of rows. Since $x + 3 = x + 1 + 1 + 1$, there will be four rows.

Step 2:   Multiply every term in each row by every term in each column. Place the product in the appropriate box of the grid.

Step 3:   Add the products in the boxes of the completed grid to find the quadratic expression.

$$x^2 + x^2 + x + x + x + x + x + x + x + 1 + 1 + 1 = 2x^2 + 7x + 3$$

Fill in the area representations of the binomial products, and write the resulting quadratic expressions on the lines provided.

1. $(2x + 1)(x + 4)$

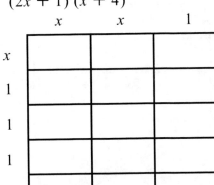

_____

3. $(x + 3)(2x + 2)$

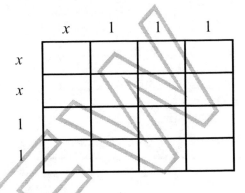

_____

2. $(3x + 1)(x + 2)$

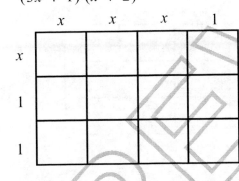

_____

4. $(2x + 1)^2$

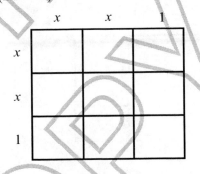

_____

Label the rows and columns of the area representations, fill in the grids, and write the resulting quadratic expressions of the binomials products.

5. $(x + 3)(x + 3)$

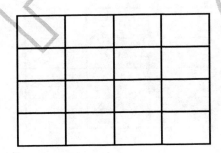

_____

6. $(x + 2)(4x + 1)$

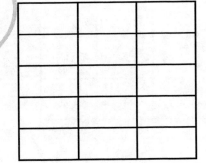

_____

7. $(2x + 1)(2x + 3)$

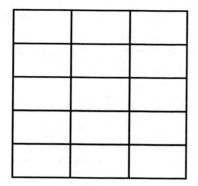

_____

8. $(x + 2)(2x + 3)$

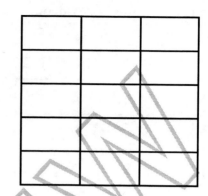

_____

Label the rows and columns of the area representations, fill in the grids, and write in the binomial products on the lines provided.

9. $x^2 + 3x + 2$

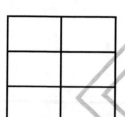

_____

10. $x^2 + 4x + 4$

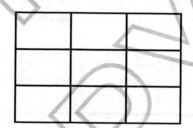

_____

Use the area representations to find the binomial and quadratic expressions, and write your answers in the lines provided.

11.

x^2	x^2	x
x	x	1
x	x	1

_____

12.

x^2	x	x
x	1	1
x	1	1
x	1	1
x	1	1

_____

138

Find the quadratic expressions by making an area representation for each of the binomial pairs.

13. $(x + 2)^2$

16. $(4x + 3)(x + 2)$

19. $(x + 1)^2$

14. $(x + 1)(x + 4)$

17. $(3x + 3)(2x + 2)$

20. $(x + 5)(3x + 1)$

15. $(x + 5)(x + 3)$

18. $(2x + 3)^2$

21. $(x + 4)^2$

Determine the areas of the rectangles or selected shaded regions using area representation. Write your answers in the lines below. (Area = length × width)

22. Find the area of the shaded region.

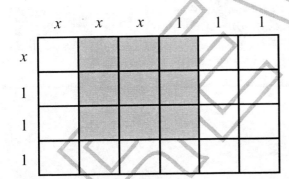

_____

23. Find the area of the rectangle.

$3x + 1$

x

_____

24. Find the area of the shaded region.

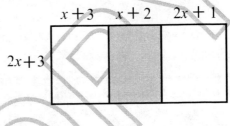

_____

25. Find the area of the rectangle.

$2x + 1$

$x + 1$

_____

MULTIPLYING TWO BINOMIALS USING THE FOIL METHOD

When you multiply two binomials such as $(x + 6)(x - 5)$, you must multiply each term in the first binomial by each term in the second binomial. The easiest way is to use the **FOIL** method. If you can remember the word **FOIL**, it can help you keep order when you multiply. The "F" stands for **first**, "O" stands for **outside**, "I" stands for **inside**, and "L" stands for **last**.

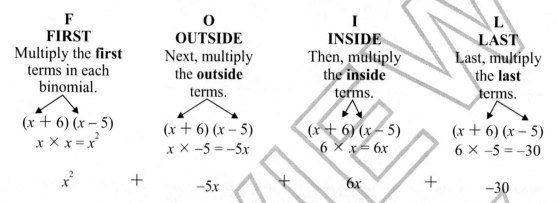

F	**O**	**I**	**L**
FIRST	**OUTSIDE**	**INSIDE**	**LAST**
Multiply the **first** terms in each binomial.	Next, multiply the **outside** terms.	Then, multiply the **inside** terms.	Last, multiply the **last** terms.
$(x + 6)(x - 5)$	$(x + 6)(x - 5)$	$(x + 6)(x - 5)$	$(x + 6)(x - 5)$
$x \times x = x^2$	$x \times -5 = -5x$	$6 \times x = 6x$	$6 \times -5 = -30$
$x^2 +$	$-5x +$	$6x +$	-30

Now, just combine like terms, $6x - 5x = x$, and write your answer.
$(x + 6)(x + 5) = x^2 + x - 30$

Note: It is customary for mathematicians to write polynomials in descending order. That means that the term with the highest number exponent comes first in a polynomial. The next highest exponent is second, and so on. When you use the **FOIL** method, the terms will always be in the customary order. You just need to combine like terms and write your answer.

Multiply the following binomials.

1. $(y - 7)(y + 3)$

2. $(2x + 4)(x + 9)$

3. $(4b - 3)(3b - 4)$

4. $(6g + 2)(g - 9)$

5. $(7k - 5)(-4k - 3)$

6. $(8v - 2)(3v + 4)$

7. $(10p + 2)(4p + 3)$

8. $(3h - 9)(-2h - 5)$

9. $(w - 4)(w - 7)$

10. $(6x + 1)(x - 2)$

11. $(5t + 3)(2t - 1)$

12. $(4y - 9)(4y + 9)$

13. $(a + 6)(3a + 5)$

14. $(3z - 8)(z - 4)$

15. $(5c + 2)(6c + 5)$

16. $(y + 3)(y - 3)$

27. $(8b - 8)(2b - 1)$

38. $(2x + 8)(2x - 3)$

17. $(2w - 5)(4w + 6)$

28. $(z + 3)(3z + 5)$

39. $(k - 1)(6k + 12)$

18. $(7x + 1)(x - 4)$

29. $(7y - 5)(y - 3)$

40. $(3w + 11)(2w + 2)$

19. $(6t - 9)(4t - 4)$

30. $(9x + 5)(3x - 1)$

41. $(8y - 10)(5y - 3)$

20. $(5b + 6)(6b + 2)$

31. $(3t + 1)(t + 10)$

42. $(6d + 13)(d - 1)$

21. $(2z + 1)(10z + 4)$

32. $(2w - 9)(8w + 7)$

43. $(7h + 3)(2h + 4)$

22. $(11w - 8)(w + 3)$

33. $(8s - 2)(s + 4)$

44. $(5n + 9)(5n - 5)$

23. $(5d - 9)(9d + 9)$

34. $(4k - 1)(8k + 9)$

45. $(6z + 5)(z - 8)$

24. $(9g + 2)(g - 2)$

35. $(h + 12)(h - 2)$

46. $(4p + 5)(2p - 9)$

25. $(4p + 7)(2p + 3)$

36. $(3x + 7)(7x + 3)$

47. $(b + 2)(5b + 7)$

26. $(m + 5)(m - 5)$

37. $(2v - 6)(2v + 6)$

48. $(9y - 3)(8y - 7)$

SIMPLIFYING EXPRESSIONS WITH EXPONENTS

EXAMPLE 1: **Simplify** $(2a + 5)^2$

When you simplify an expression such as $(2a + 5)^2$, it is best to write the expression as two binomials and use FOIL to simplify.

$(2a + 5)^2 = (2a + 5)(2a + 5)$

Using FOIL we have $4a^2 + 10a + 10a + 25 = 4a^2 + 20a + 25$

EXAMPLE 2: **Simplify** $4(3a + 2)^2$

Using order of operations, we must simplify the exponent first.
$= 4(3a + 2)(3a + 2)$
$= 4(9a^2 + 6a + 6a + 4)$
$= 4(9a^2 + 12a + 4)$ Now multiply by 4.
$= 4(9a^2 + 12a + 4) = 36a^2 + 48a + 16$

Multiply the following binomials.

1. $(y + 3)^2$

2. $7(2x + 4)^2$

3. $6(4b - 3)^2$

4. $5(6g + 2)^2$

5. $(-4k - 3)^2$

6. $3(-2h - 5)^2$

7. $-2(8v - 2)^2$

8. $(10p + 2)^2$

9. $6(-2h - 5)^2$

10. $6(w - 7)^2$

11. $2(6x + 1)^2$

12. $(9x + 2)^2$

13. $(5t + 3)^2$

14. $3(4y - 9)^2$

15. $8(a + 6)^2$

16. $4(3z - 8)^2$

17. $3(5c + 2)^2$

18. $4(3t + 9)^2$

CHAPTER 9 REVIEW

Simplify:

1. $3a^2 + 9a^2$

2. $-6z^2(z+3)$

3. $(4b^2)(5b^3)$

4. $7x^2 - 9x^2$

5. $(5p-4)-(3p+2)$

6. $-5t(3t+9)^2$

7. $3(2g+3)^2$

8. $14d^4 - 9d^4$

9. $(7w-4)(w-8)$

10. $15t^2 + 4t^2$

11. $(7c^4)(9c^2)$

12. $(9x+2)(x+5)$

13. $4y(4y^2 - 9y + 2)$

14. $(5w^6)(9w^9)$

15. $8x^3 + 12x^3$

16. $15p^5 - 11p^5$

17. $(4d+9)(2d+7)$

18. $4w(-3w^2 + 7w - 5)$

19. $24z^6 - 10z^6$

20. $-7y^3 - 8y^3$

21. $(7x^4)(7x^5)$

22. $17p^2 + 9p^2$

23. $4(6y - 5)^2$

24. $(3c^2)(6c^8)$

25. Add $2x^2 + 9x$ and $5x^2 - 8x + 2$

26. $4t(6t^2 + 4t - 6) + 8t(3t + 3)$

27. Subtract $y^2 + 4y - 6$ from $3y^2 + 7$

28. $2x(4x^2 + 6x - 3) + 4x(x + 3)$

29. $(6t - 4) - (6t^2 + t - 2)$

30. $(4x + 6) + (7x^2 - 2x + 3)$

31. Subtract $5a - 2$ from $a + 9$

32. $(-2y + 4) + (4y - 6)$

33. $2t(t + 6) - 5t(2t + 7)$

34. Add $3c - 4$ and $c^2 - 3c - 2$

35. $2b(b - 4) - (b^2 + 2b + 1)$

36. $(6k^2 + 5k) + (k^2 + k + 9)$

Chapter 10 | Factoring Polynomials

In a multiplication problem, the numbers multiplied together are called **factors**. The answer to a multiplication problem is called the **product**.

$$5 \times 4 = 20$$

factors   product

If we reverse the problem $20 = 5 \times 4$, we say we have **factored** 20 into 5×4.

In this chapter, we will factor **polynomials**.

EXAMPLE: Find the greatest common factor of $2y^3 + 6y^2$.

Step 1: Look at the whole numbers. The greatest common factor of 2 and 6 is 2. Factor the 2 out of each term.

$2(y^3 + 3y^2)$

Step 2: Look at the remaining terms, $y^3 + 3y^2$. What are the common factors of each term?

$y^3 = y \times \boxed{y \times y}$
$3y^2 = 3 \times \boxed{y \times y}$ ← common factors $= y^2$

Step 3: Factor 2 and y^2 out of each term: $2y^2(y + 3)$

Check: $2y^2(y + 3) = 2y^3 + 6y^2$

Find the greatest common factor of each of the following.

1. $6x^4 + 18x^2$

2. $14y^3 + 7y$

3. $4b^5 + 12b^3$

4. $10a^3 + 5$

5. $2y^3 + 8y^2$

6. $6x^4 - 12x^2$

7. $18y^2 - 12y$

8. $15a^3 - 25a^2$

9. $4x^3 + 16x^2$

10. $6b^2 + 21b^5$

11. $27m^3 + 18m^4$

12. $100x^4 - 25x^3$

13. $4b^4 - 12b^3$

14. $18c^2 + 24c$

15. $20y^3 + 30y^5$

16. $16x^2 - 24x^5$

17. $15a^4 - 25a^2$

18. $24b^3 + 16b^6$

19. $36y^4 + 9y^2$

20. $42x^3 + 49x$

Factoring larger polynomials with 3 or 4 terms works the same way.

EXAMPLE: $4x^5 + 16x^4 + 12x^3 + 8x^2$

Step 1: Find the greatest common factor of the whole numbers. 4 can be divided evenly into 4, 16, 12, and 8; therefore, 4 is the greatest common factor.

$$4x^5 + 16x^4 + 12x^3 + 8x^2 = 4(x^5 + 4x^4 + 3x^3 + 2x^2)$$

Step 2: Next, find the greatest common factor of the variables. x^5, x^4, x^3, and x^2 can each be divided by x^2, the lowest power of x in each term.

$$4x^5 + 16x^4 + 12x^3 + 8x^2 = 4x^2(x^3 + 4x^2 + 3x + 2)$$

Factor each of the following polynomials.

1. $5a^3 + 15a^2 + 20a$

2. $18y^4 + 6y^3 + 24y^2$

3. $12x^5 + 21x^3 + x^2$

4. $6b^4 + 3b^3 + 15b^2$

5. $14c^3 + 28c^2 + 7c$

6. $15b^4 - 5b^2 + 20b$

7. $t^3 + 3t^2 - 5t$

8. $8a^3 - 4a^2 + 12a$

9. $16b^5 - 12b^4 - 20b^2$

10. $20x^4 + 16x^3 - 24x^2 + 28x$

11. $40b^7 + 30b^5 - 50b^3$

12. $20y^4 - 15y^3 + 30y^2$

13. $4m^5 + 8m^4 + 12m^3 + 6m^2$

14. $16x^5 + 20x^4 - 12x^3 + 24x^2$

15. $18y^4 + 21y^3 - 9y^2$

16. $3n^5 + 9n^3 + 12n^2 + 15n$

17. $4d^6 - 8d^2 + 2d$

18. $10w^2 + 4w + 2$

19. $6t^3 - 3t^2 + 9t$

20. $25p^5 - 10p^3 - 5p^2$

21. $18x^4 + 9x^2 - 36x$

22. $6b^4 - 12b^2 - 6b$

23. $y^3 + 3y^2 - 9y$

24. $10x^5 - 2x^4 + 4x^2$

Find the greatest common factor of $4a^3b^2 - 6a^2b^2 + 2a^4b^3$.

Step 1:   The greatest common factor of the whole numbers is 2.

$$4a^3b^2 - 6a^2b^2 + 2a^4b^3 = 2(2a^3b^2 - 3a^2b^2 + a^4b^3)$$

Step 2:   Next, find the lowest power of each variable that is in each term. Factor them out of each term. The lowest power of a is a^2. The lowest power of b is b^2.

$$4a^3b^2 - 6a^2b^2 + 2a^4b^3 = 2a^2b^2(2a - 3 + a^2b)$$

Factor each of the following polynomials.

1. $3a^2b^2 - 6a^3b^4 + 9a^2b^3$

2. $12x^4y^3 + 18x^3y^4 - 24x^3y^3$

3. $20x^2y - 25x^3y^3$

4. $12x^2y - 20x^2y^2 + 16xy^2$

5. $8a^3b + 12a^2b + 20a^2b^3$

6. $36c^4 + 42c^3 + 24c^2 - 18c$

7. $14m^3n^4 - 28m^3n^2 + 42m^2n^3$

8. $16x^4y^2 - 24x^3y^2 + 12x^2y^2 - 8xy^2$

9. $32c^3d^4 - 56c^2d^3 + 64c^4d^2$

10. $21a^4b^3 + 27a^2b^3 + 15a^3b^2$

11. $4w^3t^2 + 6w^2t - 8wt^2$

12. $5pq^3 - 2p^2q^2 - 9p^3q$

13. $49x^3t^3 + 7xt^2 - 14xt^3$

14. $9cd^4 - 3d^4 - 6c^2d^3$

15. $12a^2b^3 - 14ab + 10ab^2$

16. $25x^4 + 10x - 20x^2$

17. $bx^3 - b^2x^2 + b^3x$

18. $4k^3a^2 + 22ka + 16k^2a^2$

19. $33w^4y^2 - 9w^3y^2 + 24w^2y^2$

20. $18x^3 - 9x^5 + 27x^2$

FACTORING BY GROUPING

Not all polynomials have a common factor in each term. In this case, they may sometimes be factored by grouping.

EXAMPLE: Factor $ab + 4a + 2b + 8$.

Step 1: Factor an a from the first two terms and a 2 from the last two terms.

$a(b + 4) + 2(b + 4)$

Now the polynomial has two terms, $a(b + 4)$ and $2(b + 4)$. Notice $(b + 4)$ is a factor of each term.

Step 2: Factor out the common factor of each term:

$ab + 4a + 2b + 8 = (b + 4)(a + 2)$.

Check: Multiply using the FOIL method to check.

$(b + 4)(a + 2) = ab + 4a + 2b + 8$

Factor the following polynomials by grouping.

1. $xy + 4x + 2y + 8$

2. $cd + 5c + 4d + 20$

3. $xy - 4x + 6y - 24$

4. $ab + 6a + 3b + 18$

5. $ab + 3a - 5b - 15$

6. $xy - 2x + 6y - 12$

7. $cd + 4c + 4d + 16$

8. $mn - 5m + 3n - 15$

9. $ab + 4a + 3b + 12$

10. $xy + 7x - 4y - 28$

11. $ab - 2a + 8b - 16$

12. $cd + 4c - 5d - 20$

13. $mn + 6m - 2n - 12$

14. $xy - 9x - 3y + 27$

15. $bc - 3b + 5c - 15$

16. $ab + a + 7b + 7$

17. $xy + 4y + 2y + 8$

18. $cd + 9c - d - 9$

19. $ab + 2a - 7b - 14$

20. $xy - 6x - 2y + 12$

21. $wz + 6z - 4w - 24$

From the previous page, you may have noticed that after finding the greatest common factor of the first two terms, whatever was left in parentheses was a factor of the second two terms. This is true of second degree (having an exponent of 2 in one of the terms) equations as well.

EXAMPLE: Factor $3x^2 + 4x - 3xy - 4y$.

Step 1: Find the greatest common factor of the first two terms. In $3x^2 + 4x$, both terms have an x. x can be factored out, so we have $x(3x + 4)$.

Step 2: Find the greatest common factor of the last two terms. In $-3xy - 4y$, both terms have a $-y$. $-y$ can be factored out, so we have $-y(3x + 4)$.

Step 3: You now have $x(3x + 4) - y(3x + 4)$. $(3x + 4)$ is a common factor in both terms, so you can factor it out.

$$3x^2 + 4x - 3xy - 4y = (3x + 4)(x - y)$$

Check: Use the FOIL method to check the answer.
$$(3x + 4)(x - y) = 3x^2 + 4x - 3xy - 4y$$

Factor the following polynomials by grouping.

1. $3ax^2 + 3ay + 4x^2 + 4y$

2. $2cx^2 + 2cy + 5y + 5x^2$

3. $y^2 + 4y - 5y - 20$

4. $3xy - 9x + 2y^2 - 6y$

5. $b^2 + 2b - 3b - 6$

6. $x^2 + bx - 2x - 2b$

7. $8c^3 + 5c^2 - 32c - 20$

8. $6y^2 + 4y + 9y + 6$

9. $3a^3 + 4a^2 + 6a + 8$

10. $5b^3 - 10b^2 + 3b - 6$

11. $a^3 + 6a - 2a^2 - 12$

12. $7x^3 + 4x^2 - 21x - 12$

13. $4y^3 - 12y^2 + 2y - 6$

14. $6x^2 + 3xy - 2x - y$

15. $2b^3 - 5b^2 + 8b - 20$

16. $3a^3 + 4a^2 - 6a - 8$

17. $10y^3 + 20y^2 + 18y + 36$

18. $27x^2 + 6x + 45x + 10$

When polynomials have only one variable, they can also be factored by grouping. First, however, they must be arranged in descending order. In the example below, you could try to factor it by grouping without rearranging it in descending order, but you will see it does not work.

EXAMPLE: $2n^3 - 8 + 8n^2 - 2n$

Step 1: Arrange in descending order (exponents go from highest to lowest).

$$2n^3 - 8 + 8n^2 - 2n = 2n^3 + 8n^2 - 2n - 8$$

Step 2: Factor by grouping the first two terms and the last two terms.

$$2n^3 + 8n^2 - 2n - 8 = 2n^2(n + 4) - 2(n + 4) = (2n^2 - 2)(n + 4)$$

Check: Check using the FOIL method.

$$(2n^2 - 2)(n + 4) = 2n^3 + 8n^2 - 2n - 8$$

It is correct. The terms are the same as the terms in the original problem above. Only the order of the terms is different.

Factor the following polynomials by grouping. Be sure to arrange terms in descending order first.

1. $a^3 - 3 - 3a^2 + a$

2. $3c^2 - 4c + c^3 - 12$

3. $x^3 - 28 - 4x^2 + 7x$

4. $-8 + y^3 - y + 8y^2$

5. $b^3 - 15 - 5b^2 + 3b$

6. $d^3 + 20 - 4d - 5d^2$

7. $-3y^2 - 18 + y^3 + 6y$

8. $x^3 - 2x + 5x^2 - 10$

9. $-2y^2 - 3y + 6 + y^3$

10. $6a^2 - 3a - 18 + a^3$

11. $b^3 - 5 + b - 5b^2$

12. $c^3 - 14 - 7c + 2c^2$

13. $3d^2 - 4d - 12 + d^3$

14. $12 + a^3 + 6a + 2a^2$

15. $x^3 - 20 + 4x^2 - 5x$

16. $y^3 - 8y - 8 + y^2$

17. $b^3 - 6 - 3b^2 + 2b$

18. $-7 - c + c^3 + 7c^2$

FINDING THE NUMBERS

The next kind of factoring we will do requires thinking of two numbers with a certain sum and a certain product.

EXAMPLE: Which two numbers have a sum of 8 and a product of 12? In other words, what pair of numbers would answer both equations?

$$\underline{\hspace{1cm}} + \underline{\hspace{1cm}} = 8 \text{ and } \underline{\hspace{1cm}} \times \underline{\hspace{1cm}} = 12$$

You may think $4 + 4 = 8$, but 4×4 does not equal 12.
Or you may think $7 + 1 = 8$, but 7×1 does not equal 12.

$6 + 2 = 8$ and $6 \times 2 = 12$, so 6 and 2 are the pair of numbers that will work in both equations.

For each problem below, find one pair of numbers that will solve both equations.

1. $\underline{\hspace{1cm}} + \underline{\hspace{1cm}} = 14$ and $\underline{\hspace{1cm}} \times \underline{\hspace{1cm}} = 40$

2. $\underline{\hspace{1cm}} + \underline{\hspace{1cm}} = 10$ and $\underline{\hspace{1cm}} \times \underline{\hspace{1cm}} = 21$

3. $\underline{\hspace{1cm}} + \underline{\hspace{1cm}} = 18$ and $\underline{\hspace{1cm}} \times \underline{\hspace{1cm}} = 81$

4. $\underline{\hspace{1cm}} + \underline{\hspace{1cm}} = 12$ and $\underline{\hspace{1cm}} \times \underline{\hspace{1cm}} = 20$

5. $\underline{\hspace{1cm}} + \underline{\hspace{1cm}} = 7$ and $\underline{\hspace{1cm}} \times \underline{\hspace{1cm}} = 12$

6. $\underline{\hspace{1cm}} + \underline{\hspace{1cm}} = 8$ and $\underline{\hspace{1cm}} \times \underline{\hspace{1cm}} = 15$

7. $\underline{\hspace{1cm}} + \underline{\hspace{1cm}} = 10$ and $\underline{\hspace{1cm}} \times \underline{\hspace{1cm}} = 25$

8. $\underline{\hspace{1cm}} + \underline{\hspace{1cm}} = 14$ and $\underline{\hspace{1cm}} \times \underline{\hspace{1cm}} = 48$

9. $\underline{\hspace{1cm}} + \underline{\hspace{1cm}} = 12$ and $\underline{\hspace{1cm}} \times \underline{\hspace{1cm}} = 36$

10. $\underline{\hspace{1cm}} + \underline{\hspace{1cm}} = 17$ and $\underline{\hspace{1cm}} \times \underline{\hspace{1cm}} = 72$

11. $\underline{\hspace{1cm}} + \underline{\hspace{1cm}} = 15$ and $\underline{\hspace{1cm}} \times \underline{\hspace{1cm}} = 56$

12. $\underline{\hspace{1cm}} + \underline{\hspace{1cm}} = 9$ and $\underline{\hspace{1cm}} \times \underline{\hspace{1cm}} = 18$

13. $\underline{\hspace{1cm}} + \underline{\hspace{1cm}} = 13$ and $\underline{\hspace{1cm}} \times \underline{\hspace{1cm}} = 40$

14. $\underline{\hspace{1cm}} + \underline{\hspace{1cm}} = 16$ and $\underline{\hspace{1cm}} \times \underline{\hspace{1cm}} = 63$

15. $\underline{\hspace{1cm}} + \underline{\hspace{1cm}} = 10$ and $\underline{\hspace{1cm}} \times \underline{\hspace{1cm}} = 16$

16. $\underline{\hspace{1cm}} + \underline{\hspace{1cm}} = 8$ and $\underline{\hspace{1cm}} \times \underline{\hspace{1cm}} = 16$

17. $\underline{\hspace{1cm}} + \underline{\hspace{1cm}} = 9$ and $\underline{\hspace{1cm}} \times \underline{\hspace{1cm}} = 20$

18. $\underline{\hspace{1cm}} + \underline{\hspace{1cm}} = 13$ and $\underline{\hspace{1cm}} \times \underline{\hspace{1cm}} = 36$

19. $\underline{\hspace{1cm}} + \underline{\hspace{1cm}} = 15$ and $\underline{\hspace{1cm}} \times \underline{\hspace{1cm}} = 50$

20. $\underline{\hspace{1cm}} + \underline{\hspace{1cm}} = 11$ and $\underline{\hspace{1cm}} \times \underline{\hspace{1cm}} = 30$

Now that you have mastered positive numbers, take up the challenge of finding pairs of negative numbers or pairs where one number is negative and one is positive.

EXAMPLE: Which two numbers have a sum of −3 and a product of −40? In other words, what pair of numbers would answer both equations?

$$\underline{\hspace{1cm}} + \underline{\hspace{1cm}} = -3 \text{ and } \underline{\hspace{1cm}} \times \underline{\hspace{1cm}} = -40$$

It is faster to look at the factors of 40 first. 8 and 5 and 10 and 4 are possibilities. 8 and 5 have a difference of 3, and in fact, $5 + (-8) = -3$ and $5 \times (-8) = -40$. This pair of numbers, 5 and −8, will satisfy both equations.

For each problem below, find one pair of numbers that will solve both equations.

1. $\underline{\hspace{1cm}} + \underline{\hspace{1cm}} = -2$ and $\underline{\hspace{1cm}} \times \underline{\hspace{1cm}} = -35$

2. $\underline{\hspace{1cm}} + \underline{\hspace{1cm}} = 4$ and $\underline{\hspace{1cm}} \times \underline{\hspace{1cm}} = -5$

3. $\underline{\hspace{1cm}} + \underline{\hspace{1cm}} = 4$ and $\underline{\hspace{1cm}} \times \underline{\hspace{1cm}} = -12$

4. $\underline{\hspace{1cm}} + \underline{\hspace{1cm}} = -6$ and $\underline{\hspace{1cm}} \times \underline{\hspace{1cm}} = 8$

5. $\underline{\hspace{1cm}} + \underline{\hspace{1cm}} = 3$ and $\underline{\hspace{1cm}} \times \underline{\hspace{1cm}} = -40$

6. $\underline{\hspace{1cm}} + \underline{\hspace{1cm}} = 10$ and $\underline{\hspace{1cm}} \times \underline{\hspace{1cm}} = -11$

7. $\underline{\hspace{1cm}} + \underline{\hspace{1cm}} = 6$ and $\underline{\hspace{1cm}} \times \underline{\hspace{1cm}} = -27$

8. $\underline{\hspace{1cm}} + \underline{\hspace{1cm}} = 8$ and $\underline{\hspace{1cm}} \times \underline{\hspace{1cm}} = -20$

9. $\underline{\hspace{1cm}} + \underline{\hspace{1cm}} = -5$ and $\underline{\hspace{1cm}} \times \underline{\hspace{1cm}} = -24$

10. $\underline{\hspace{1cm}} + \underline{\hspace{1cm}} = -3$ and $\underline{\hspace{1cm}} \times \underline{\hspace{1cm}} = -28$

11. $\underline{\hspace{1cm}} + \underline{\hspace{1cm}} = -2$ and $\underline{\hspace{1cm}} \times \underline{\hspace{1cm}} = -48$

12. $\underline{\hspace{1cm}} + \underline{\hspace{1cm}} = -1$ and $\underline{\hspace{1cm}} \times \underline{\hspace{1cm}} = -20$

13. $\underline{\hspace{1cm}} + \underline{\hspace{1cm}} = -3$ and $\underline{\hspace{1cm}} \times \underline{\hspace{1cm}} = 2$

14. $\underline{\hspace{1cm}} + \underline{\hspace{1cm}} = 1$ and $\underline{\hspace{1cm}} \times \underline{\hspace{1cm}} = -30$

15. $\underline{\hspace{1cm}} + \underline{\hspace{1cm}} = -7$ and $\underline{\hspace{1cm}} \times \underline{\hspace{1cm}} = 12$

16. $\underline{\hspace{1cm}} + \underline{\hspace{1cm}} = 6$ and $\underline{\hspace{1cm}} \times \underline{\hspace{1cm}} = -16$

17. $\underline{\hspace{1cm}} + \underline{\hspace{1cm}} = 5$ and $\underline{\hspace{1cm}} \times \underline{\hspace{1cm}} = -24$

18. $\underline{\hspace{1cm}} + \underline{\hspace{1cm}} = -4$ and $\underline{\hspace{1cm}} \times \underline{\hspace{1cm}} = 4$

19. $\underline{\hspace{1cm}} + \underline{\hspace{1cm}} = -1$ and $\underline{\hspace{1cm}} \times \underline{\hspace{1cm}} = -42$

20. $\underline{\hspace{1cm}} + \underline{\hspace{1cm}} = -6$ and $\underline{\hspace{1cm}} \times \underline{\hspace{1cm}} = 8$

FACTORING TRINOMIALS

In the previous chapter, you multiplied binomials (two terms) together, and the answer was a trinomial (three terms).

For example, $(x + 6)(x - 5) = x^2 + x - 30$

Now, you need to practice factoring a trinomial into two binomials.

EXAMPLE 1:   Factor $x^2 + 6x + 8$

Step 1:   When the trinomial is in descending order as in the example above, you need to find a pair of numbers in which the sum of the two numbers equals the number in the second term, while the product of the two numbers equals the third term. In the above example, find the pair of numbers that has a sum of 6 and a product of 8.

$$\underline{} + \underline{} = 6 \quad \text{and} \quad \underline{} \times \underline{} = 8$$

The pair of numbers that satisfy both equations is 4 and 2.

Step 2:   Use the pair of numbers in the binomials.

The factors of $x^2 + 6x + 8$ are $(x + 4)(x + 2)$

Check: To check, use the FOIL method.

$(x + 4)(x + 2) = x^2 + 4x + 2x + 8 = x^2 + 6x + 8$

Notice, when the second term and the third term of the trinomial are both positive, both numbers in the solution pair are positive.

EXAMPLE 2:   Factor $x^2 - x - 6$ Find the pair of numbers where ...

the sum is −1 and the product is −6

$$\underline{} + \underline{} = -1 \quad \text{and} \quad \underline{} \times \underline{} = -6$$

The pair of numbers that satisfies both equations is 2 and −3.

The factors of $x^2 - x - 6$ are $(x + 2)(x - 3)$.

Notice, if the third term is negative, one number in the solution pair is positive, and the other number is negative.

EXAMPLE 3:   Factor $x^2 - 7x + 12$ Find the pair of numbers where ...

the sum is −7 and the product is 12.

$$\underline{} + \underline{} = -7 \quad \text{and} \quad \underline{} \times \underline{} = 12$$

The pair of numbers that satisfies both equations is −3 and −4.

The factors of $x^2 - 7x + 12$ are $(x - 3)(x - 4)$.

Notice, if the second term of a trinomial is negative and the third term is positive, both numbers in the solution pair are negative.

Find the factors of the following trinomials.

1. $x^2 - x - 2$

2. $y^2 + y - 6$

3. $w^2 + 3w - 4$

4. $t^2 + 5t + 6$

5. $x^2 + 2x - 8$

6. $k^2 - 4k + 3$

7. $t^2 + 3t - 10$

8. $x^2 - 3x - 4$

9. $y^2 - 5y + 6$

10. $y^2 + y - 20$

11. $a^2 - a - 6$

12. $b^2 - 4b - 5$

13. $c^2 - 5c - 14$

14. $c^2 - c - 12$

15. $d^2 + d - 6$

16. $x^2 - 3x - 28$

17. $y^2 + 3y - 18$

18. $a^2 - 9a + 20$

19. $b^2 - 2b - 15$

20. $c^2 + 7c - 8$

21. $t^2 - 11t + 30$

22. $w^2 + 13w + 36$

23. $m^2 - 2m - 48$

24. $y^2 + 14y + 49$

25. $x^2 + 7x + 10$

26. $a^2 - 7a + 6$

27. $d^2 - 6d - 27$

Sometimes a trinomial has a greatest common factor which must be factored out first.

EXAMPLE : Factor $4x^2 + 8x - 32$

Step 1: Begin by factoring out the greatest common factor, 4.

$$4(x^2 + 2x - 8)$$

Step 2: Factor by finding a pair of numbers whose sum is 2 and product is −8.
4 and −2 will work, so

$$4(x^2 + 2x - 8) = 4(x + 4)(x - 2)$$

Check: Multiply to check. $4(x + 4)(x - 2) = 4x^2 + 8x - 32$

Factor the following trinomials. Be sure to factor out the greatest common factor first.

1. $2x^2 + 6x + 4$

2. $3y^2 - 9y + 6$

3. $2a^2 + 2a - 12$

4. $4b^2 + 28b + 40$

5. $3y^2 - 6y - 9$

6. $10x^2 + 10x - 200$

7. $5c^2 - 10c - 40$

8. $6d^2 + 30d - 36$

9. $4x^2 + 8x - 60$

10. $6a^2 - 18a - 24$

11. $5b^2 + 40b + 75$

12. $3c^2 - 6c - 24$

13. $2x^2 - 18x + 28$

14. $4y^2 - 20y + 16$

15. $7a^2 - 7a - 42$

16. $6b^2 - 18b - 60$

17. $11d^2 + 66d + 88$

18. $3x^2 - 24x + 45$

In the following problems, instead of factoring out just a whole number first, you need to factor out a whole number with a variable or a whole number with a variable and an exponent. Study the following two examples.

EXAMPLE 1: $4a^3 - 4a^2 - 24a = 4a(a^2 - a - 6) = 4a(a - 3)(a + 2)$

EXAMPLE 2: $y^4 + 3y^3 - 4y^2 = y^2(y^2 + 3y - 4) = y^2(y - 1)(y + 4)$

Factor the following trinomials.

1. $x^4 - x^3 - 12x^2$

2. $3c^3 - 6c^2 - 24c$

3. $5b^3 + 10b^2 - 40b$

4. $3y^4 - 9y^3 - 12y^2$

5. $2x^5 + 8x^4 - 10x^3$

6. $6d^3 + 24d^2 + 24d$

7. $2y^3 - 16y^2 + 32y$

8. $6b^4 - 18b^3 - 60b^2$

9. $a^3 - 3a^2 - 4a$

10. $4x^4 + 4x^3 - 24x^2$

11. $y^5 - y^4 - 42y^3$

12. $b^4 + 11b^3 + 24b^2$

13. $4c^3 - 4c^2 - 48c$

14. $11a^4 + 33a^3 + 22a^2$

15. $2x^5 + 2x^4 - 112x^3$

16. $10d^3 - 70d^2 - 180d$

17. $4y^6 + 4y^5 - 24y^4$

18. $2a^3 - 14a^2 + 20a$

19. $6b^5 - 24b^4 + 18b^3$

20. $x^6 + 2x^5 + x^4$

21. $5d^4 - 35d^3 + 50d^2$

22. $a^3 + 3a^2 - 54a$

23. $3y^3 - 42y^2 + 147y$

24. $8x^3 + 24x^2 + 16x$

Some trinomials have a whole number in front of the first term that cannot be factored out of the trinomial. The trinomial can still be factored.

EXAMPLE : Factor $2x^2 + 5x - 3$

Step 1: To get a product of $2x^2$, one factor must begin with $2x$ and the other with x.

$(2x \quad) (x \quad)$

Step 2: Now think: What two numbers give a product of -3? The two possibilities are 3 and -1 or -3 and 1. We know they could be in any order so there are 4 possible arrangements.

$(2x + 3) (x - 1)$
$(2x - 3) (x + 1)$
$(2x + 1) (x - 3)$
$(2x - 1) (x + 3)$

Step 3: Multiply each possible answer until you find the arrangement of the numbers that works. Multiply the outside terms and the inside terms and add them together to see which one will equal $5x$.

$(2x + 3) (x - 1) = 2x^2 + x - 3$
$(2x - 3) (x + 1) = 2x^2 - x - 3$
$(2x + 1) (x - 3) = 2x^2 - 5x - 3$
$(2x - 1) (x + 3) = 2x^2 + 5x - 3$ ← —————— This arrangement works, so:

The factors of $2x^2 + 5x - 3$ are $(2x - 1) (x + 3)$.

Alternative: You can do some of the multiplying in your head. For the above example, ask yourself the following question: What two numbers give a product of -3 and give a sum of 5 (the whole number in the second term) when one of the numbers is first multiplied by 2 (the whole number in front of the first term)? The pair of numbers, -1 and 3, have a product of -3 and a sum of 5 when the 3 is first multiplied by 2. Therefore, the 3 will go opposite the factor with the $2x$ so that when the terms are multiplied, you get -5.

You can use this method to at least narrow down the possible pairs of numbers when you have several from which to choose.

Factor the following trinomials.

1. $3y^2 + 14y + 8$

2. $5a^2 + 24a - 5$

3. $7b^2 + 30b + 8$

4. $2c^2 - 9c + 9$

5. $2y^2 - 7y - 15$

6. $3x^2 + 4x + 1$

7. $7y^2 + 13y - 2$

8. $11a^2 + 35a + 6$

9. $5y^2 + 17y - 12$

10. $3a^2 + 4a - 7$

11. $2a^2 + 3a - 20$

12. $5b^2 - 13b - 6$

13. $3y^2 - 4y - 32$

14. $2x^2 - 17x + 36$

15. $11x^2 - 29x - 12$

16. $5c^2 + 2c - 16$

17. $7y^2 - 30y + 27$

18. $2x^2 - 3x - 20$

19. $5b^2 + 24b - 5$

20. $7d^2 + 18d + 8$

21. $3x^2 - 20x + 25$

22. $2a^2 - 7a - 4$

23. $5m^2 + 12m + 4$

24. $9y^2 - 5y - 4$

25. $2b^2 - 13b + 18$

26. $7x^2 + 31x - 20$

27. $3c^2 - 2c - 21$

There can be more than 4 possible arrangements to try if the whole number in the first term can be factored two or more ways. Factoring this kind of trinomial can only be done by trial and error.

EXAMPLE : Factor $6x^2 - 5x - 4$

The 6 can be factored as 6×1 or as 2×3.
The last term, -4, can be factored as 2×-2, -2×2, 4×-1, or -4×1.
The possible combinations are:

$$(6x - 4)(x + 1) = 6x^2 + 2x - 4$$

$$(6x + 4)(x - 1) = 6x^2 - 2x - 4$$

$$(6x - 2)(x + 2) = 6x^2 + 10x - 4$$

$$(6x + 2)(x - 2) = 6x^2 - 10x - 4$$

$$(6x - 1)(x + 4) = 6x^2 + 23x - 4$$

$$(6x + 1)(x - 4) = 6x^2 - 23x - 4$$

$$(2x - 4)(3x + 1) = 6x^2 - 10x - 4$$

$$(2x + 4)(3x - 1) = 6x^2 + 10x - 4$$

$$(2x - 2)(3x + 2) = 6x^2 - 2x - 4$$

$$(2x + 2)(3x - 2) = 6x^2 + 2x - 4$$

$$(2x - 1)(3x + 4) = 6x^2 + 5x - 4$$

$$(2x + 1)(3x - 4) = 6x^2 - 5x - 4$$

You can see from these wrong guesses that changing the signs of the factors only changes the sign of the middle number. It will not give you the -5 you are looking for. Knowing this can eliminate half the guesses.

This arrangement works so:

the factors of $6x^2 - 5x - 4$ are $(2x + 1)(3x - 4)$

If you find the factors that give you the right middle number for the middle term, but the sign is wrong, then switch the signs of your factors.

159

Find the factors of the following trinomials.

1. $4a^2 + 29a + 30$

2. $6x^2 + 17x - 3$

3. $6y^2 - y - 2$

4. $20a^2 + a - 12$

5. $8b^2 - 7b - 18$

6. $4x^2 - 4x - 35$

7. $10y^2 + 29y - 21$

8. $12a^2 - 8a - 7$

9. $12c^2 + 17c - 5$

10. $14a^2 + 31a + 15$

11. $6x^2 + 29x + 28$

12. $21y^2 + 13y + 2$

13. $27h^2 - 39h - 10$

14. $30a^2 + 37a - 4$

15. $20w^2 + 19w + 3$

16. $36x^2 + 19x - 6$

17. $36y^2 - 36y + 5$

18. $42b^2 - 59b + 20$

19. $10x^2 - 3x - 18$

20. $12x^2 + 29x - 8$

21. $4y^2 + 4y - 24$

22. $10b^2 + 42b + 8$

23. $8a^2 - 34a + 30$

24. $12c^2 - 13c - 4$

25. $10x^2 - 14x - 12$

26. $18d^2 + 36d + 10$

27. $9y^2 - 76y + 32$

160

FACTORING TRINOMIALS WITH TWO VARIABLES

Some trinomials have two variables with exponents. These trinomials can still be factored.

EXAMPLE : Factor $x^2 + 5xy + 6y^2$

Step 1: Notice there is an x^2 in the 1st term and a y^2 in the last term. When you see two different terms that are squared, you know there has to be an x and a y in each factor:

$(x \quad y)(x \quad y)$

Step 2: Now think: What are two numbers whose sum is 5 and product is 6? You see that 3 and 2 will work. Put 3 and 2 in the factors:

$(x + 3y)(x + 2y)$

Check: Multiply to check. $(x + 3y)(x + 2y) = x^2 + 3xy + 2xy + 6y^2 = x^2 + 5xy + 6y^2$

Factor the following trinomials.

1. $a^2 + 6ab + 8b^2$

2. $x^2 + 3xy - 4y^2$

3. $c^2 - 2cd - 15d^2$

4. $g^2 + 7gh + 10h^2$

5. $a^2 - 5ab + 6b^2$

6. $c^2 - cd - 30d^2$

7. $x^2 + 5xy - 24y^2$

8. $a^2 - 4ab + 4b^2$

9. $c^2 - 11cd + 30d^2$

10. $x^2 - 6xy + 8y^2$

11. $g^2 - gh - 42h^2$

12. $a^2 - ab - 20b^2$

13. $x^2 + 12xy + 32y^2$

14. $c^2 + 3cd - 40d^2$

15. $x^2 + 6xy - 27y^2$

16. $a^2 - 2ab - 48b^2$

17. $c^2 - 3cd - 28d^2$

18. $x^2 + xy - 6y^2$

FACTORING THE DIFFERENCE OF TWO SQUARES

The product of a term and itself is called a **perfect square**.

25 is a perfect square because $5 \times 5 = 25$
49 is a perfect square because $7 \times 7 = 49$

Any variable with an even exponent is a perfect square.

y^2 is a perfect square because $y \times y = y^2$
y^4 is a perfect square because $y^2 \times y^2 = y^4$

When two terms that are both perfect squares are subtracted, factoring those terms is very easy. To factor the difference of perfect squares, you use the square root of each term, a plus sign in the first factor, and a minus sign in the second factor.

EXAMPLE 1 : Factor $4x^2 - 9$ The example has two terms which are both perfect squares, and the terms are subtracted.

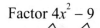

Step 1: $(2x \quad 3\,)(2x \quad 3\,)$ Find the square root of each term. Use the square roots in each of the factors.

Step 2: 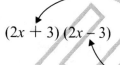 $(2x + 3)(2x - 3)$ Use a plus sign in one factor and a minus sign in the other factor.

Check: Multiply to check. $(2x + 3)(2x - 3) = 4x^2 - 6x + 6x - 9 = 4x^2 - 9$

The inner and outer terms add to zero.

EXAMPLE 2: Factor $81y^4 - 1$.

Step 1: 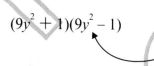 $(9y^2 + 1)(9y^2 - 1)$ Factor like the example above. Notice, the second factor is also the difference of two perfect squares.

Step 2: $(9y^2 + 1)(3y + 1)(3y - 1)$ Factor the second term further.
Note: You cannot factor the *sum* of two perfect squares.

Check: Multiply in reverse to check your answer.
$(9y^2 + 1)(3y + 1)(3y - 1) = (9y^2 + 1)(9y^2 - 3y + 3y - 1) = (9y^2 + 1)(9y^2 - 1)$
$(9y^2 + 1)(9y^2 - 1) = 81y^4 - 9y^2 + 9y^2 - 1 = 81y^4 - 1$

Factor the following differences of perfect squares.

1. $64x^2 - 49$

2. $4y^4 - 25$

3. $9a^4 - 4$

4. $25c^4 - 9$

5. $64y^2 - 9$

6. $x^4 - 16$

7. $49x^2 - 4$

8. $4d^2 - 25$

9. $9a^2 - 16$

10. $100y^4 - 49$

11. $c^4 - 36$

12. $36x^2 - 25$

13. $25x^2 - 4$

14. $9x^4 - 64$

15. $49x^2 - 100$

16. $16x^2 - 81$

17. $9y^4 - 1$

18. $64c^2 - 25$

19. $25d^2 - 64$

20. $36a^4 - 49$

21. $16x^4 - 16$

22. $b^2 - 25$

23. $c^4 - 144$

24. $9y^2 - 4$

25. $81x^4 - 16$

26. $4b^2 - 36$

27. $9w^2 - 9$

28. $64a^2 - 25$

29. $49y^2 - 121$

30. $x^6 - 9$

CHECKLIST FOR FACTORING POLYNOMIALS

When factoring polynomials, follow these steps:

1. Check to see if it has a greatest common factor that can be factored out first.

    $$4x^2 + 4xy - 24y^2 = 4(x^2 + xy - 6y^2)$$

2. See if it is the difference of two squares. Remember, the sum of two squares cannot be factored.

3. If a polynomial has more than 3 terms, try factoring by grouping.

4. If none of the above are possibilities, factor by trial and error.

EXAMPLE 1:   Factor $50y^2 - 18$

First, factor out the greatest common factor.

$$2(25y^2 - 9)$$

Second, factor the difference of 2 squares.

$$2(5y + 3)(5y - 3)$$

EXAMPLE 2:   Factor $24xy - 30y - 8x + 10$

First, factor out the greatest common factor.

$$2(12xy - 15y - 4x + 5)$$

Second, factor by grouping.

$$2 \times [3y(4x - 5) - 1(4x - 5)]$$

$$2(3y - 1)(4x - 5)$$

In example 2, you could factor by grouping first. You would get $(4x - 5)(6y - 2)$. Then, you could factor out 2 from the second factor to get $2(4x - 5)(3y - 1)$, the same answer as above.

EXAMPLE 3:   Factor $4a^2 + 26a + 12$

First, factor out the greatest common factor.

$$2(2a^2 + 13a + 6)$$

Now, factor by trial and error.

$$2(2a + 1)(a + 6)$$

Factor completely the following polynomials.

1. $20a^2 + 10a - 10$

2. $2y^2 - 6y - 20$

3. $2x^2 + 2x - 40$

4. $3x^2 + 3bx - 6x - 6b$

5. $8b^2 - 20b - 12$

6. $4ab + 16a + 8b + 32$

7. $2ab + 6a - 10b - 30$

8. $12c^2 + 18c - 12$

9. $9a^2 - 36$

10. $3x^2 + 6x + 3$

11. $6x^2 + 18x + 12$

12. $36x^2 - 9$

13. $2x^2 - 18x + 40$

14. $2xy + 8x + 6y + 24$

15. $10y^2 + 14y - 12$

16. $16x^2 - 16$

17. $5b^2 + 20b - 25$

18. $4d^2 - 64$

19. $12a^2 + 36a - 48$

20. $36a^4 - 9$

21. $8x^2 + 4xy - 12x - 6y$

22. $16b^4 - 81$

23. $15y^2 + 30y - 15xy - 30x$

24. $2a^3 + 2a^2 - 4a$

25. $3w^2y + 15wy - 6w^3 - 30w^2$

26. $8x^3 - 2x$

27. $6b^2 + 5b - 25$

28. $4x^2y + 8x - 4xy^2 - 8y$

29. $5x^4 + 20x^2$

30. $6a^2x + 15ax - 6ax^2 - 15a^2$

SIMPLIFYING ALGEBRAIC RATIOS

Sometimes algebraic expressions are written as ratios. We will use what we learned so far in this chapter to factor the terms in the numerator and the denominator when possible, then simplify the ratio.

EXAMPLE : Simplify $\dfrac{c^2 - 25}{c^2 + 5c}$

Step 1: The numerator is the difference of two perfect squares, so it can be easily factored as in the previous section. Use the square root of each of the terms in the parentheses, with a plus sign in one and a minus sign in the other.

$$c^2 - 25 = (c - 5)(c + 5)$$

Step 2: Find the greatest common factor in the denominator and factor it out. In this case, it is the variable c.

$$c^2 + 5c = c\,(c + 5)$$

Step 3: Simplify $\dfrac{c^2 - 25}{c^2 + 5b} = \dfrac{(c - 5)(c + 5)}{c\,(c + 5)} = \dfrac{(c - 5)}{c}$

Simplify the following algebraic ratios. Check for perfect squares and common factors.

1. $\dfrac{25x^2 - 4}{5x^2 - 2x}$

2. $\dfrac{64c^2 - 25}{8c^2 + 5c}$

3. $\dfrac{36a^2 - 49}{6a^2 - 7a}$

4. $\dfrac{x^2 - 9}{x^2 + 3x}$

5. $\dfrac{9a^2 - 16}{3a^2 - 4a}$

6. $\dfrac{16x^2 - 81}{4x^2 - 9x}$

7. $\dfrac{49x^2 - 100}{7x^2 + 10x}$

8. $\dfrac{x^4 - 16}{x^2 + 2x}$

9. $\dfrac{4y^2 - 36}{2y^2 + 6y}$

10. $\dfrac{81y^4 - 16}{9y^2 + 4}$

11. $\dfrac{25x^4 - 225}{5x^2 + 15}$

12. $\dfrac{3y^3 + 9}{y^9 - 9}$

CHAPTER 10 REVIEW

Factor the following polynomials completely.

1. $8x - 18$

2. $6x^2 - 18x$

3. $16b^3 + 8b$

4. $15a^3 + 40$

5. $20y^6 - 12y^4$

6. $5a - 15a^2$

7. $4y^2 - 36$

8. $25a^4 - 49b^2$

9. $3ax + 3ay + 4x + 4y$

10. $ax - 2x + ay - 2y$

11. $2bx + 2x - 2by - 2y$

12. $2b^2 - 2b - 12$

13. $7x^3 + 14x - 3x^2 - 6$

14. $3a^3 + 4a^2 + 9a + 12$

15. $27y^2 + 42y - 5$

16. $12b^2 + 25b - 7$

17. $c^2 + cd - 20d^2$

18. $x^2 - 4xy - 21y^2$

19. $6y^2 + 30y + 36$

20. $2b^2 + 6b - 20$

21. $16b^4 - 81d^4$

22. $9w^2 - 54w - 63$

23. $m^2p^2 - 5mp + 2m^2p - 10m$

24. $12x^2 + 27x$

25. $2xy - 36 + 8y - 9x$

26. $2a^4 - 32$

27. $21c^2 + 41c + 10$

28. $x^2 - y + xy - x$

29. $2b^3 - 24 + 16b - 3b^2$

30. $5 - 2a - 25a^2 + 10a^3$

USING FACTORING

In the previous chapter, we factored polynomials such as $y^2 - 4y - 5$ into two factors:

$$y^2 - 4y - 5 = (y + 1)(y - 5)$$

In this chapter, we learn that any equation that can be put in the form $ax^2 + bx + c = 0$ is a quadratic equation if a, b, and c are real numbers and $a \neq 0$. $ax^2 + bx + c = 0$ is the standard form of a quadratic equation. To solve these equations, follow the steps below.

EXAMPLE: Solve $y^2 - 4y - 5 = 0$.

Step 1: Factor the left side of the equation.

$$y^2 - 4y - 5 = 0$$
$$(y + 1)(y - 5) = 0$$

Step 2: If the product of these two factors equals zero, then the two factors individually must be equal to zero. Therefore, to solve, we set each factor equal to zero.

$$
\begin{array}{ll}
(y + 1) = 0 & (y - 5) = 0 \\
\underline{\quad -1 \quad -1} & \underline{\quad +5 \quad +5} \\
y = -1 & y = 5
\end{array}
$$

The equation has two solutions: $y = -1$ and $y = 5$.

Check: To check, substitute each solution into the original equation.
When $y = -1$, the equation becomes:
$$(-1)^2 - (4)(-1) - 5 = 0$$
$$1 + 4 - 5 = 0$$
$$0 = 0$$

When $y = 5$, the equation becomes:
$$5^2 - (4)(5) - 5 = 0$$
$$25 - 20 - 5 = 0$$
$$0 = 0$$

Both solutions produce true statements.
The solution set for the equation is $\{-1, 5\}$. The solutions of a quadratic equation are also called the zeros of the function. The graph of a quadratic equation is a parabola. When a parabola is graphed, the zeros are the x-intercepts. In the example shown above, the x-intercepts of the graph will be at $(-1, 0)$ and $(5, 0)$.

Solve each of the following quadratic equations by factoring and setting each factor equal to zero. Check by substituting answers back in the original equation.

1. $x^2 + x - 6 = 0$

2. $y^2 - 2y - 8 = 0$

3. $a^2 + 2a - 15 = 0$

4. $y^2 - 5y + 4 = 0$

5. $b^2 - 9b + 14 = 0$

6. $x^2 - 3x - 4 = 0$

7. $y^2 + y - 20 = 0$

8. $d^2 + 6d + 8 = 0$

9. $y^2 - 7y + 12 = 0$

10. $x^2 - 3x - 28 = 0$

11. $a^2 - 5a + 6 = 0$

12. $b^2 + 3b - 10 = 0$

13. $a^2 + 7a - 8 = 0$

14. $c^2 + 3c + 2 = 0$

15. $x^2 - x - 42 = 0$

16. $a^2 + a - 6 = 0$

17. $b^2 + 7b + 12 = 0$

18. $y^2 + 2y - 15 = 0$

19. $a^2 - 3a - 10 = 0$

20. $d^2 + 10d + 16 = 0$

21. $x^2 - 4x - 12 = 0$

Quadratic equations that have a whole number and a variable in the first term are solved the same way as the previous page. Factor the trinomial, and set each factor equal to zero to find the solution set.

EXAMPLE: Solve $2x^2 + 3x - 2 = 0$

$(2x - 1)(x + 2) = 0$

Set each factor equal to zero and solve:

$$2x - 1 = 0$$
$$\underline{+1 \quad +1}$$
$$\frac{2x}{2} = \frac{1}{2}$$
$$x = \tfrac{1}{2}$$

$$x + 2 = 0$$
$$\underline{-2 \quad -2}$$
$$x = -2$$

The solution set is $\{\tfrac{1}{2}, -2\}$

Solve the following quadratic equations.

1. $3y^2 + 12y + 32 = 0$

2. $5c^2 - 2c - 16 = 0$

3. $7d^2 + 18d + 8 = 0$

4. $3a^2 - 10a - 8 = 0$

5. $11x^2 - 31x - 6 = 0$

6. $5b^2 + 17b + 6 = 0$

7. $3x^2 - 11x - 20 = 0$

8. $5a^2 + 47a - 30 = 0$

9. $2c^2 - 5c - 25 = 0$

10. $2y^2 + 11y - 21 = 0$

11. $5a^2 + 23a - 42 = 0$

12. $3d^2 + 11d - 20 = 0$

13. $3x^2 - 10x + 8 = 0$

14. $7b^2 + 23b - 20 = 0$

15. $9a^2 - 58a + 24 = 0$

16. $4c^2 - 25c - 21 = 0$

17. $8d^2 + 53d + 30 = 0$

18. $4y^2 - 29y + 30 = 0$

19. $8a^2 + 37a - 15 = 0$

20. $3x^2 - 41x + 26 = 0$

21. $8b^2 + 2b - 3 = 0$

Sometimes quadratic equations are not in standard form. They are not already set equal to zero. These equations must first be put in standard form in order to solve.

EXAMPLE 1: $y^2 - 5y = -4$

Step 1:   Add 4 to both sides so the equation will be set equal to 0.

$$\begin{array}{r} y^2 - 5y = -4 \\ +4 +4 \\ \hline y^2 - 5y + 4 = 0 \end{array}$$

Step 2:   Factor the left side of the equation.

$$y^2 - 5y + 4 = 0$$
$$(y - 4)(y - 1) = 0$$

Step 3:   Set each factor equal to 0 and solve.

$$\begin{array}{r} y - 4 = 0 \\ +4 +4 \\ \hline y = 4 \end{array} \qquad \begin{array}{r} y - 1 = 0 \\ +1 +1 \\ \hline y = 1 \end{array}$$

The solution set is $\{4, 1\}$.

EXAMPLE 2: $y^2 - 11y + 10 = -14$

Step 1:   Add 14 to both sides so the equation will be set equal to 0.

$$\begin{array}{r} y^2 - 11y + 10 = -14 \\ +14 +14 \\ \hline y^2 - 11y + 24 = 0 \end{array}$$

Step 2:   Factor: $(y - 8)(y - 3) = 0$

Step 3:   Set each factor equal to 0 and solve for y.

$$y - 8 = 0 \qquad\qquad y - 3 = 0$$
$$y = 8 \qquad\qquad\qquad y = 3$$

The solution set is $\{8, 3\}$.

EXAMPLE 3: $x^2 = x + 20$

Step 1:   Add $-x - 20$ to both sides so the equation will equal to zero.

$$\begin{array}{r} x^2 = x + 20 \\ \underline{-x - 20 \quad -x - 20} \\ x^2 - x - 20 = 0 \end{array}$$

Step 2:   Factor the left side of the equation.

$$x^2 - x - 20 = 0$$
$$(x + 4)(x - 5) = 0$$

Step 3:   Set each factor equal to 0 and solve.

$$\begin{array}{r} x + 4 = 0 \\ \underline{-4 \ -4} \\ x = -4 \end{array} \qquad \begin{array}{r} x - 5 = 0 \\ \underline{+5 +5} \\ x = 5 \end{array}$$

The solution set is $\{-4, 5\}$.

EXAMPLE 4: $8x^2 = -42x - 10$

Step 1:   Add $42x + 10$ to both sides so the equation will equal to zero.
to

$$\begin{array}{r} 8x^2 = -42x - 10 \\ \underline{+42x + 10 \quad +42x + 10} \\ 8x^2 + 42x + 10 = 0 \end{array}$$

Step 2:   Factor the left side of the equation.

$$8x^2 + 42x + 10 = 0$$
$$(8x + 2)(x + 5) = 0$$

Step 3:   Set each factor equal to 0 and solve.

$$\begin{array}{r} 8x + 2 = 0 \\ \underline{-2 \ -2} \\ \frac{8x}{8} = \frac{-2}{8} \end{array} \qquad \begin{array}{r} x + 5 = 0 \\ \underline{-5 \ -5} \\ x = -5 \end{array}$$

$$x = -\frac{1}{4}$$

The solution set is $\{-\frac{1}{4}, -5\}$.

Solve each of the quadratic equations below. Put each in standard form first by setting the equation equal to zero.

1. $x^2 - x = 12$

2. $y^2 + 2y = 15$

3. $b^2 - 4b - 2 = 10$

4. $c^2 - 11c = -28$

5. $a^2 + 2a = 35$

6. $y^2 - 10y + 5 = -11$

7. $x^2 + 10x = -24$

8. $d^2 - 7d - 1 = 17$

9. $4x^2 = 4x + 15$

10. $10b^2 = 37b + 12$

11. $2y^2 = -9y - 18$

12. $c^2 = -4c + 21$

13. $t^2 = -10t - 16$

14. $4x^2 = -8x + 5$

15. $3y^2 = 10y - 3$

16. $4b^2 = 3b + 1$

17. $a^2 - 5a + 8 = 14$

18. $b^2 - b = 20$

19. $c^2 + 9c = -14$

20. $y^2 - 6y - 3 = 13$

21. $x^2 - 3x = 18$

22. $d^2 - 5d + 1 = 37$

23. $x^2 + 2x = 24$

24. $y^2 - 11y - 4 = -28$

25. $5a^2 = 13a - 6$

26. $c^2 = 5c + 36$

27. $t^2 = 11t - 24$

28. $6x^2 = -11x - 3$

29. $y^2 = 2y + 35$

30. $p^2 = 7p + 18$

31. $w^2 = 14w - 48$

32. $m^2 = -10m - 25$

33. $x^2 - 4x - 10 = 35$

34. $a^2 + 2a = 8$

35. $b^2 - 7b + 2 = 10$

36. $y^2 + y + 7 = 13$

37. $k^2 + 14k = -40$

38. $p^2 + 7p + 8 = 16$

39. $x^2 - 3x - 5 = -7$

40. $y^2 - 3y = 18$

41. $5x^2 = -27x - 10$

42. $6y^2 = -11y + 7$

43. $a^2 = 8a - 12$

44. $b^2 = -13b - 42$

45. $k^2 = -16k - 63$

46. $4m^2 = -9m + 9$

47. $y^2 = 2y + 8$

48. $6x^2 = 7x - 2$

SOLVING THE DIFFERENCE OF TWO SQUARES

To solve the difference of two squares, first factor. Then set each factor equal to zero.

EXAMPLE: $25x^2 - 36 = 0$

Step 1: Factor the left hand side of the equation.

$$25x^2 - 36 = 0$$
$$(5x + 6)(5x - 6) = 0$$

Step 2: Set each factor equal to zero and solve.

$$5x + 6 = 0 \qquad\qquad 5x - 6 = 0$$
$$\underline{-6 \quad -6} \qquad\qquad \underline{+6 \quad +6}$$
$$\frac{5x}{5} = \frac{-6}{5} \qquad\qquad \frac{5x}{5} = \frac{6}{5}$$

$$x = \frac{-6}{5} \qquad\qquad x = \frac{6}{5}$$

Check: Substitute each solution in the equation to check.

for $x = -\dfrac{6}{5}$:

$$25x^2 - 36 = 0$$

$$25\left(-\frac{6}{5}\right)\left(-\frac{6}{5}\right) - 36 = 0 \leftarrow \text{Substitute } \frac{-6}{5} \text{ for } x.$$

$$25\left(\frac{36}{25}\right) - 36 = 0 \leftarrow \text{Cancel the 25's.}$$

$$36 - 36 = 0 \leftarrow \text{A true statement. } x = \frac{-6}{5} \text{ is a solution.}$$

for $x = \dfrac{6}{5}$:

$$25x^2 - 36 = 0$$

$$25\left(\frac{6}{5}\right)\left(\frac{6}{5}\right) - 36 = 0 \leftarrow \text{Substitute } \frac{6}{5} \text{ for } x.$$

$$25\left(\frac{36}{25}\right) - 36 = 0 \leftarrow \text{Cancel the 25's.}$$

$$36 - 36 = 0 \leftarrow \text{A true statement. } x = \frac{6}{5} \text{ is a solution.}$$

The solution set is $\{-\frac{6}{5}, \frac{6}{5}\}$.

Find the solution sets for the following.

1. $25a^2 - 16 = 0$

2. $c^2 - 36 = 0$

3. $9x^2 - 64 = 0$

4. $100y^2 - 49 = 0$

5. $4b^2 - 81 = 0$

6. $d^2 - 25 = 0$

7. $9x^2 - 1 = 0$

8. $16a^2 - 9 = 0$

9. $36y^2 - 1 = 0$

10. $36y^2 - 25 = 0$

11. $d^2 - 16 = 0$

12. $64b^2 - 9 = 0$

13. $81a^2 - 4 = 0$

14. $64y^2 - 25 = 0$

15. $4c^2 - 49 = 0$

16. $x^2 - 81 = 0$

17. $49b^2 - 9 = 0$

18. $a^2 - 64 = 0$

19. $9x^2 - 1 = 0$

20. $4y^2 - 9 = 0$

21. $t^2 - 100 = 0$

22. $16k^2 - 81 = 0$

23. $81a^2 - 4 = 0$

24. $36b^2 - 16 = 0$

SOLVING PERFECT SQUARES

When the square root of a constant, variable, or polynomial results in a constant, variable, or polynomial without irrational numbers, the expression is a **perfect square**. Some examples are 49, x^2, and $(x-2)^2$.

EXAMPLE 1:   Solve the perfect square for x. $(x-5)^2 = 0$

Step 1:   Take the square root of both sides.
$\sqrt{(x-5)^2} = \sqrt{0}$
$(x-5) = 0$

Step 2:   Solve the equation.
$(x-5) = 0$
$x - 5 + 5 = 0 + 5$
$x = 5$

EXAMPLE 2:   Solve the perfect square for x. $(x-5)^2 = 64$

Step 1:   Take the square root of both sides.
$\sqrt{(x-5)^2} = \sqrt{64}$
$(x-5) = \pm 8$
$(x-5) = 8$ and $(x-5) = -8$

Step 2:   Solve the two equations.
$(x-5) = 8$ and $(x-5) = -8$
$x - 5 + 5 = 8 + 5$ and $x - 5 + 5 = -8 + 5$
$x = 13$ **and** $x = -3$

Solve the perfect square for x.

1. $(x-5)^2 = 0$

2. $(x+1)^2 = 0$

3. $(x+11)^2 = 0$

4. $(x-4)^2 = 0$

5. $(x-1)^2 = 0$

6. $(x+8)^2 = 0$

7. $(x+3)^2 = 4$

8. $(x-5)^2 = 16$

9. $(x-10)^2 = 100$

10. $(x+9)^2 = 9$

11. $(x-4.5)^2 = 25$

12. $(x+7)^2 = 36$

13. $(x+2)^2 = 49$

14. $(x-1)^2 = 4$

15. $(x+8.9)^2 = 49$

16. $(x-6)^2 = 81$

17. $(x-12)^2 = 121$

18. $(x+2.5)^2 = 64$

COMPLETING THE SQUARE

"Completing the square" is another way of factoring a quadratic equation. To complete the square, the equation must be converted into a perfect square.

EXAMPLE 1:   Solve $x^2 - 10x + 9 = 0$ by completing the square.

Completing the square:

Step 1:   The first step is to get the constant on the other side of the equation. Subtract 9 from both sides:

$x^2 - 10x + 9 - 9 = -9$

$x^2 - 10x = -9$

Step 2:   Determine the coefficient of the x. The coefficient in this example is 10. Substitute the coefficient of x into the equation $(\frac{p}{2})^2$.

$(10 \div 2)^2 = 5^2 = 25$

Step 3:   Add the $(\frac{p}{2})^2$ value, 25, to both sides:

$x^2 - 10x + 25 = -9 + 25$

$x^2 - 10x + 25 = 16$

Step 4:   Now factor the $x^2 - 10x + 25$ into a perfect square:

$(x - 5)^2 = 16$

Solving the perfect square:

Step 5:   Take the square root of both sides.

$\sqrt{(x-5)^2} = \sqrt{16}$

$(x - 5) = \pm 4$

$(x - 5) = 4$ and $(x - 5) = -4$

Step 6:   Solve the two equations.

$(x - 5) = 4$ and $(x - 5) = -4$

$x - 5 + 5 = 4 + 5$ and $x - 5 + 5 = -4 + 5$

$\mathbf{x = 9}$ and $\mathbf{x = 1}$

Solve for x by completing the square.

1. $x^2 + 2x - 3 = 0$ 6. $x^2 - 4x = 0$ 11. $x^2 - 16x + 60 = 0$

2. $x^2 - 8x + 7 = 0$ 7. $x^2 + 12x + 27 = 0$ 12. $x^2 - 8x - 48 = 0$

3. $x^2 + 6x - 7 = 0$ 8. $x^2 + 2x - 24 = 0$ 13. $x^2 + 24x + 44 = 0$

4. $x^2 - 16x - 36 = 0$ 9. $x^2 + 12x - 85 = 0$ 14. $x^2 + 6x + 5 = 0$

5. $x^2 - 14x + 49 = 0$ 10. $x^2 - 8x + 15 = 0$ 15. $x^2 - 11x + 5.25 = 0$

USING THE QUADRATIC FORMULA

You may be asked to use the quadratic formula to solve an algebra problem known as a **quadratic equation**. The equation should be in the form $ax^2 + bx + c = 0$.

EXAMPLE: Using the quadratic formula, find x in the following equation: $x^2 - 8x = -7$.

Step 1: Make sure the equation is set equal to 0.

$$\begin{array}{r} x^2 - 8x = -7 \\ +7 = +7 \\ \hline x^2 - 8x + 7 = 0 \end{array}$$

The quadratic formula, $\dfrac{-b \pm \sqrt{b^2 - 4ac}}{2a}$, will be given to you on your formula sheet with your test.

Step 2:

In the formula, a is the number x^2 is multiplied by, b is the number x is multiplied by, and c is the last term of the equation. For the equation in the example, $x^2 - 8x + 7$, $a = 1$, $b = -8$, and $c = 7$. When we look at the formula we notice a $\pm$ sign. This means there will be two solutions to the equation, one when we use the plus sign and one when we use the minus sign. Substituting the numbers from the problem into the formula, we have:

$$\frac{8 + \sqrt{8^2 - (4)(1)(7)}}{2(1)} = 7 \qquad \text{And} \qquad \frac{8 - \sqrt{8^2 - (4)(1)(7)}}{2(1)} = 1$$

The solutions are (7, 1)

For each of the following equations, use the quadratic formula to find two solutions.

1. $x^2 + x - 6 = 0$

2. $y^2 - 2y - 8 = 0$

3. $a^2 + 2a - 15 = 0$

4. $y^2 - 5y + 4 = 0$

5. $b^2 - 9b + 14 = 0$

6. $x^2 - 3x - 4 = 0$

7. $y^2 + y - 20 = 0$

8. $d^2 + 6d + 8 = 0$

9. $y^2 - 7y + 12 = 0$

10. $x^2 - 3x - 28 = 0$

11. $a^2 - 5a + 6 = 0$

12. $b^2 + 3b - 10 = 0$

13. $a^2 + 7a - 8 = 0$

14. $c^2 + 3c + 2 = 0$

15. $x^2 - x - 42 = 0$

16. $a^2 + a - 6 = 0$

17. $b^2 + 7b + 12 = 0$

18. $y^2 + 2y - 15 = 0$

19. $a^2 - 3a - 10 = 0$

20. $d^2 + 10d + 16 = 0$

21. $x^2 - 4x - 12 = 0$

CHAPTER 11 REVIEW

Factor and solve each of the following quadratic equations.

1. $16b^2 - 25 = 0$

2. $a^2 - a - 30 = 0$

3. $x^2 - x = 6$

4. $100x^2 - 49 = 0$

5. $81y^2 = 9$

6. $y^2 = 21 - 4y$

7. $y^2 - 7y + 8 = 16$

8. $6x^2 + x - 2 = 0$

9. $3y^2 + y - 2 = 0$

10. $b^2 + 2b - 8 = 0$

11. $4x^2 + 19x - 5 = 0$

12. $8x^2 = 6x + 2$

13. $2y^2 - 6y - 20 = 0$

14. $-6x^2 + 7x - 2 = 0$

15. $y^2 + 3y - 18 = 0$

Using the quadratic formula, find both solutions for the variable.

16. $x^2 + 10x - 11 = 0$

17. $y^2 - 14y + 40 = 0$

18. $b^2 + 9b + 18 = 0$

19. $y^2 - 12y - 13 = 0$

20. $a^2 - 8a - 48 = 0$

21. $x^2 + 2x - 63 = 0$

Using Formulas

SUBSTITUTING NUMBERS IN FORMULAS

You will also be expected to know how to substitute numbers for variables (letters) in formulas.

EXAMPLE: Area of a parallelogram: $A = b \times h$

Find the area of the parallelogram if $b = 20$ cm and $h = 10$ cm

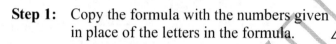

Step 1: Copy the formula with the numbers given in place of the letters in the formula.

$$A = 20 \times 10$$

Step 2: Solve the problem. $A = 20 \times 10 = 200$. Therefore, $A = 200$ cm². Solve the following problems using the formulas given.

1. The volume of a rectangular pyramid is determined by using the following formula:

$$V = \frac{lwh}{3}$$

l = Length of the base of the pyramid
w = Width of the base of the pyramid
h = Height of the pyramid

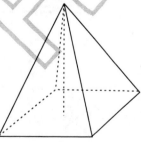

Find the volume of the pyramid if $l = 6$ in, $w = 6$ in, and $h = 11$ in.

2. Find the volume of a cone with a radius of 30 inches and a height of 60 inches using the formula:

$$V = \frac{1}{3}\pi r^2 h \qquad \pi = 3.14$$

3. Lumber is measured by the following formula:

$$\text{Number of board feet} = \frac{LWT}{12}$$

L = Length of the board in feet
W = Width of the board in feet
T = Thickness of the board in feet

Find the number of board feet if $L = 14$ feet, $W = 8$ feet, and $T = 6$ feet

4. The perimeter of a square is figured by the formula $P = 4s$ where

P = Perimeter (the distance around the outside)
s = length of one side

Find the perimeter if $s = 6$.

5. What is the circumference of a circle with a diameter of 8 cm?

$C = \pi d$ $\qquad$ $\pi = \textbf{3.14}$

---

6. Find the area of the trapezoid.

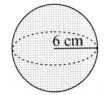

$A = \dfrac{1}{2} h\,(a + b)$

$a = 11$ in
$b = 23$ in
$h = 18$ in

---

7. Find the volume of a sphere with a radius of 6 cm. $\pi = 3.14$

$V = \dfrac{4}{3} \pi r^3$

6 cm

---

8. Find the area of the following ellipse given by the equation: $A = \pi ab$

$\pi = 3.14$
$a = 2$ cm
$b = 4$ cm

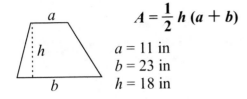

---

9. The formula for changing from degrees Fahrenheit to degrees Celsius is:

$$C = \dfrac{5\,(F - 32)}{9}$$

If it is 68°F outside, how many degrees Celsius is it?

---

10. Find the volume. $V = \dfrac{4}{3} \pi r^3$

$\pi = \textbf{3.14}$

8 cm

---

11. Louise has a cone-shaped mold to make candles. The diameter of the base is 10 cm, and it is 13 cm tall. How many cubic centimeters of liquid wax will it hold?

$\pi = \textbf{3.14}$

$V = \dfrac{1}{3} \pi r^2 h$

---

12. To convert from degrees Celsius to degrees Fahrenheit, use the following formula:

$$F = \dfrac{9C}{5} + 32$$

If it is 15° C outside, what is the temperature in degrees Fahrenheit?

---

13. Lumber is measured with the following formula:

$$\textbf{Number of board feet} = \dfrac{LWT}{12}$$

L = Length of the board in feet
W = Width of the board in feet
T = Thickness of the board in feet

Find the number of board feet if

L= 12 feet, W = 6 feet, and
T = 6 feet

FINDING THE DISTANCE BETWEEN TWO POINTS

To find the distance between any two points on a Cartesian plane, use the following formula:

$$d = \sqrt{(y_2 - y_1)^2 + (x_2 - x_1)^2}$$

EXAMPLE: Find the distance between $(-2, 1)$ and $(3, -4)$.

Plugging the values from the ordered pairs into the formula, we find:

$$d = \sqrt{(-4 - 1)^2 + [3 - (-2)]^2}$$

$$d = \sqrt{(-5)^2 + (5)^2}$$

$$d = \sqrt{25 + 25} = \sqrt{50}$$

To simplify, we look for perfect squares that are factors of 50.
$50 = 25 \times 2$. Therefore,

$$d = \sqrt{25} \times \sqrt{2}$$

$$d = 5\sqrt{2}$$

Find the distance between the following pairs of points using the distance formula above.

1. $(6, -1)\ (5, 2)$

2. $(-4, 3)\ (2, -1)$

3. $(10, 2)\ (6, -1)$

4. $(-2, 5)\ (-4, 3)$

5. $(8, -2)\ (3, -9)$

6. $(2, -2)\ (8, 1)$

7. $(3, 1)\ (5, 5)$

8. $(-2, -1)\ (3, 4)$

9. $(5, -3)\ (-1, -5)$

10. $(6, 5)\ (3, -4)$

11. $(-1, 0)\ (-9, -8)$

12. $(-2, 0)\ (-6, 6)$

13. $(2, 4)\ (8, 10)$

14. $(-10, -5)\ (2, -7)$

15. $(-3, 6)\ (1, -1)$

MANIPULATING FORMULAS AND EQUATIONS

Sometimes you are given a formula such as $A = l \times w$ (A = area, l = length, and w = width) and you need to solve for w. For example: The area of a playground is 4500 square feet. The length is 600 feet. What is the width of the playground? Starting with $A = l \times w$, you need to solve for w. You need to have w on one side of the equation and all the other variables on the other.

$$A = l \times w \Rightarrow \frac{A}{l} = \frac{\cancel{l} \times w}{\cancel{l}} \Rightarrow \frac{A}{l} = w \Rightarrow w = \frac{A}{l}$$ You have solved $A = l \times w$ for w.

Solve each of the following formulas and equations for the given variable.

1. $C = 2\pi r$ for r.

2. $I = PRT$ for r.

3. $V = \pi r^2 h$ for π.

4. $A = \frac{1}{2}bh$ for h.

5. $d = 4a + 3c$ for c

6. $h = 6a + 9c^2$ for a

7. $y = 4xz$ for z

8. $5t = 9y + 22$ for y

9. $17 - 9m = n - 23$ for n

10. $7x + 4 = \frac{9y}{4}$ for y

11. $8 + 2a = 5b - 6$ for b

12. $A = s^2$ for s

13. $a^2 + b^2 = c^2$ for a

14. $I = PRT$ for P

15. $x = 4a + 7$ for a

16. $9 - 5y = 6x + 2$ for x

17. $D = rt$ for r

18. $A = lw$ for w

19. $a^2 + b^2 = c^2$ for b^2

20. $C = 2dr$ for r

21. $V = \pi r^2 h$ for h

22. $V = \frac{1}{3}bh$ for b

23. $A = \pi r^2$ for r

24. $S = 4\pi r^2$ for r

25. $y = \frac{1}{4}x + 5$ for x

26. $x = -\frac{1}{5}y - 3$ for y

27. $a = \frac{b}{3}$ for b

28. $c = 3d + \frac{2}{5}$ for d

29. $g = \frac{2}{3}h - 2$ for h

30. $A = s^2$ for s

31. $F = \frac{9}{5}c + 32$ for c

32. $y = mx + b$ for m

COMPOUND INTEREST

The formula for finding compound interest is:

$$A = P(1+\frac{n}{k})^{kt}$$

A = ending balance after interest is added on
P = principle amount invested
k = how many times per year interest is compounded
n = rate

1. Lisa invested $1000 into an account that pays 6% interest compounded annually. If this account is for her newborn, how much will the account be worth on his 21st birthday, which is exactly 21 years from now?

2. Mr. Dumple wants to open up a savings account. He has looked at two different banks. Bank 1 is offering a rate of 5% compounded daily. Bank 2 is offering an account that has a rate of 8% but is only compounded semi-yearly. Mr. Dumple puts $5000 and wants to take it out for his retirement in 10 years. Which bank will give him the most money back?

EXAMPLE: What is the ending balance if $500 is invested at 8% compounded quarterly for one year?

$$A = 500 \, (1+\frac{.08}{4})^{4 \cdot 1}$$

$$A = 500 \, (1.02)^{4}$$

$$A = 500 \, (1.08^{2})$$

$$A = \$541$$

Complete the chart below with the missing information.

	Beginning Principal	Interest Rate	Compounded	Duration	Ending Balance
1.	$1,000	9%	Quarterly	Six Months	
2.	$2,500	5%	Semiannually	One Year	
3.	$1,700	6%	Quarterly	Six Months	
4.	$3,200	11%	Quarterly	Six Months	
5.	$6,300	10%	Semiannually	One Year	
6.	$3,700	2%	Monthly	Two Months	
7.	$5,200	3%	Quarterly	Six Months	
8.	$7,500	8%	Yearly	Two Years	
9.	$4,250	4%	Semiannually	One Year	
10.	$5,900	1%	Monthly	Two Months	

CHAPTER 12 REVIEW

1. Seven students formed a study group. At the first meeting, they all exchanged phone numbers. Using the formula, $n \times (n - 1)$, where n = number of students, determine how many phone numbers were exchanged.

2. A shirt originally priced at $36 is put on sale Wednesday for 20% off. By Saturday it had not sold and was marked 40% off the already reduced price. Which formula below will help you determine the new price?

   A. $.60\,x$ C. $x - (.60\,x)$
   B. $.40\,x$ D. $.60\,[x - (.20x)]$

3. 1 kg = 2.2 pounds
   If a truck can carry 2,000 pounds, how many kilograms can it carry?

4. Dan needs to know the volume of a sphere that has a 22 inch diameter. Using $\pi = \frac{22}{7}$, find the volume of the sphere. $V = \frac{4}{3}\pi r^3$

5. Angela wants to know how much 1,000 in a savings account will earn in 1 year at 3% interest. Use I = PRT

6. If Juan drove 512 miles in 8 hours and 12 minutes, what was his average speed? Use D = RT.

7. Andy has 10 more jellybeans than his friend Charlie, but half as many as Robert. Which of the following is a formula for how many jellybeans Robert has?

   A. $R = C + 10$ C. $R = A + \frac{1}{2}C$
   B. $R = 2C + 20$ D. $R = 2A + 10$

8. When building a pipe organ, the length in meters of each pipe is $\frac{1}{2}$ the wavelength of the note it produces. Find the length of a pipe needed to produce an A note with frequency 440.

   $fl = v$
   f = frequency
   l = wavelength
   v = speed of sound (about 335 m/s)

9. For the spreadsheet below, using the formula $3(B^2 + A)$, fill in the values for Column C.

	A	B	C
1	2	3	
2	4	6	
3	5	2	

10. In the spreadsheet below, using the formula $5A^2 - 2B$, fill in the values for Column C.

	A	B	C
1	2	3	
2	4	6	
3	5	2	

Using the Quadratic Equation, find both solutions for x.

11. $x^2 + 10x - 11 = 0$

12. $x^2 - 14x + 40 = 0$

13. $x^2 + 9x + 18 = 0$

14. $x^2 - 12x - 13 = 0$

Chapter 13 | Graphing and Writing Equations

This chapter helps students with Standard 3, Algebra and Functions.

GRAPHING LINEAR EQUATIONS

In addition to graphing ordered pairs, the Cartesian plane can be used to graph the solution set for an equation. Any equation with two variables that are both to the first power is called a **linear equation**. The graph of a linear equation will always be a straight line.

EXAMPLE 1: Graph the solution set for $x + y = 7$.

Step 1: Make a list of some pairs of numbers that will work in the equation.

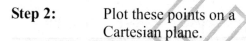

$$\frac{x + y = 7}{}$$
$$4 + 3 = 7 \qquad (4, 3)$$
$$-1 + 8 = 7 \qquad (-1, 8)$$
$$5 + 2 = 7 \qquad (5, 2)$$
$$0 + 7 = 7 \qquad (0, 7)$$

ordered pair solutions

Step 2: Plot these points on a Cartesian plane.

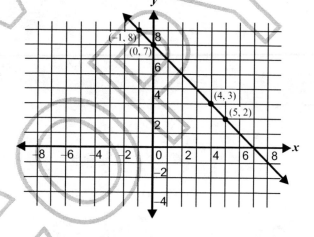

Step 3: By passing a line through these points, we graph the solution set for $x + y = 7$.

This means that every point on this line is a solution to the equation $x + y = 7$. For example, $(1, 6)$ is a solution; and therefore, the line passes through the point $(1, 6)$.

Make a table of solutions for each linear equation below. Then plot the ordered pair solutions on graph paper. Draw a line through the points. **(If one of the points does not line up, you have made a mistake.)**

1. $x + y = 6$

2. $y = x + 1$

3. $y = x - 2$

4. $x + 2 = y$

5. $x - 5 = y$

6. $x - y = 0$

EXAMPLE 2: Graph the equation $y = 2x - 5$.

Step 1: This equation has 2 variables, both to the first power, so we know the graph will be a straight line.

Substitute some numbers for x or y to find pairs of numbers that satisfy the equation. For the above equation, it will be easier to substitute values of x in order to find the corresponding value for y. Record the values for x and y in a table.

x	y
0	−5
1	−3
2	−1
3	1

If x is 0, y would be −5
If x is 1, y would be −3
If x is 2, y would be −1
If x is 3, y would be 1

Step 2: Graph the ordered pairs, and draw a line through the points.

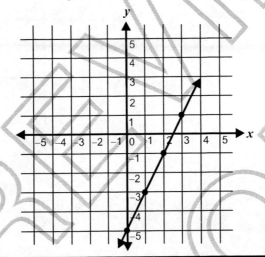

Find pairs of numbers that satisfy the equations below, and graph the line on graph paper.

1. $y = -2x + 2$

2. $2x - 2 = y$

3. $-x + 3 = y$

4. $y = x + 1$

5. $4x - 2 = y$

6. $y = 3x - 3$

7. $x = 4y - 3$

8. $2x = 3y + 1$

9. $x + 2y = 4$

GRAPHING HORIZONTAL AND VERTICAL LINES

The graph of some equations is a horizontal or a vertical line.

EXAMPLE 1: $y = 3$

Step 1: Make a list of ordered pairs that satisfy the equation $y = 3$. This can also be written in slope-intercept form as $y = 0x + 3$.

x	y
0	3
1	3
2	3
3	3

No matter what value of x you choose, y is always 3.

Step 2: Plot these points on a Cartesian plane, and draw a line through the points.

The graph is a horizontal line.

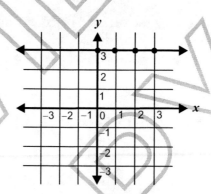

EXAMPLE 2: $2x + 3 = 0$

Step 1: For these equations with only one variable, find what x equals first.

$$2x + 3 = 0$$
$$2x = -3$$
$$x = \frac{-3}{2}$$

Step 2: No matter which value of y you choose, the value of x does not change.

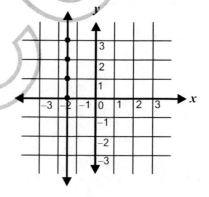

The graph is a vertical line.

Find pairs of numbers that satisfy the equations below, and graph the line on graph paper.

1. $2y + 2 = 0$

2. $x = -4$

3. $3x = 3$

4. $y = 5$

5. $4x - 2 = 0$

6. $2x - 6 = 0$

7. $4y = 1$

8. $5x + 10 = 0$

9. $3y + 12 = 0$

10. $x + 1 = 0$

11. $2y - 8 = 0$

12. $3x = -9$

13. $x = -2$

14. $6y - 2 = 0$

15. $5x - 5 = 0$

16. $2y - 4 = 0$

17. $2y - 2 = 0$

18. $3x + 1 = 0$

19. $4y = -2$

20. $-2y = 6$

21. $-4x = -8$

22. $3y = -6$

23. $x = 2$

24. $4y = 8$

FINDING THE INTERCEPTS OF A LINE

The x-intercept is the point where the graph of a line crosses the x-axis. The y-intercept is the point where the graph of a line crosses the y-axis.

> **To find the x-intercept, set $y = 0$**
> **To find the y-intercept, set $x = 0$**

EXAMPLE: Find the x- and y-intercepts of the line $6x + 2y = 18$.

Step 1:   To find the x-intercept, set $y = 0$.

$$6x + 2(0) = 18$$
$$\frac{6x}{6} = \frac{18}{6}$$
$$x = 3 \qquad \text{The } x\text{-intercept is at the point } (3, 0).$$

Step 2:   To find the y-intercept, set $x = 0$.

$$6(0) + 2y = 18$$
$$\frac{2y}{2} = \frac{18}{2}$$
$$y = 9 \qquad \text{The } y\text{-intercept is at the point } (0, 9).$$

Step 3:   You can now use the two intercepts to graph the line.

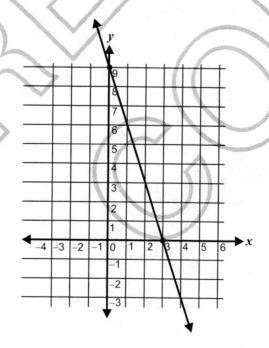

For each of the following equations, find both the *x* and the *y* intercepts of the line. For extra practice, draw each of the lines on graph paper.

1. $6x - 2y = 6$

2. $2x + 4y = 8$

3. $4x + 3y = 12$

4. $x - 3y = -4$

5. $8x + 3y = 8$

6. $5x - 4y = 10$

7. $-2x - 2y = 6$

8. $-6x + 4y = 12$

9. $6x - 2y = -6$

10. $-5x - 5y = 15$

11. $9x - 6y = -18$

12. $6x + 6y = 18$

13. $-3x - 6y = 21$

14. $8x + 3y = -8$

15. $-3x + 9y = 9$

16. $12x + 6y = 24$

17. $x - 2y = -4$

18. $-2x - 4y = 8$

19. $5x + 4y = 15$

20. $12x + 18y = 60$

21. $7x - 14y = 21$

22. $5x + 10y = 15$

23. $-12x + 16y = 48$

24. $33x - 11y = -33$

25. $-2x - 6y = -8$

26. $14x + 3y = 21$

27. $10x - 5y = 20$

28. $10x + 15y = 30$

29. $-18x + 27y = 54$

30. $21x + 42y = 63$

UNDERSTANDING SLOPE

The **slope** of a line refers to how steep a line is. When we graph a line using ordered pairs, we can easily determine the slope. Slope is often represented by the letter *m*.

$$\text{The formula for slope of a line is: } m = \frac{y_2 - y_1}{x_2 - x_1} \text{ or } \frac{\text{rise}}{\text{run}}$$

EXAMPLE 1: What is the slope of the following line that passes through the ordered pairs $(-4, -3)$ and $(1, 3)$?

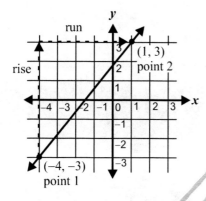

y_2 is 3, the *y*-coordinate of point 2.

y_1 is -3, the *y*-coordinate of point 1.

x_2 is 1, the *x*-coordinate of point 2.

x_1 is -4, the *x*-coordinate of point 1.

Use the formula for slope given above. $\quad m = \dfrac{3 - (-3)}{1 - (-4)} = \dfrac{6}{5}$

The slope is $\frac{6}{5}$. This shows us that we can go up 6 (rise) and over 5 to the right (run) to find another point on the line.

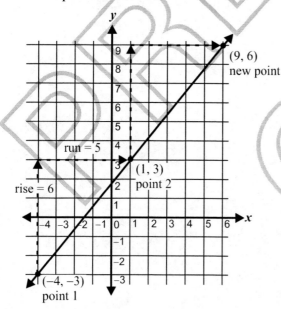

EXAMPLE 2: Find the slope of a line through the points (–2, 3) and (1, –2). It doesn't matter which pair we choose for point 1 and point 2. The answer is the same.

let point 1 be (–2, 3)
let point 2 be (1, –2)

$$\text{slope} = \frac{(y_2 - y_1)}{(x_2 - x_1)} = \frac{-2 - 3}{1 - (-2)} = \frac{-5}{3}$$

When the slope is negative, the line will slant left. For this example, the line will go **down** 5 units and then over 3 to the **right**.

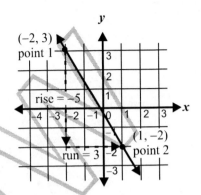

EXAMPLE 3: What is the slope of a line that passes through (1, 1) and (3, 1)?

$$\text{slope} = \frac{1 - 1}{3 - 1} = \frac{0}{2} = 0$$

When $y_2 - y_1 = 0$, the slope will equal 0, and the line will be horizontal.

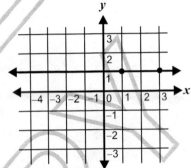

EXAMPLE 4: What is the slope of a line that passes through (2, 1) and (2, –3)?

$$\text{Slope} = \frac{-3 - 1}{2 - 2} = \frac{4}{0} = \text{undefined}$$

When $x_2 - x_1 = 0$, the slope is undefined, and the line will be vertical.

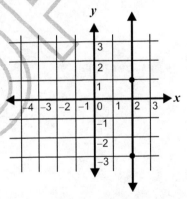

The following lines summarize what we know about slope.

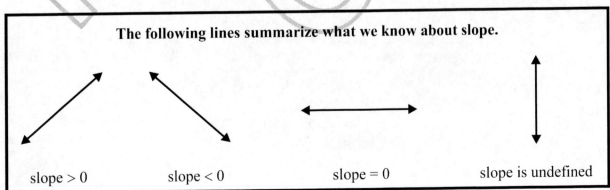

slope > 0   slope < 0   slope = 0   slope is undefined

Find the slope of the line that goes through the following pairs of points. Use the formula slope = $\dfrac{y_2 - y_1}{x_2 - x_1}$. Then, using graph paper, graph the line through the two points, and label the rise and the run. (See Examples 1 and 2.)

1. (2, 3) (4, 5)

2. (1, 3) (2, 5)

3. (−1, 2) (4, 1)

4. (1, −2) (4, −2)

5. (3, 0) (3, 4)

6. (3, 2) (−1, 8)

7. (4, 3) (2, 4)

8. (2, 2) (1, 5)

9. (3, 4) (1, 2)

10. (3, 2) (3, 6)

11. (6, −2) (3, −2)

12. (1, 2) (3, 4)

13. (−2, 1) (−4, 3)

14. (5, 2) (4, −1)

15. (1, −3) (−2, 4)

16. (2, −1) (3, 5)

17. (2, 4) (5, 3)

18. (5, 2) (2, 5)

19. (4, 5) (6, 6)

20. (2, 1) (−1, −3)

SLOPE - INTERCEPT FORM OF A LINE

An equation that contains two variables, each to the first degree, is a **linear equation**. The graph for a linear equation is a straight line. To put a linear equation in slope-intercept form, solve the equation for y. This form of the equation shows the slope and the y-intercept. Slope-intercept form follows the pattern of $y = mx + b$. The "m" represents slope, and the "b" represents the y-intercept. The y-intercept is the point at which the line crosses the y-axis.

When the slope of a line is not 0, the graph of the equation shows a **direct variation** between y and x. When y increases, x increases in a certain proportion. The proportion stays constant. The constant is called the **slope** of the line.

EXAMPLE: Put the equation $2x + 3y = 15$ in slope-intercept form. What is the slope of the line? What is the y-intercept? Graph the line.

Step 1: Solve for y:

$$2x + 3y = 15$$
$$\underline{-2x \qquad\qquad -2x}$$
$$\frac{3y}{3} = \frac{-2x}{3} + \frac{15}{3}$$

slope-intercept form: $\quad y = \dfrac{-2}{3}x + 5$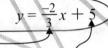

The slope is $\dfrac{-2}{3}$ and the y-intercept is 5

Step 2: Knowing the slope and the y-intercept, we can graph the line.

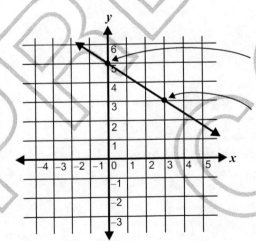

The y-intercept is 5, so the line passes through the point (0, 5) on the y-axis.

The slope is $\dfrac{-2}{3}$, so go down 2 and over 3 to get a second point.

195

Put each of the following equations in slope-intercept form by solving for y. On your graph paper, graph the line using the slope and y-intercept.

1. $4x - y = 5$

2. $2x + 4y = 16$

3. $3x - 2y = 10$

4. $x + 3y = -12$

5. $6x + 2y = 0$

6. $8x - 5y = 10$

7. $-2x + y = 4$

8. $-4x + 3y = 12$

9. $-6x + 2y = 12$

10. $x - 5y = 5$

11. $3x - 2y = -6$

12. $3x + 4y = 2$

13. $-x = 2 + 4y$

14. $2x = 4y - 2$

15. $6x - 3y = 9$

16. $4x + 2y = 8$

17. $6x - y = 4$

18. $-2x - 4y = 8$

19. $5x + 4y = 16$

20. $6 = 2y - 3x$

VERIFY THAT A POINT LIES ON A LINE

To know whether or not a point lies on a line, substitute the coordinates of the point into the formula for the line. If the point lies on the line, the equation will be true. If the point does not lie on the line, the equation will be false.

EXAMPLE 1: Does the point $(5, 2)$ lie on the line given by the equation $x + y = 7$?

Solution: Substitute 5 for x and 2 for y in the equation. $5 + 2 = 7$. Since this is a true statement, the point $(5, 2)$ does lie on the line $x + y = 7$.

EXAMPLE 2: Does the point $(0, 1)$ lie on the line given by the equation $5x + 4y = 16$?

Solution: Substitute 0 for x and 1 for y in the equation. $5x + 4y = 16$.
Does $5(0) + 4(1) = 16$? No, it equals 4, not 16. Therefore, the point $(0, 1)$ is not on the line given by the equation $5x + 4y = 16$.

For each point below, state whether or not it lies on the line given by the equation that follows the point coordinates.

1. $(2, 4)$ $6x - y = 8$

2. $(1, 1)$ $6x - y = 5$

3. $(3, 8)$ $-2x + y = 2$

4. $(9, 6)$ $-2x + y = 0$

5. $(3, 7)$ $x - 5y = -32$

6. $(0, 5)$ $-6x - 5y = 3$

7. $(2, 4)$ $4x + 2y = 16$

8. $(9, 1)$ $3x - 2y = 29$

9. $(6, 8)$ $6x - y = 28$

10. $(-2, 3)$ $x + 2y = 4$

11. $(4, -1)$ $-x - 3y = -1$

12. $(-1, -3)$ $2x + y = 1$

196

GRAPHING A LINE KNOWING A POINT AND SLOPE

If you are given a point of a line and the slope of a line, the line can be graphed.

EXAMPLE 1: Given that line *l* has a slope of $\frac{4}{3}$ and contains the point $(2, -1)$, graph the line.

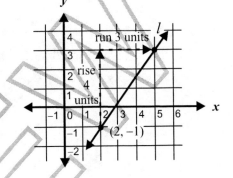

Step 1: Plot and label the point $(2, -1)$ on a Cartesian plane.

Step 2: The slope, *m*, is $\frac{4}{3}$, so the rise is 4 and the run is 3. From the point $(2, -1)$, count 4 units up and 3 units to the right.

Step 3: Draw the line through the two points.

EXAMPLE 2: Given a line that has a slope of $\frac{-1}{4}$ and passes through the point $(-3, 2)$, graph the line.

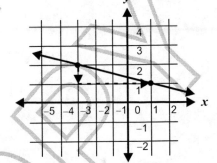

Step 1: Plot the point $(-3, 2)$.

Step 2: Since the slope is negative, go **down** 1 unit and over 4 to get a second point.

Step 3: Graph the line through the two points.

Graph the line on your own graph paper for each of the following problems. First, plot the point. Then use the slope to find a second point. Draw the line formed from the point and the slope.

1. $(2, -2), m = \frac{3}{4}$ 6. $(-2, 1), m = \frac{4}{3}$ 11. $(-2, -3), m = \frac{2}{3}$ 16. $(3, -3), m = \frac{-3}{4}$

2. $(3, -4), m = \frac{1}{2}$ 7. $(-4, -2), m = \frac{1}{2}$ 12. $(4, -1), m = \frac{-1}{3}$ 17. $(-2, 5), m = \frac{1}{3}$

3. $(1, 3), m = \frac{-1}{3}$ 8. $(1, -4), m = \frac{3}{4}$ 13. $(-1, 5), m = \frac{2}{5}$ 18. $(-2, -3), m = \frac{-3}{4}$

4. $(2, -4), m = 1$ 9. $(2, -1), m = \frac{-1}{2}$ 14. $(-2, 3), m = \frac{3}{4}$ 19. $(4, -3), m = \frac{2}{3}$

5. $(3, 0), m = \frac{-1}{2}$ 10. $(5, -2), m = \frac{1}{4}$ 15. $(4, 4), m = \frac{-1}{2}$ 20. $(1, 4), m = \frac{-1}{2}$

7.3.8

197

FINDING THE EQUATION OF A LINE USING TWO POINTS OR A POINT AND SLOPE

If you can find the slope of a line and know the coordinates of one point, you can write the equation for the line. You know the formula for the slope of a line is:

$$m = \frac{y_2 - y_1}{x_2 - x_1} \quad \text{or} \quad \frac{y_2 - y_1}{x_2 - x_1} = m$$

Using algebra, you can see that if you multiply both sides of the equation by $x_2 - x_1$, you get:

$$y - y_1 = m(x - x_1) \quad \longleftarrow \quad \text{point-slope form of an equation}$$

EXAMPLE: Write the equation of the line passing through the points $(-2, 3)$ and $(1, 5)$.

Step 1: First, find the slope of the line using the two points given.

$$m = \frac{y_2 - y_1}{x_2 - x_1} = \frac{5 - 3}{1 - (-2)} = \frac{2}{3}$$

Step 2: Pick one of the points to use in the point-slope equation. For point $(-2, 3)$, we know $x_1 = -2$ and $y_1 = 3$, and we know $m = \frac{2}{3}$. Substitute these values into the point-slope form of the equation.

$$y - y_1 = m(x - x_1)$$

$$y - 3 = \frac{2}{3}[x - (-2)]$$

$$y - 3 = \frac{2}{3}x + \frac{4}{3}$$

$$y = \frac{2}{3}x + \frac{13}{3}$$

Use the point-slope formula to write an equation for each of the following lines.

1. $(1, -2)$, $m = 2$

2. $(-3, 3)$, $m = \frac{1}{3}$

3. $(4, 2)$, $m = \frac{1}{4}$

4. $(5, 0)$, $m = 1$

5. $(3, -4)$, $m = \frac{1}{2}$

6. $(-1, 4)$ $(2, -1)$

7. $(2, 1)$ $(-1, -3)$

8. $(-2, 5)$ $(-4, 3)$

9. $(-4, 3)$ $(2, -1)$

10. $(3, 1)$ $(5, 5)$

11. $(-3, 1)$, $m = 2$

12. $(-1, -2)$, $m = \frac{4}{3}$

13. $(2, -5)$, $m = -2$

14. $(-1, 3)$, $m = \frac{1}{3}$

15. $(0, -2)$, $m = -\frac{3}{2}$

7.3.8

198

SLOPES OF PERPENDICULAR LINES

Once you have found the slope of a line, determining the slope of a line perpendicular to it is done in two simple steps:

1. multiply the slope by -1
2. invert the slope

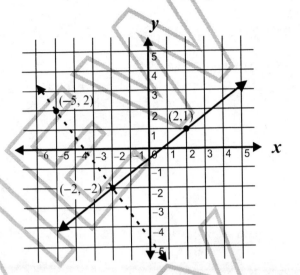

EXAMPLE: The solid line on the graph at right has a slope of $\frac{3}{4}$.

Step 1: $(-1)\frac{3}{4} = -\frac{3}{4}$.

Step 2: Taking the inverse of the negative slope, $-\frac{3}{4}$, gives $-\frac{4}{3}$.

The dotted line on the graph at right has a slope of $-\frac{4}{3}$ and is perpendicular to the solid line with a slope of $\frac{3}{4}$.

---

1. Find the slope of a line perpendicular to the following line, and draw it in as a dotted line:

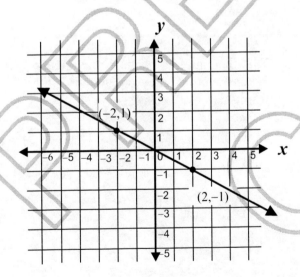

2. Find the slope of a line perpendicular to lines with the following slopes.

A. $-\frac{4}{5}$

B. $\frac{3}{2}$

C. $\frac{5}{7}$

D. $-\frac{6}{5}$

E. $-\frac{7}{9}$

F. $\frac{1}{4}$

G. $-\frac{4}{1}$

EQUATIONS OF PERPENDICULAR LINES

Now that we know how to calculate the slope of a line running perpendicular to a given line, we can easily find the equation of this perpendicular line. Writing the equation for a line perpendicular to another line involves three steps:

1. Find the slope of the perpendicular line.
2. Choose one point on the first line.
3. Use the point-slope form to write the equation.

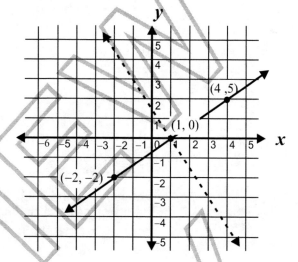

EXAMPLE: The solid line on the graph at right has a slope of $\frac{2}{3}$. Write the equation of a line perpendicular to the solid line.

Step 1: Find the slope of the perpendicular line. Multiply the slope by -1 and then invert it: $\frac{2}{3} \rightarrow -\frac{3}{2}$
The slope of the line perpendicular, shown as a dotted line on the graph at right, is $-\frac{3}{2}$.

Step 2: Choose one point on the first line. We will use $(1, 0)$ in this example. The point $(-2, -2)$ or $(4, 5)$ could also be used.

Step 3: Use the point-slope formula, $(y - y_1) = m(x - x_1)$, to write the equation of the perpendicular line. Remember, we chose $(1, 0)$ as our point.
So, $(y - 0) = -\frac{3}{2}(x - 1)$. Simplified, $y = -\frac{3}{2}x + \frac{3}{2}$.

---

1. Find the slope of the line perpendicular to the solid line shown below, and draw the perpendicular as a dotted line. Use one point on the solid line and the calculated slope to find the equation of the perpendicular line.

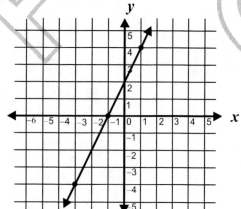

Find the equation of the perpendicular line using the point and slope given and the formula $(y - y_1) = m(x - x_1)$.

2. $(2, 1), 5$

3. $(3, 2), 2$

4. $(-2, 1), -3$

5. $(-4, 2), -\frac{1}{2}$

6. $(-1, 4), 1$

7. $(3, 3), \frac{2}{3}$

8. $(5, -1), -1$

9. $(\frac{1}{2}, \frac{3}{4}), 4$

10. $(\frac{2}{3}, \frac{3}{4}), -\frac{1}{6}$

11. $(7, -2), -\frac{1}{8}$

12. $(5, 0), \frac{4}{5}$

13. $(-3, -3), -\frac{7}{3}$

14. $(\frac{1}{4}, 4), \frac{1}{2}$

15. $(0, 6), -\frac{1}{9}$

16. $(-\frac{1}{3}, -2), \frac{1}{3}$

17. $(0, \frac{1}{4}), \frac{5}{6}$

CHANGING THE SLOPE OR *Y*-INTERCEPT OF A LINE

When the slope and/or the *y*-intercept of a linear equation changes, the graph of the line will also change.

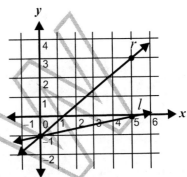

Figure 1

EXAMPLE 1: Consider line *l* shown in figure 1 at right. What happens to the graph of the line if the slope is changed to $\frac{4}{5}$?

Step 1: Determine the *y*-intercept of the line. For line *l*, it can easily be seen from the graph that the *y*-intercept is at the point $(0, -1)$.

Step 2: Find the slope of the line using two points that the line goes through: $(0, -1)$ and $(5, 0)$.

$$m = \frac{y_2 - y_1}{x_2 - x_1} = \frac{0 - (-1)}{5 - 0} = \frac{1}{5}$$

Step 3: Write the equation of line *l* in slope-intercept form:
$y = mx + b$.
$y = \frac{1}{5}x - 1$

Step 4: Rewrite the equation of the line using a slope of $\frac{4}{5}$, and then graph the line.
Equation of new line: $y = \frac{4}{5}x - 1$
The graph of the new line is labeled line *r* and is shown in Figure 1.
A line with slope $\frac{4}{5}$ is steeper than a line with slope $\frac{1}{5}$.

Note: The greater the numerator, or "rise," of the slope, the steeper the line will be. The greater the denominator, or "run," of the slope, the flatter the line will be.

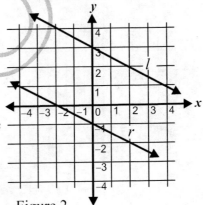

Figure 2

EXAMPLE 2: Consider line *l* shown in figure 2 at right. The equation of the line is $y = -\frac{1}{2}x + 3$. What happens to the graph of the line if the *y*-intercept is changed to -1?

Step 1: Rewrite the equation of the line replacing the *y*-intercept with -1.
Equation of new line: $y = -\frac{1}{2}x - 1$.

Step 2: Graph the new line. Line *r* in figure 2 is the graph of the equation $y = -\frac{1}{2}x - 1$.
Since both lines *l* and *r* have the same slope, they are parallel. Line *r*, with a *y*-intercept of -1, sits below line *l*, with a *y*-intercept of 3.

Put each pair of the following equations in slope-intercept form. Write P if the lines are parallel and NP if the lines are not parallel.

1. $y = x + 1$ _____
    $2y - 2x = 6$

2. $2x + y = 6$ _____
    $2x = 8 - y$

3. $x + 5y = 0$ _____
    $5y + 5 = x$

4. $y = 3 - \frac{1}{3}x$ _____
    $3y + x = -6$

5. $x = 2y$ _____
    $-x = -2y + 14$

6. $y = x + 2$ _____
    $-y = x + 4$

7. $y = 4 - \frac{1}{4}x$ _____
    $3x + 4y = 4$

8. $x + y = 5$ _____
    $5 - y = 2x$

9. $x - 4y = 0$ _____
    $4y = x - 8$

Consider the line (l) shown on each of the following graphs, and write the equation of the line in the space provided. Then, on the same graph, graph the line (r) for which the equation is given. Write how the slope and y-intercept of line l compare to the slope and y-intercept of line r for each graph.

10.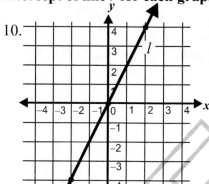

line l: _____
line r: ____$y = -2x$____
slopes: _____
y-intercepts: _____

12.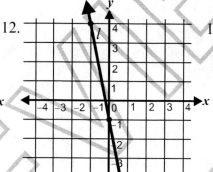

line l: _____
line r: ____$y = -3x - 1$____
slopes: _____
y-intercepts: _____

14.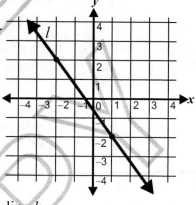

line l: _____
line r: ____$y = \frac{1}{4}x - 2$____
slopes: _____
y-intercepts: _____

11.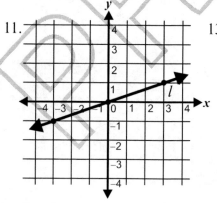

line l: _____
line r: ____$y = \frac{1}{3}x + 2$____
slopes: _____
y-intercepts: _____

13.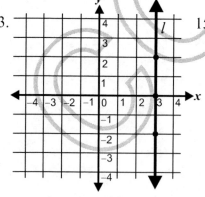

line l: _____
line r: ____$y = -3$____
slopes: _____
y-intercepts: _____

15.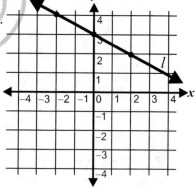

line l: _____
line r: ____$y = -\frac{1}{2}x - 3$____
slopes: _____
y-intercepts: _____

202

WRITING AN EQUATION FROM DATA

Data is often written in a two column format. If the increases or decreases in the ordered pairs are at a constant rate, then a linear equation for the data can be found.

EXAMPLE: **Write an equation for the following set of data.**

Dan set his car on cruise control and noted the distance he went every 5 minutes.

Minutes in operation (x)	Distance traveled (y)
5	28,490 miles
10	28,494 miles

Step 1: Write two ordered pairs in the form (minutes, distance) for Dan's driving, (5, 28490), and (10, 28494), and find slope.

Step 2: Use the ordered pairs to write the equation in the form $y = mx + b$.
Place the slope, m, that you found and one of the pairs of points as x_1 and y_1 in the following formula:

$$y - y_1 = m(x - x_1)$$
$$y - 28490 = \tfrac{4}{5}(x - 5)$$
$$y - 28490 + 28490 = \tfrac{4}{5}x - 4 + 28490$$
$$y = \tfrac{4}{5}x + 28486$$

It doesn't matter which pair of points you use; the answer will be the same.

Write an equation for each of the following sets of data, assuming the relationship is linear.

1.
Doug's Doughnut Shop

Year in Business	Total Sales
1	$55,000
4	$85,000

4.
Jim's Depreciation on His Jet Ski

Years	Value
1	$4,500
6	$2,500

2.
Gwen's Green Beans

Days Growing	Height in Inches
2	5
6	12

5.
Stepping on the Brakes

Seconds	MPH
2	51
5	18

3.
At the Gas Pump

Gallons Purchased	Total Cost
5	$6.00
7	$8.40

6.
Stepping on the Accelerator

Seconds	MPH
4	35
7	62

GRAPHING LINEAR DATA

Many types of data are related by a constant ratio. This type of data is linear. The slope of the line described by linear data is the ratio between the data. Plotting linear data with a constant ratio can be helpful in finding additional values.

EXAMPLE 1: A department store prices socks per pair. Each pair of socks costs $0.75. Plot pairs of socks versus price on a Cartesian plane.

Step 1: Since the price of the socks is constant, you know that one pair of socks costs $0.75, 2 pairs of socks cost $1.50, 3 pairs of socks cost $2.25, and so on. Make a list of a few points.

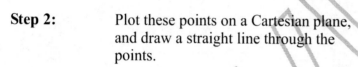

Pair(s) x	Price y
1	.75
2	1.50
3	2.25

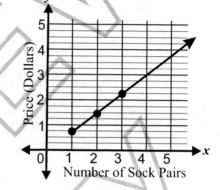

Step 2: Plot these points on a Cartesian plane, and draw a straight line through the points.

EXAMPLE 2: What is the slope of the data? What does the slope describe?

Solution: You can determine the slope either by the graph or by the data points. For this data, the slope is .75. Remember, slope is rise/run. For every $0.75 going up the y-axis, you go across one pair of socks on the x-axis. The slope describes the price per pair of socks.

EXAMPLE 3: Use the graph created in Example 1 to answer the following questions. How much would 5 pairs of socks cost? How many pairs of socks could you purchase for $3.00? Extending the line gives useful information about the price of additional pairs of socks.

Solution 1: The line that represents 5 pairs of socks intersects the data line at $3.75 on the y-axis. Therefore, 5 pairs of socks would cost $3.75.

Solution 2: The line representing the value of $3.00 on the y-axis intersects the data line at 4 on the x-axis. Therefore, $3.00 will buy exactly 4 pairs of socks.

Use the information given to make a line graph for each set of data, and answer the questions related to each graph.

1. The diameter of a circle versus the circumference of a circle is a constant ratio. Use the data given below to graph a line to fit the data. Extend the line, and use the graph to answer the next question.

Circle

Diameter	Circumference
4	12.56
5	15.70

2. Using the graph of the data in question 1, estimate the circumference of a circle that has a diameter of 3 inches.

3. If the circumference of a circle is 3 inches, about how long is the diameter?

4. What is the slope of the line you graphed in question 1?

5. What does the slope of the line in question 4 describe?

6. The length of a side on a square and the perimeter of a square are a constant ratio. Use the data below to graph this relationship.

Square

Length of side	Perimeter
2	8
3	12

7. Using the graph from question 6, what is the perimeter of a square with a side that measures 4 inches?

8. What is the slope of the line graphed in question 6?

9. Conversions are often constant ratios. For example, converting from pounds to ounces follows a constant ratio. Use the data below to graph a line that can be used to convert pounds to ounces.

Measurement Conversion

Pounds	Ounces
2	32
4	64

10. Use the graph from question 9 to convert 40 ounces to pounds.

11. What does the slope of the line graphed for question 9 represent?

12. Graph the data below and create a line that shows converting weeks to days.

Time

Weeks	Days
1	7
2	14

13. About how many days are in $2\frac{1}{2}$ weeks?

14. Graph a data line that converts feet to inches.

15. Using the graph in question 14, how many inches are in 4.5 feet?

16. What is the slope of the line converting feet to inches?

17. An electronics store sells DVDs for $25 each. Graph a data line showing total cost versus number of DVDs purchased.

18. Using the graph in question 17, how many DVDs could be purchased for $150?

IDENTIFYING GRAPHS OF LINEAR EQUATIONS

Match each equation below with the graph of the equation.

A: $x = y$

B: $x = -4$

C: $y = -x$

D: $y = 4$

E: $x = \frac{1}{2}y$

F: $y = -4$

G: $x = -2y$

H: $4x = y$

I: $-2x = y$

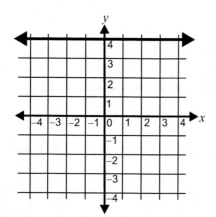

1. _____

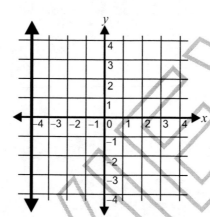

2. _____

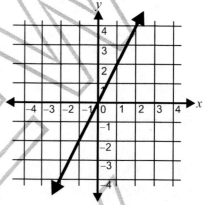

3. _____

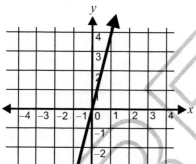

4. _____

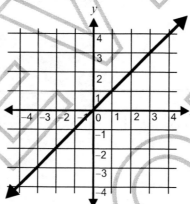

5. _____

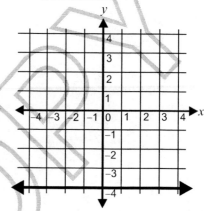

6. _____

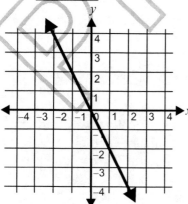

7. _____

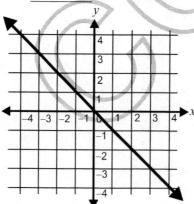

8. _____

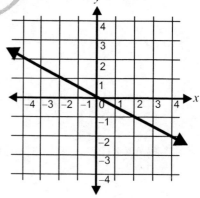

9. _____

Match each equation below with the graph of the equation.

A: $y = 4x$

B: $y = -4x$

C: $4x + y = 4$

D: $x - 2y = 6$

E: $y = 3x - 1$

F: $2x + 3y = 6$

G: $y = 3x + 2$

H: $x + 2y = 6$

I: $y = x - 3$

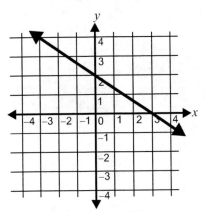

1. _____

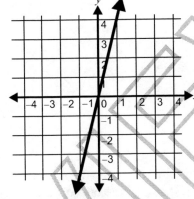

2. _____

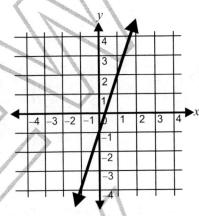

3. _____

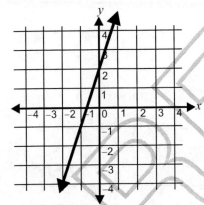

4. _____

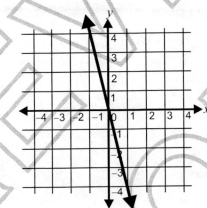

5. _____

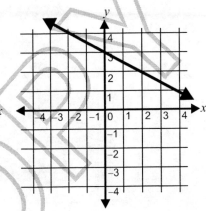

6. _____

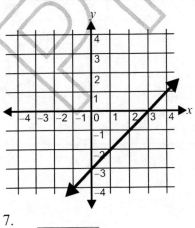

7. _____

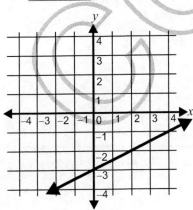

8. _____

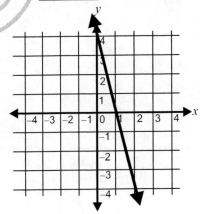

9. _____

Copyright © American Book Company

GRAPHING NON-LINEAR EQUATIONS

Equations may involve variables which are squared or cubed (raised to the second or third power). The best way to find values for the x and y variables in an equation is to plug one number into x, and then find the corresponding value for y just like you did at the beginning of this chapter. Then, plot the points and draw a line through the points.

EXAMPLE 1: $y = x^2$

Step 1: Make a table and find several values for x and y.

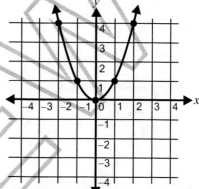

x	y
-2	4
-1	1
0	0
1	1
2	4

Step 2: Plot the points, and draw a curve through the points. Notice the shape of the curve. This type of curve is called a **parabola**. Equations with one squared term will be parabolas.

EXAMPLE 2: $y = x^3$

Step 1: Make a table and find several values for x and y.

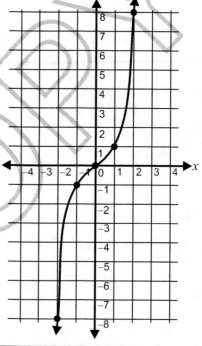

x	y
-2	-8
-1	-1
0	0
1	1
2	8

Step 2: Plot the points, and draw a curve through the points. Equations with one cubed term will always have this shape of curve.

Graph the equations below on a Cartesian plane.

1. $y = 2x^2$
2. $y = 3 - x^3$
3. $y = x^2 - 2$

4. $y = 2x^3$
5. $y = x^2 + 3$
6. $y = x^3 - 2$

7. $y = 3x^2 - 5$
8. $y = x^3 + 1$
9. $y = -x^2$

10. $y = -x^3$
11. $y = 2x^2 - 1$
12. $y = 2 - 2x^3$

1. Graph the solution set for the linear equation: $x - 3 = y$.

    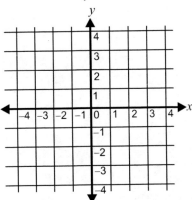

2. Which of the following is not a solution of $3x = 5y - 1$?

    A. $(3, 2)$
    B. $(7, 4)$
    C. $(\frac{1}{3}, 0)$
    D. $(-2, -1)$

3. $(-2, 1)$ is a solution for which of the following equations?

    A. $y + 2x = 4$
    B. $-2x + 1 = 5$
    C. $x + 2y = -4$
    D. $2x - y = -5$

4. Graph the equation $2x - 4 = 0$.

    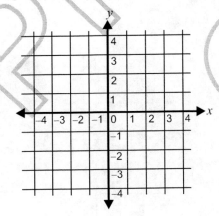

5. What is the slope of the line that passes through the points $(5, 3)$ and $(6, 1)$?

6. What is the slope of the line that passes through the points $(-1, 4)$ and $(-6, -2)$?

7. What is the x-intercept for the following equation?
    $$6x - y = 30$$

8. What is the y-intercept for the following equation?
    $$4x + 2y = 28$$

9. Graph the equation $3y = 9$.

    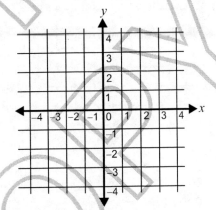

10. Write the following equation in slope-intercept form:
    $$3x = -2y + 4$$

11. What is the slope of the line $y = -\frac{1}{2}x + 3$?

12. What is the x-intercept of the line $y = 5x + 6$?

13. What is the y-intercept of the line $y - \frac{2}{3}x + 2 = 0$?

14. Graph the line which has a slope of 2 and a *y*-intercept of –3.

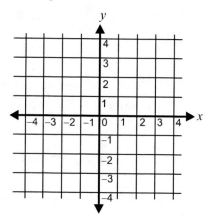

15. Graph the line which has a slope of –2 and a *y*-intercept of –3.

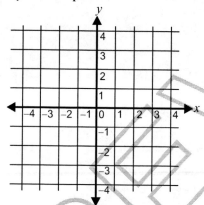

16. Which of the following points does ***not*** lie on the line $y = 3x - 2$?

A. (0, –2)
B. (1, 1)
C. (–1, 5)
D. (2, 4)

17. Which of the following points lies on the line $2y = -x + 1$?

A. $(\frac{1}{2}, 0)$
B. $(2, -\frac{1}{2})$
C. (0, 1)
D. (–1, –1)

18. Graph the equation $-x = 6 + 2y$.

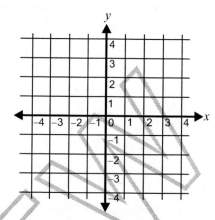

19. Find the equation of the line which contains the point (0, 2) and has a slope of $\frac{3}{4}$.

20. Which is the graph of $x - 3y = 6$?

A.

B.

C.

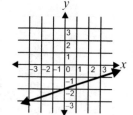

D.

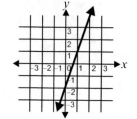

21. Which of the following is the graph of the line which has a slope of −2 and a y-intercept of (0, 3)?

A.
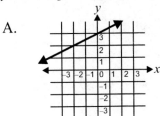

B.
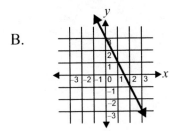

C.
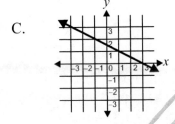

D.
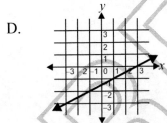

22. Given that a line contains the point (2, 3) and has a slope of $-\frac{1}{2}$, graph the line.

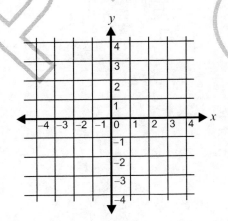

23. Paulo turned on the oven to preheat it. After one minute, the oven temperature was 200°. After 2 minutes, the oven temperature was 325°.

Oven Temperature

Minutes	Temperature
1	200°
2	325°

Assuming the oven temperature rose at a constant rate, write an equation that fits the data.

24. Write an equation that fits the data given below. Assume the data is linear.

Plumber Charges per Hour

Hour	Charge
2	$170
3	$220

25. What is the name of the curve described by the equation $y = 2x^2 - 1$?

26. Graph the following equation:

$$y = \frac{1}{2}x^3$$

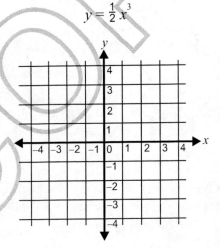

27. The data given below show conversions between miles per hour and kilometers per hour. Based on this data, graph a conversion line on the Cartesian plane below.

Speed

MPH	KPH
5	8
10	16

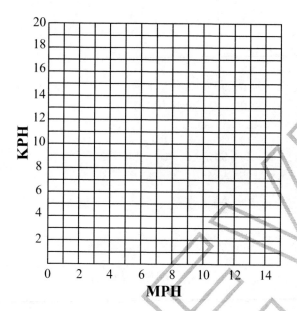

28. What would be the approximate conversion of 9 mph to kph?

29. What would be the approximate conversion of 13 kph to mph?

30. A bicyclist travels 12 mph downhill. Approximately how many kph is the bicyclist traveling?

31. Use the data given below to graph the interest versus the interest rate on $80.00 in one year.

$80.00 Principal

Interest Rate	Interest–1 year
5%	$4.00
10%	$8.00

32. About how much interest would accrue in one year at an 8% interest rate?

33. What is the slope of the line describing interest versus interest rate?

34. What information does the slope give in problem 33?

35. The graph of the line $y = 3x - 1$ is shown below. On the same graph, draw the line $y = -\frac{1}{3}x - 1$.

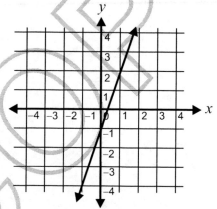

36. Which of the following statements is an accurate comparison of the lines $y = 3x - 1$ and $y = -\frac{1}{3}x - 1$?

   A. Only their y-intercepts are different.
   B. Only their slopes are different.
   C. Both their y-intercepts and their slopes are different.
   D. There is no difference between these two lines.

Chapter 14 | Graphing Inequalities

In the previous chapter, you learned to graph linear equations. In this chapter, you will learn to graph linear inequalities.

EXAMPLE 1: In the previous chapter, you would graph the equation $x = 3$ as:

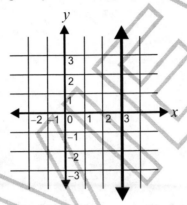

In this chapter, we graph inequalities such as $x > 3$ (read x is greater than 3). To show this, we use a broken line since the points on the line $x = 3$ are not included in the solution. We shade all points greater than 3.

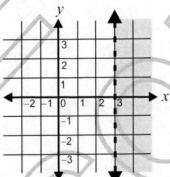

When we graph $x \geq 3$ (read x is greater than or equal to 3), we use a solid line because the points on the line $x = 3$ are included in the graph.

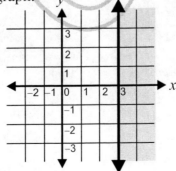

213

Graph the following inequalities on your own graph paper.

1. $y < 2$
2. $x \geq 4$
3. $y \geq 1$
4. $x < -1$
5. $y \geq -2$
6. $x \leq -4$
7. $x > -3$
8. $y \leq 3$

9. $x \leq 5$
10. $y > -5$
11. $x \geq 3$
12. $y < -1$
13. $x \leq 0$
14. $y > -1$
15. $y \leq 4$
16. $x \geq 0$

17. $y \geq 3$
18. $x < 4$
19. $x \leq -2$
20. $y < -2$
21. $y \geq -4$
22. $x \geq -1$
23. $y \leq 5$
24. $x < -3$

EXAMPLE 2: Graph $x + y \geq 3$

Step 1: First, we graph $x + y \geq 3$ by changing the inequality to an equality. Think of ordered pairs that will satisfy the equation $x + y = 3$. Then, plot the points, and draw the line.

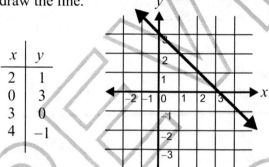

x	y
2	1
0	3
3	0
4	-1

This divides the Cartesian plane into 2 half-planes, $x + y \geq 3$ and $x + y \leq 3$. One half-plane is above the line, and the other is below the line.

Step 2: To determine which side of the line to shade, first choose a test point. If the point you choose makes the inequality true, then the point is on the side you shade. If the point you choose does not make the inequality true, then shade the side that does not contain the test point.

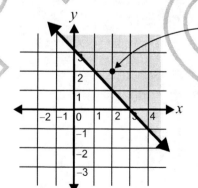

For our test point, let's choose (2, 2). Substitute (2, 2) into the inequality.

$x + y \geq 3$
$2 + 2 \geq 3$

$4 \geq 3$ is true, so shade the side that includes this point.

Use a solid line because of the $\geq$ sign.

Graph the following inequalities on your own graph paper.

1. $x + y \leq 4$

2. $x + y \geq 3$

3. $x \geq 5 - y$

4. $x \leq 1 + y$

5. $x - y \geq -2$

6. $x < y + 4$

7. $x + y < -1$

8. $x - y \leq 0$

9. $x \geq y + 2$

10. $x < -y + 1$

11. $-x + y > 1$

12. $-x - y < -2$

For more complex inequalities, it is easier to graph by first changing the inequality to an equality and second, putting the equation in slope-intercept form.

EXAMPLE: $2x + 4y \leq 8$

Step 1:   Change the inequality to an equality.
$2x + 4y = 8$

Step 2:   Put the equation in slope-intercept form by solving the equation for y.

$$2x + 4y = 8$$
$$\underline{-2x \qquad\qquad -2x}$$
$$\frac{4y}{4} = \frac{-2x}{4} + \frac{8}{4}$$
$$y = -\frac{1}{2}x + 2$$

Step 3:   Graph the line. If the inequality is $<$ or $>$, use a dotted line. If the inequality is $\leq$ or $\geq$, use a solid line. For this example, we should use a solid line.

Step 4:   Determine which side of the line to shade. Pick a point like $(0, 0)$ to see if it is true in the inequality.

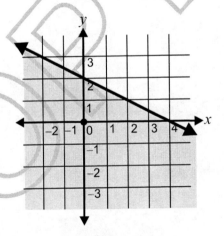

$2x + 4y \leq 8$, so substitute $(0, 0)$.
Is $0 + 0 \leq 8$? Yes, $0 \leq 8$, so shade the side of the line that includes the point $(0, 0)$.

Graph the following inequalities on your own graph paper.

1. $2x + y \geq 1$

2. $3x - y \leq 3$

3. $x + 3y > 12$

4. $4x - 3y < 12$

5. $y \geq 3x + 1$

6. $x - 2y > -2$

7. $x \leq y + 4$

8. $x + y < -1$

9. $-4y \geq 2x + 1$

10. $x \leq 4y - 2$

11. $3x - y \geq 4$

12. $y \geq 2x - 5$

13. $x + 7y < 1$

14. $-2y < 4x - 1$

15. $y > 4x + 1$

CHAPTER 14 REVIEW

Graph the following inequalities on a Cartesian plane using your own graph paper.

1. $x \geq 4$

2. $x < -3$

3. $y \leq 2$

4. $y \geq -1$

5. $2y \geq 8$

6. $3x < 6$

7. $-4y \geq 12$

8. $-2x \leq 4$

9. $2x + 5y \geq 10$

10. $-2y + 4x < 8$

11. $-2x - 3y > 9$

12. $5y > -10x + 5$

13. $y \leq 2x - 6$

14. $-2x + y < 1$

15. $3x - 4y > 8$

16. $y + 5 \leq x$

17. $2x - y \geq -1$

18. $y < -5$

19. $x + y > -2$

20. $3y < 2x + 6$

21. $x \leq -2$

22. $y \geq x + 2$

23. $2x + y < 5$

24. $3 + y > x$

25. $y - 2x \leq 3$

26. $2y + 6 < x$

27. $2x - y \geq 4$

28. $y \leq 3x$

29. $y > 4x + 1$

30. $2y < 3$

SYSTEMS OF EQUATIONS

Two linear equations considered at the same time are called a **system** of linear equations. The graph of a linear equation is a straight line. The graphs of two linear equations can show that the lines are **parallel**, **intersecting**, or **collinear**. Two lines that are **parallel** will never intersect and have no ordered pairs in common. If two lines are **intersecting**, they have one point in common, and in this chapter, you will learn to find the ordered pair for that one point. If the graph of two linear equations is the same line, the lines are said to be **collinear**.

If you are given a system of two linear equations, and you put both equations in slope-intercept form, you can immediately tell if the graph of the lines will be **parallel**, **intersecting**, or **collinear**.

If two linear equations have the same slope and the same y-intercept, then they are both equations for the same line. They are called **collinear** or **coinciding** lines. A line is made up of an infinite number of points extending infinitely far in two directions. Therefore, collinear lines have an infinite number of points in common.

EXAMPLE:
$$2x + 3y = -3$$
$$4x + 6y = -6$$

Or in slope intercept form the equations become :
$$y = -\frac{2}{3}x - 1$$
$$y = -\frac{2}{3}x - 1$$

We notice that both slopes are equal to $-\frac{2}{3}$

If two linear equations have the same slope but different y-intercepts, they are **parallel** lines. Parallel lines never touch each other, so they have no points in common.

If two linear equations have different slopes, then they are intersecting lines and share exactly one point in common.

The chart below summarizes what we know about the graphs of two equations in slope-intercept form.

y-Intercepts	Slopes	Graphs	Number of Solutions
same	same	collinear	infinite
different	same	distinct parallel lines	none (they never touch)
same or different	different	intersecting lines	exactly one

For the pairs of equations below, put each equation in slope-intercept form, and tell whether the graphs of the lines will be collinear, parallel, or intersecting.

1. $x - y = -1$ _____
   $-x + y = 1$

2. $x - 2y = 4$ _____
   $-x + 2y = 6$

3. $y - 2 = x$ _____
   $x + 2 = y$

4. $x = y - 1$ _____
   $-x = y - 1$

5. $2x + 5y = 10$ _____
   $4x + 10y = 20$

6. $x + y = 3$ _____
   $x - y = 1$

7. $2y = 4x - 6$ _____
   $-6x + y = 3$

8. $x + y = 5$ _____
   $2x + 2y = 10$

9. $2x = 3y - 6$ _____
   $4x = 6y - 6$

10. $2x - 2y = 2$ _____
    $3y = -x + 5$

11. $x = -y$ _____
    $x = 4 - y$

12. $2x = y$ _____
    $x + y = 3$

13. $x = y + 1$ _____
    $y = x + 1$

14. $x - 2y = 4$ _____
    $-2x + 4y = -8$

15. $2x + 3y = 4$ _____
    $-2x + 3y = 4$

16. $2x - 4y = 1$ _____
    $-6x + 12y = 3$

17. $-3x + 4y = 1$ _____
    $6x + 8y = 2$

18. $x + y = 2$ _____
    $5x + 5y = 10$

19. $x + y = 4$ _____
    $x - y = 4$

20. $y = -x + 3$ _____
    $x - y = 1$

218

FINDING COMMON SOLUTIONS FOR INTERSECTING LINES

When two lines intersect, they share exactly one point in common.

EXAMPLE: $3x + 4y = 20$ and $4x + 2y = 12$

Put each equation in slope-intercept form.

$$3x + 4y = 20 \qquad\qquad 2y - 4x = 12$$
$$4y = -3x + 20 \qquad\qquad 2y = 4x + 12$$
$$y = -\tfrac{3}{4}x + 5 \qquad\qquad y = 2x + 6$$

slope-intercept form

Straight lines with different slopes are **intersecting lines**. Look at the graph of the lines on the same Cartesian plane.

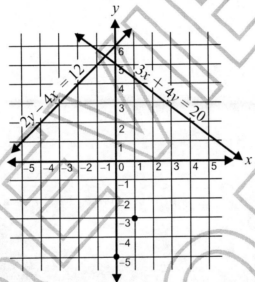

You can see from looking at the graph that the intersecting lines share one point in common. However, it is hard to tell from looking at the graph what the coordinates are for the point of intersection. To find the exact point of intersection, you can use the **substitution method** to solve the system of equations algebraically.

SOLVING SYSTEMS OF EQUATIONS BY SUBSTITUTION

You can solve systems of equations algebraically by using the substitution method.

EXAMPLE: Find the point of intersection of the following two equations:

Equation 1: $x - y = 3$
Equation 2: $2x + y = 9$

Step 1: Solve one of the equations for x or y. Let's choose to solve equation 1 for x.
Equation 1: $x - y = 3$
$x = y + 3$

Step 2: Substitute the value of x from equation 1 in place of x in equation 2.
Equation 2: $2x + y = 9$
$2(y + 3) + y = 9$
$2y + 6 + y = 9$
$3y + 6 = 9$
$3y = 3$
$y = 1$

Step 3: Substitute the solution for y back in equation 1 and solve for x.
Equation 1: $x - y = 3$
$x - 1 = 3$
$x = 4$

Step 4: The solution set is (4, 1). Substitute in one or both of the equations to check.

Equation 1: $x - y = 3$ Equation 2: $2x + y = 9$
$4 - 1 = 3$ $2(4) + 1 = 9$
$3 = 3$ $8 + 1 = 9$
$9 = 9$

The point (4, 1) is common to both equations. This is the **point of intersection**.

For each of the following pairs of equations, find the point of intersection, the common solution, using the substitution method.

1. $x + 2y = 8$
   $2x - 3y = 2$

2. $x - y = -5$
   $x + y = 1$

3. $x - y - 4$
   $x + y = 2$

4. $x - y = -1$
   $x + y = 9$

5. $-x + y = 2$
   $x + y = 8$

6. $x + 4y = 10$
   $x + 5y = 12$

7. $2x + 3y = 2$
   $4x - 9y = -1$

8. $x + 3y = 5$
   $x - y = 1$

9. $-x = y - 1$
   $x = y - 1$

10. $x - 2y = 2$
    $2y + x = -2$

11. $5x + 2y = 1$
    $2x + 4y = 10$

12. $3x - y = 2$
    $5x + y = 6$

13. $2x + 3y = 3$
    $4x + 5y = 5$

14. $x - y = 1$
    $-x - y = 1$

15. $x = y + 3$
    $y = 3 - x$

SOLVING SYSTEMS OF EQUATIONS BY ADDING OR SUBTRACTING

You can solve systems of equations algebraically by adding or subtracting an equation from another equation or system of equations.

EXAMPLE 1:   Find the point of intersection of the following two equations:
Equation 1: $x + y = 10$
Equation 2: $-x + 4y = 5$

Step 1:   Eliminate one of the variables by adding the two equations together. Since the x has the same coefficient in each equation, but opposite signs, it will cancel nicely by adding.

$$\begin{array}{r} x + y = 10 \\ + \; (-x + 4y = 5) \\ \hline 0 + 5y = 15 \\ 5y = 15 \\ y = 3 \end{array}$$

Add each like term together.
Simplify.
Divide both sides by 5.

Step 2:   Substitute the solution for y back into an equation, and solve for x.
Equation 1: $x + y = 10$ Substitute 3 for y.
$\qquad\qquad x + 3 = 10$ Subtract 3 from both sides.
$\qquad\qquad\quad x = 7$

Step 3:   The solution set is (7, 3). Substitute in both of the equations to check.

Equation 1: $x + y = 10$ Equation 2: $-x + 4y = 5$
$\qquad\qquad 7 + 3 = 10$ $\qquad\qquad\qquad -(7) + 4(3) = 5$
$\qquad\qquad\quad 10 = 10$ $\qquad\qquad\qquad\qquad -7 + 12 = 5$
$\qquad\qquad\qquad\qquad\qquad\qquad\qquad\qquad\qquad\qquad 5 = 5$

The point (7, 3) is the point of intersection.

EXAMPLE 2:   Find the point of intersection of the following two equations:
Equation 1: $3x - 2y = -1$
Equation 2: $-4y = -x - 7$

Step 1:   Put the variables on the same side of each equation. Take equation 2 out of y-intercept form.

$\qquad\qquad\qquad -4y = -x - 7$ Add x to both sides.
$\qquad\qquad x - 4y = -x + x - 7$ Simplify.
$\qquad\qquad x - 4y = -7$

Step 2:   Add the two equations together to cancel one variable. Since each variable has the same sign and different coefficients, we have to multiply one equation by a negative number so one of the variables will cancel. Equation 1's y variable has a coefficient of 2, and if multiplied by -2, the y will have the same variable as the y in equation 2, but a different sign. This will cancel nicely when added.

$\qquad\qquad -2(3x - 2y = -1)$ Multiply by -2.
$\qquad\qquad -6x + 4y = 2$

Step 3: Add the two equations.

$$-6x + 4y = 2$$
$$+ \quad (x - 4y = -7)$$

Add equation 2 to equation 1.

$$-5x + 0 = -5$$ Simplify.

$$-5x = -5$$ Divide both sides by -5.

$$x = 1$$

Step 4: Substitute the solution for x back into an equation and solve for y.

Equation 1: $3x - 2y = -1$ Substitute 1 for x.

$3(1) - 2y = -1$ Simplify.

$3 - 2y = -1$ Subtract 3 from both sides.

$3 - 3 - 2y = -1 - 3$ Simplify.

$-2y = -4$ Divide both sides by -2.

$y = 2$

Step 5: The solution set is $(1, 2)$. Substitute in both equations to check.

Equation 1: $3x - 2y = -1$ Equation 2: $-4y = -x - 7$

$3(1) - 2(2) = -1$ $-4(2) = -1 - 7$

$3 - 4 = -1$ $-8 = -8$

$-1 = -1$

The point (1, 2) is the point of intersection.

For each of the following pairs of equations, find the point of intersection by adding the 2 equations together. Remember you might need to change the coefficients and/or signs of the variables before adding.

1. $x + 2y = 8$
   $-x - 3y = 2$

2. $x - y = 5$
   $2x + y = 1$

3. $x - y = -1$
   $x + y = 9$

4. $3x - y = -1$
   $x + y = 13$

5. $-x + 4y = 2$
   $x + y = 8$

6. $x + 4y = 10$
   $x + 7y = 16$

7. $2x - y = 2$
   $4x - 9y = -3$

8. $x + 3y = 13$
   $5x - y = 1$

9. $-x = y - 1$
   $x = y - 1$

10. $x - y = 2$
    $2y + x = 5$

11. $5x + 2y = 1$
    $4x + 8y = 20$

12. $3x - 2y = 14$
    $x - y = 6$

13. $2x + 3y = 3$
    $3x + 5y = 5$

14. $x - 4y = 6$
    $-x - y = -1$

15. $x = 2y + 3$
    $y = 3 - x$

GRAPHING SYSTEMS OF INEQUALITIES

Systems of inequalities are best solved graphically. Look at the following example.

EXAMPLE: Sketch the solution set of the following system of inequalities:
$$y > -2x - 1 \text{ and } y \leq 3x$$

Step 1: Graph both inequalities on a Cartesian plane.

Step 2: Shade the portion of the graph that represents the solution set to each inequality.

Step 3: Any shaded region that overlaps is the solution set of both inequalities.

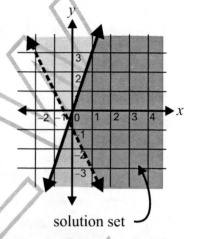

solution set

Graph the following systems of inequalities on your own graph paper. Shade and identify the solution set for both inequalities.

1. $2x + 2y \geq -4$
   $3y < 2x + 6$

2. $7x + 7y \leq 21$
   $8x < 6y - 24$

3. $9x + 12y < 36$
   $34x - 17y > 34$

4. $-11x - 22y \geq 44$
   $-4x + 2y \leq 8$

5. $24x < 72 + 36y$
   $11x + 22y \leq -33$

6. $15x - 60 < 30y$
   $20x + 10y < 40$

7. $-12x + 24y > -24$
   $10x < -5y + 15$

8. $y \geq 2x + 2$
   $y < -x - 3$

9. $3x + 4y \geq 12$
   $y > -3x + 2$

10. $-3x \leq 6 + 2y$
    $y \geq -x - 2$

11. $2x - 2y \leq 4$
    $3x + 3y \leq -9$

12. $-x \geq -2y - 2$
    $-2x - 2y > 4$

CHAPTER 15 REVIEW

For each pair of equations below, tell whether the graphs of the lines will be collinear, parallel, or intersecting.

1. $y = 4x + 1$
   $y = 4x - 3$

2. $y - 4 = x$
   $2x + 8 = 2y$

3. $x + y = 5$
   $x - y = -1$

4. $2y - 3x = 6$
   $4y = 6x + 8$

5. $5y = 3x - 7$
   $4x - 3y = -7$

6. $2x - 2y = 2$
   $y - x = -1$

Find the common solution for each of the following pairs of equations, using the substitution method.

7. $x - y = 2$
   $x + 4y = -3$

8. $x + y = 1$
   $x + 3y = 1$

9. $-4y = -2x + 4$
   $-x = -2y - 2$

10. $2x + 8y = 20$
    $5y = 12 - x$

11. $x = y - 3$
    $-x = y + 3$

12. $-2x + y = -3$
    $x - y = 9$

Graph the following systems of inequalities on your own graph paper. Identify the solution set to both inequalities.

13. $x + 2y \geq 2$
    $2x - y \leq 4$

14. $20x + 10y \leq 40$
    $3x + 2y \geq 6$

15. $6x + 8y \leq -24$
    $-4x + 8y \geq 16$

16. $14x - 7y \geq -28$
    $3x + 4y \leq -12$

17. $2y \geq 6x + 6$
    $2x - 4y \geq -4$

18. $9x - 6y \geq 18$
    $3y \geq 6x - 12$

Find the point of intersection for each pair of equations by adding and/or subtracting the two equations.

19. $2x + y = 4$
    $3x - y = 6$

20. $x + 2y = 3$
    $x + 5y = 0$

21. $x + y = 1$
    $y = x + 7$

22. $2x + 4y = 5$
    $3x + 8y = 9$

23. $2x - 2y = 7$
    $3x - 5y = \frac{5}{2}$

24. $x - 3y = -2$
    $y = -\frac{1}{3}x + 4$

RELATIONS

A **relation** is a set of ordered pairs. The set of the first members of each ordered pair is called the **domain** of the relation. The set of the second members of each ordered pair is called the **range**.

EXAMPLE: State the domain and range of the following relation.

$$\{(2, 4), (3, 7), (4, 9), (6, 11)\}$$

Solution: **Domain:** $\{2, 3, 4, 6\}$ the first member of each ordered pair
Range: $\{4, 7, 9, 11\}$ the second member of each ordered pair

State the domain and range for each relation.

1. $\{(2, 5), (9, 12), (3, 8), (6, 7)\}$

2. $\{(12, 4), (3, 4), (7, 12), (26, 19)\}$

3. $\{(4, 3), (7, 14), (16, 34), (5, 11)\}$

4. $\{(2, 45), (33, 43), (98, 9), (43, 61), (67, 54)\}$

5. $\{(78, 14), (29, 67), (84, 49), (16, 18), (98, 46)\}$

6. $\{(-8, 16), (23, -7), (-4, -9), (16, -8), (-3, 6)\}$

7. $\{(-7, -4), (-3, 16), (-4, 17), (-6, -8), (-8, 12)\}$

8. $\{(-1, -2), (3, 6), (-7, 14), (-2, 8), (-6, 2)\}$

9. $\{(0, 9), (-8, 5), (3, 12), (-8, -3), (7, 18)\}$

10. $\{(58, 14), (44, 97), (74, 32), (6, 18), (63, 44)\}$

11. $\{(-7, 0), (-8, 10), (-3, 11), (-7, -32), (-2, 57)\}$

12. $\{(18, 34), (22, 64), (94, 36), (11, 18), (91, 45)\}$

When given an equation in two variables, the **domain** is the set of x values that satisfies the equation. The **range** is the set of y values that satisfies the equation.

EXAMPLE: Find the range of the equation $3x = y + 2$ for the domain $\{-1, 0, 1, 2, 3\}$.

Solution: Solve the equation for each value of x given. The result, the y values, will be the range.

Given:

x	y
-1	
0	
1	
2	
3	

Solution:

x	y
-1	-5
0	-2
1	1
2	4
3	7

The range is $\{-5, -2, 1, 4, 7\}$.

Find the range of each relation for the given domain.

Relation	Domain	Range
1. $y = 5x$	$\{1, 2, 3, 4\}$	
2. $y = |x|$	$\{-3, -2, -1, 0, 1\}$	
3. $y = 3x + 2$	$\{0, 1, 3, 4\}$	
4. $y = -|x|$	$\{-2, -1, 0, 1, 2\}$	
5. $y = -2x + 1$	$\{0, 1, 3, 4\}$	
6. $y = 10x - 2$	$\{-2, -1, 0, 1, 2,\}$	
7. $y = 3|x| + 1$	$\{-2, -1, 0, 1, 2,\}$	
8. $y - x = 0$	$\{1, 2, 3, 4\}$	
9. $y - 2x = 0$	$\{1, 2, 3, 4\}$	
10. $y = 3x - 1$	$\{0, 1, 3, 4\}$	
11. $y = 4x + 2$	$\{0, 1, 3, 4\}$	
12. $y = 2|x| - 1$	$\{-2, -1, 0, 1, 2,\}$	

FUNCTIONS

Some relations are also **functions**. A relation is a function if **for every element in the domain, there is exactly one element in the range**. In other words, for each value for x there is only one unique value for y.

EXAMPLE 1: $\{(2, 4), (2, 5), (3, 4)\}$ is **NOT** a function because in the first pair, 2 is paired with 4, and in the second pair, 2 is paired with 5. The 2 can be paired with only one number to be a function. In this example, the x value of 2 has more than one value for y: 4 and 5.

EXAMPLE 2: $\{(1, 2), (3, 2), (5, 6)\}$ **IS** a function. Each first number is paired with only one second number. The 2 is repeated as a second number, but the relation remains a function.

Determine whether the ordered pairs of numbers below represent a function. Write "F" if it is a function. Write "NF" if it is not a function.

1. $\{(-1, 1), (-3, 3), (0, 0), (2, 2)\}$ _____

2. $\{(-4, -3), (-2, -3), (-1, -3), (2, -3)\}$ _____

3. $\{(5, -1), (2, 0), (2, 2), (5, 3)\}$ _____

4. $\{(-3, 3), (0, 2), (1, 1), (2, 0)\}$ _____

5. $\{(-2, -5), (-2, -1), (-2, 1), (-2, 3)\}$ _____

6. $\{(0, 2), (1, 1), (2, 2), (4, 3)\}$ _____

7. $\{(4, 2), (3, 3), (2, 2), (0, 3)\}$ _____

8. $\{(-1, -1), (-2, -2), (3, -1), (3, 2)\}$ _____

9. $\{(2, -2), (0, -2), (-2, 0), (1, -3)\}$ _____

10. $\{(2, 1), (3, 2), (4, 3), (5, -1)\}$ _____

11. $\{(-1, 0), (2, 1), (2, 4), (-2, 2)\}$ _____

12. $\{(1, 4), (2, 3), (0, 2), (0, 4)\}$ _____

13. $\{(0, 0), (1, 0), (2, 0), (3, 0)\}$ _____

14. $\{(-5, -1), (-3, -2), (-4, -9), (-7, -3)\}$ _____

15. $\{(8, -3), (-4, 4), (8, 0), (6, 2)\}$ _____

16. $\{(7, -1), (4, 3), (8, 2), (2, 8)\}$ _____

17. $\{(4, -3), (2, 0), (5, 3), (4, 1)\}$ _____

18. $\{(2, -6), (7, 3), (-3, 4), (2, -3)\}$ _____

19. $\{(1, 1), (3, -2), (4, 16), (1, -5)\}$ _____

20. $\{(5, 7), (3, 8), (5, 3), (6, 9)\}$ _____

FUNCTION NOTATION

Function notation is used to represent relations which are functions. Some commonly used letters to represent functions include f, g, h, F, G, and H.

EXAMPLE 1: $f(x) = 2x - 1$; find $f(-3)$

Find $f(-3)$ means replace x with -3 in the relation $2x - 1$.

$f(-3) = 2(-3) - 1$
$f(-3) = -6 - 1 = -7$

Solution: $f(-3) = -7$

EXAMPLE 2: $g(x) = 4 - 2x^2$: find $g(2)$

$g(2) = 4 - 2(2)^2 = 4 - 2(4) = 4 - 8 = -4$

Solution: $g(2) = -4$

Find solutions for each of the following.

1. $F(x) = 2 + 3x^2$; find $F(3)$

2. $f(x) = 4x + 6$; find $f(-4)$

3. $H(x) = 6 - 2x^2$; find $H(-1)$

4. $g(x) = -3x + 7$; find $g(-3)$

5. $f(x) = -5 + 4x$; find $f(7)$

6. $G(x) = 4x^2 + 4$; find $G(0)$

7. $f(x) = 7 - 6x$; find $f(-4)$

8. $h(x) = 2x^2 + 10$; find $h(5)$

9. $F(x) = 7 - 5x$; find $F(2)$

10. $f(x) = -4x^2 + 5$; find $f(-2)$

INDEPENDENT AND DEPENDENT VARIABLES

As stated previously, a relation is a function if for every element in the domain there is exactly one element in the range. The domain values are generally known, and the range values are determined by solving the function. As each domain value is applied to the function, only one range value will result. The variable that is used to represent the domain values is called the **independent variable** because it is not dependent on any other value. The variable that is used to represent the range values is called the **dependent variable** because its value will be determined by its corresponding domain value.

EXAMPLE: Mrs. Alexander assigned to her students an open book quiz containing 35 questions to be completed at home. Those students who returned the completed quiz by the due date would receive 30 points for turning the assignment in on time and 2 points for each correct answer. A student's grade on the open book quiz can be expressed as the function $f(a) = 30 + 2a$, where a represents the number of correct answers. Identify the independent and dependent variables in this function.

Solution: The independent variable in this problem is a, the number of correct answers because it is not dependent on any other value in the function.

The dependent variable in this problem is the grade, $f(a)$, because it is dependent on the number of correct answers. The dependent variable could have also been assigned a variable such as G, T, or y. Using function notation clearly illustrates in the algebraic sentence that the dependent variable is a function of the independent variable.

Identify the independent and dependent variables in the following functions.

1. A local bookstore is encouraging its customers to drop off used books to be given to schools, libraries, and other community organizations. They are offering to anyone who drops off books a special hard cover edition of *Oliver Twist* for $25.95 minus $.10 for each used book. The cost for the special edition of *Oliver Twist* can be expressed as $G(u) = \$25.95 - \$.10u$.

    Independent variable _____ Dependent variable _____

2. Claudia is planning a surprise birthday party for her best friend. To make sure that she has enough food, she is ordering 1 sub sandwich for every person who is coming to the party plus an additional 10 sub sandwiches. The number of sandwiches Claudia is ordering can be written algebraically as follows: $s = 10 + p$.

    Independent variable _____ Dependent variable _____

3. John and Mike are brothers who are training for their school swim team. John has been swimming longer than Mike and is able to swim more laps. For every lap that Mike swims John swims 3, and the number of laps that John swims can be expressed as $j = 3m$.

    Independent variable _____ Dependent variable _____

Write a function for each of the following word problems. Identify the independent and dependent variables.

1. All delivery drivers at Victor's Pizza Pub are hired to work 5 hour shifts. For each shift worked, a delivery driver gets paid $40 plus $2 for every pizza delivered. Write a function that expresses a delivery driver's earnings for one shift.

2. At 8:00 am the temperature outside was $50°$. As the morning progressed, the temperature rose by $3°$ every hour. Write a function that describes the temperature at any given hour after 8:00 am.

3. Austin wanted to borrow $325 from his father to buy a new mountain bike. His father agreed to loan him the money if Austin would pay off the debt by doing odd jobs around the house and in the yard earning a wage of $6.50/hour. Write a function that will help Austin calculate how much debt he has left to pay his father.

4. A new shopping center is leasing store space at a monthly rate of $3.00/ft^2. Each individual store will be 20 ft wide, but the length will vary. Write a function that expresses the monthly lease rate of any individual store.
Remember that *Area = Length × Width*.

5. Every year a professional baseball player gives $10,000 to a national research fund. He also gives $1500 for every home run he hits. Express the baseball player's contributions as an algebraic sentence.

6. The local natural gas company charges a monthly usage fee of $25. In addition, each household is charged $.67 per therm of natural gas used during the month. Write a function that a homeowner could use to calculate his/her monthly gas bill.

7. Boy Scout Troop 575 is planning an exciting summer mountain adventure. To raise money for the trip the boys are selling tins of popcorn. Each member of the troop must pay $400 for the trip. For each case of popcorn a Boy Scout sells, $10 will be applied toward his trip fees. Write a formula that describes the amount of money a Boy Scout must pay out of pocket.

8. Oak Hills High School is putting on a spring musical. Tickets are being sold for $6.50 per person. The drama club at Oak Hills gets $\frac{1}{3}$ of the total ticket sales to use for future programs. Write a function that expresses how much money the drama club will receive from the spring musical.

9. Josie rented a car for one day from a company that charges $30 per day plus $.20 per mile. What function would Josie use to calculate her total bill before taxes?

10. Hannah wanted to participate in a yard sale being sponsored by her school. She would have to pay $5.00 to rent the space for her items and then would receive 65% of the money her items generated. The remaining 35% would be given to a local charity. Write a function that expresses Hannah's net profit.

CHAPTER 16 REVIEW

1. What is the domain of the following relation?

$\{(-1, 2), (2, 5), (4, 9), (6, 11)\}$

2. What is the range of the following relation?

$\{(0, -2), (-1, -4), (-2, 6), (-3, -8)\}$

3. Find the range of the relation $y = 5x$ for the domain $\{0, 1, 2, 3, 4\}$.

4. Find the range of the relation $y = 2(x - 4)$ for the domain $\{2, 3, 4, 5, 6\}$.

5. Find the range of the following relation for the domain $\{0, 2, 6, 8, 10\}$.

$$y = \frac{3(x - 4)}{2}$$

6. Find the range of the following relation for the domain $\{-8, -3, 7, 12, 17\}$.

$$y = \frac{3(x - 2)}{5}$$

7. Find the range of the following relation for the domain $\{-8, -4, 0, 4, 8\}$.

$$y = 10 - 2x$$

8. Find the range of the following relation for the domain $\{-7, -1, 2, 5, 8\}$.

$$y = \frac{4 + x}{3}$$

9. Trent sells computers and other electronic devices for Computer Town. He receives $300 per week and 60% of his total sales. Write a function that expresses Trent's weekly earnings. Identify the independent and dependent variables.

Function _____

Independent variable _____

Dependent variable _____

For each of the following relations given in questions 11-15, write F if it is a function and NF if it is not a function.

10. $\{(1, 2), (2, 2), (3, 2)\}$ _____

11. $\{(-1, 0), (0, 1), (1, 2), (2, 3)\}$ _____

12. $\{(2, 1), (2, 2), (2, 3)\}$ _____

13. $\{(1, 7), (2, 5), (3, 6), (2, 4)\}$ _____

14. $\{(0, -1), (-1, -2), (-2, -3), (-3, -4)\}$ _____

For questions 15-20, find the range of the following functions for the given value of the domain.

15. For $g(x) = 2x^2 - 4x$, find $g(-1)$

16. For $h(x) = 3x(x - 4)$, find $h(3)$

17. For $f(n) = \frac{1}{n + 3}$, find $f(4)$

18. For $G(n) = \frac{2 - n}{2}$, find $G(8)$

19. For $H(x) = 2x(x - 1)$, find $H(4)$

20. For $f(x) = 7x^2 + 3x - 2$, find $f(2)$

PERIMETER

The **perimeter** is the distance around a polygon. To find the perimeter, add the lengths of the sides.

EXAMPLES:

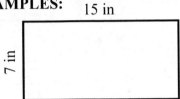

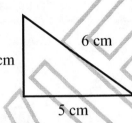

$P = 7 + 15 + 7 + 15$
$P = 44$ in

$P = 4 + 6 + 5$
$P = 15$ cm

$P = 8 + 15 + 20 + 12 + 10$
$P = 65$ ft

Find the perimeter of the following polygons.

1.

8 in

5 in

_____

4.

13 cm

15 cm

10 cm

_____

7.

6 ft

6 ft

_____

10.

8 cm

1 cm

_____

2.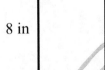

3 ft

2 ft   2 ft

5 ft   5 ft

_____

5.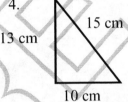

12 in

8 in   8 in

16 in

_____

8.

25 cm

20 cm   22 cm

10 cm

13 cm

_____

11.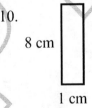

7 ft   6 ft

8 ft

_____

3.

32 cm

29 cm

33 cm

29 cm

10 cm

35 cm

_____

6.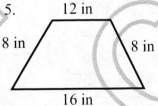

9 ft   7 ft

5 ft   6 ft

_____

9.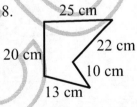

4 in

4 in   4 in

4 in   4 in

4 in   4 in

4 in

_____

12.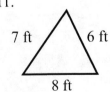

7 cm

28 cm

30 cm

24 cm

_____

AREA OF SQUARES AND RECTANGLES

Area - area is always expressed in square units, such as in^2, cm^2, ft^2, and m^2.

The area, (A), of squares and rectangles equals length (ℓ) times width (w). **$A = \ell\, w$**

EXAMPLE:

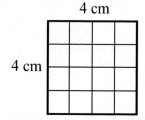
4 cm

A = ℓ w
A = 4 × 4
A = 16 cm²

If a square has an area of 16 cm^2, it means that it will take 16 squares that are 1 cm on each side to cover the area of a square that is 4 cm on each side.

Find the area of the following squares and rectangles, using the formula $A = \ell w$.

1.
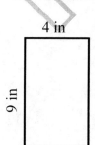

_____

2.
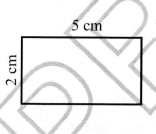

_____

3.
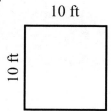

_____

4.
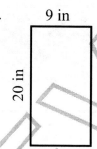

_____

5.
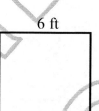

_____

6.
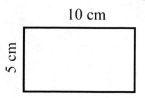

_____

7.
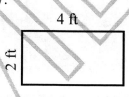

_____

8.
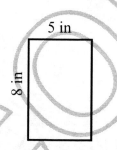

_____

9.
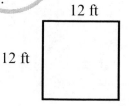

_____

10.

_____

11.
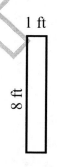

_____

12.
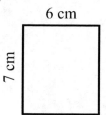

_____

AREA OF TRIANGLES

EXAMPLE: Find the area of the following triangle.

The formula for the area of a triangle is as follows:

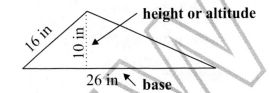

$$A = \frac{1}{2} \times b \times h$$

A = **area**
b = **base**
h = **height or altitude**

Step 1: Insert measurements from the triangle into the formula: $A = \frac{1}{2} \times 26 \times 10$

Step 2: Cancel and multiply. $A = \frac{1}{\cancel{2}} \times \frac{\cancel{26}^{13}}{1} \times \frac{10}{1} = 130 \text{ in}^2$

Note: **Area is always expressed in square units such as in^2, ft^2, cm^2, or m^2.**

Find the area of the following triangles..

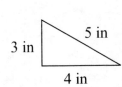

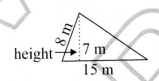

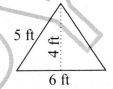

1. _____ in^2 4. _____ cm^2 7. _____ m^2 10. _____ ft^2

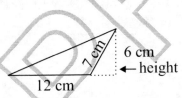

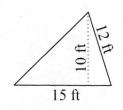

2. _____ cm^2 5. _____ ft^2 8. _____ in^2 11. _____ ft^2

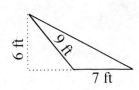

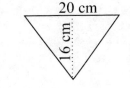

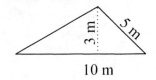

3. _____ ft^2 6. _____ cm^2 9. _____ ft^2 12. _____ m^2

235

AREA OF TRAPEZOIDS AND PARALLELOGRAMS

EXAMPLE 1: Find the area of the following parallelogram.

The formula for the area of a parallelogram is: $A = bh$.

A = **area**
b = **base**
h = **height**

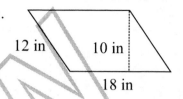

Step 1: Insert measurements from the parallelogram into the formula: $A = 18 \times 10$.

Step 2: Multiply. $18 \times 10 = 180$ in^2

EXAMPLE 2: Find the area of the following trapezoid.

The formula for the area of a trapezoid is $A = \frac{1}{2} h (b_1 + b_2)$. A trapezoid has two bases that are parallel to each other. When you add the length of the two bases together and then multiply by $\frac{1}{2}$, you find their average length.

A = **area**
b = **base**
h = **height**

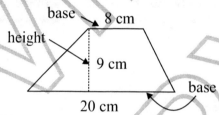

Insert measurements from the trapezoid into the formula and solve -
$\frac{1}{2} \times 9 \ (8 + 20) = 126$ cm^2.

Find the area of the following parallelograms and trapezoids.

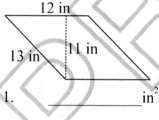

1. _____ in^2

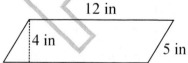

2. _____ in^2

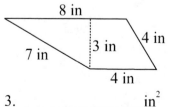

3. _____ in^2

4. _____ cm^2

5. _____ in^2

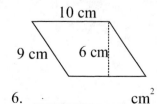

6. _____ cm^2

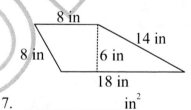

7. _____ in^2

8. _____ cm^2

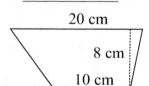

9. _____ cm^2

PARTS OF A CIRCLE

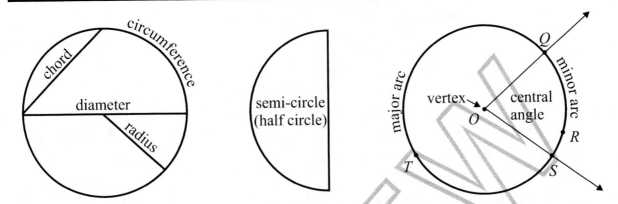

A **central angle** of a circle has the center of the circle as its vertex. The rays of a central angle each contain a radius of the circle. ∠QOS is a central angle.

The points Q and S separate the circle into **arcs**. The arc lies on the circle itself. It does not include any points inside or outside the circle. $\overarc{QRS}$, or $\overarc{QS}$, is a **minor arc** because it is less than a semicircle. A minor arc can be named by 2 or 3 points. $\overarc{QTS}$ is a **major arc** because it is more than a semicircle. A major arc must be named by 3 points.

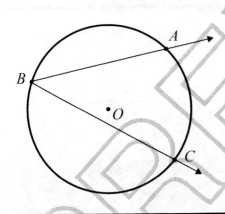

An **inscribed angle** is an angle whose vertex lies on the circle and whose sides contain **chords** of the circle. ∠ABC is an inscribed angle.

These angles are not inscribed.

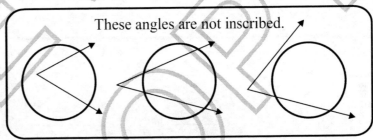

Refer to the figure on the right, and answer the following questions.

1. Identify the 2 line segments that are chords of the circle but not diameters. _____

2. Identify the largest major arc of the circle that contains point S. _____

3. Identify the vertex of the circle. _____

4. Identify the inscribed angle. _____

5. Identify the central angle. _____

6. Identify the line segment that is a diameter of the circle. _____

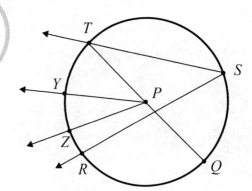

CIRCUMFERENCE

Circumference, *C* - the distance around the outside of a circle
Diameter, *d* - a line segment passing through the center of a circle from one side to the other
Radius, *r* - a line segment from the center of a circle to the edge of the circle
Pi, π - the ratio of the circumference of a circle to its diameter $\pi = 3.14$ or $\pi = \dfrac{22}{7}$

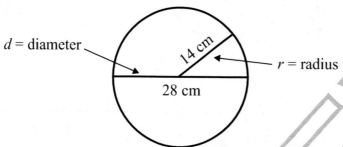

The formula for the circumference of a circle is $C = 2\pi r$ or $C = \pi d$. (The formulas are equal because the diameter is equal to twice the radius, $d = 2r$.)

EXAMPLE:

Find the circumference of the circle above.

$C = \pi d$ Use $= 3.14$
$C = 3.14 \times 28$
$C = 87.92$ cm

EXAMPLE:

Find the circumference of the circle above.

$C = 2\pi r$
$C = 2 \times 3.14 \times 14$
$C = 87.92$ cm

**Use the formulas given above to find the circumference of the following circles.
Use π = 3.14.**

1. 8 in

2. 14 ft

3. 2 cm

4. 6 m

5. 8 ft

$C =$ _____ $C =$ _____ $C =$ _____ $C =$ _____ $C =$ _____

**Use the formulas given above to find the circumference of the following circles.
Use π = $\dfrac{22}{7}$.**

6. 3 ft

7. 12 in

8. 6 m

9. 5 cm

10. 16 in

$C =$ _____ $C =$ _____ $C =$ _____ $C =$ _____ $C =$ _____

238

AREA OF A CIRCLE

The formula for the area of a circle is $A = \pi r^2$. The area is how many square units of measure would fit inside a circle.

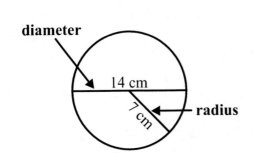

$$\pi = \frac{22}{7} \qquad \text{or} \qquad \pi = 3.14$$

EXAMPLE: Find the area of the circle, using both values for π.

Let $\pi = \frac{22}{7}$

$A = \pi r^2$

$A = \frac{22}{7} \times 7^2$

$A = \frac{22}{\cancel{7}} \times \frac{\cancel{49}}{1} = 154 \text{ cm}^2$

Let $\pi = 3.14$

$A = \pi r^2$

$A = 3.14 \times 7^2$

$A = 3.14 \times 49 = 153.86 \text{ cm}^2$

Find the area of the following circles. Remember to include units.

$\pi = 3.14 \qquad \pi = \frac{22}{7}$

1.

5 in

$A = \underline{\hspace{1cm}}$ $A = \underline{\hspace{1cm}}$

2.
16 ft

$A = \underline{\hspace{1cm}}$ $A = \underline{\hspace{1cm}}$

3.
8 cm

$A = \underline{\hspace{1cm}}$ $A = \underline{\hspace{1cm}}$

4.

3 m

$A = \underline{\hspace{1cm}}$ $A = \underline{\hspace{1cm}}$

Fill in the chart below. Include appropriate Units.

	Radius	Diameter	Area $\pi = 3.14$	$\pi = \frac{22}{7}$
5.	9 ft			
6.		4 in		
7.	8 cm			
8.		20 ft		
9.	14 m			
10.		18 cm		
11.	12 ft			
12.		6 in		

TWO-STEP AREA PROBLEMS

Solving the problems below will require two steps. You will need to find the area of two figures, and then either add or subtract the two areas to find the answer. **Carefully read the EXAMPLES below.**

EXAMPLE 1:

Find the area of the living room below.

Figure 1

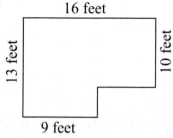

Step 1: Complete the rectangle as in Figure 2, and compute the area as if it were a complete rectangle.

Figure 2

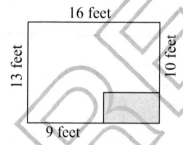

A = length × width
$A = 16 \times 13$
$A = 208 \text{ ft}^2$

Step 2: Figure the area of the shaded part.

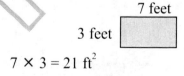

$7 \times 3 = 21 \text{ ft}^2$

Step 3: Subtract the area of the shaded part from the area of the complete rectangle.

$208 - 21 = 187 \text{ ft}^2$

EXAMPLE 2:

Find the area of the shaded sidewalk.

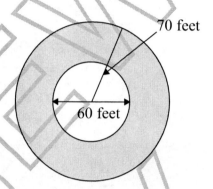

Step 1: Find the area of the outside circle.
$\pi = 3.14$
$A = 3.14 \times 70 \times 70$
$A = 15,386 \text{ ft}^2$

Step 2: Find the area of the inside circle.
$\pi = 3.14$
$A = 3.14 \times 30 \times 30$
$A = 2,826 \text{ ft}^2$

Step 3: Subtract the area of the inside circle from the area of the outside circle.

$15,386 - 2,826 = 12,560 \text{ ft}^2$

Find the area of the following figures.

1.

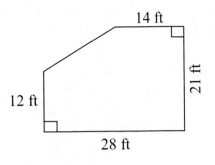

14 ft

21 ft

12 ft

28 ft

_____

2.

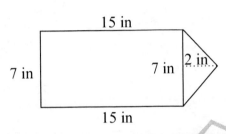

15 in

7 in   7 in 2 in

15 in

_____

3.   What is the area of the shaded circle?
Use π = 3.14, and round the answer to
the nearest whole number.

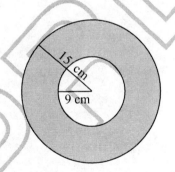

15 cm

9 cm

_____

4.

1 ft

5 ft

4 ft

18 ft

_____

5.   What is the area of the rectangle that is
shaded? Use π = 3.14 and round to the
nearest whole number.

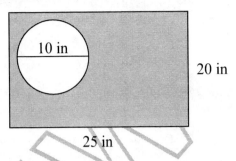

10 in

20 in

25 in

_____

6.   What is the area of the shaded part?

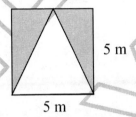

5 m

5 m

_____

7.   What is the area of the shaded part?

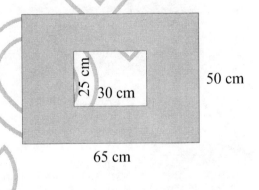

25 cm   30 cm   50 cm

65 cm

_____

24 m

8. 6 m

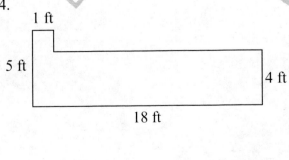

12 m

12 m

_____

241

MEASURING TO FIND PERIMETER AND AREA

Using a ruler, measure the perimeter of the following figures, and calculate the area. Be careful to measure in the correct unit. To calculate the area of some of the figures, you will have to use the two-step approach.

1.

$P = $ _____ cm

$A = $ _____ cm^2

4.

$P = $ _____ in

$A = $ _____ in^2

7.

$P = $ _____ cm

$A = $ _____ cm^2

2.

$P = $ _____ in

$A = $ _____ in^2

5.

$P = $ _____ in

$A = $ _____ in^2

8.

$P = $ _____ cm

$A = $ _____ cm^2

3.

$P = $ _____ in

$A = $ _____ in^2

6.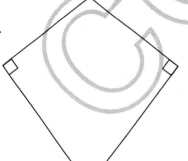

$P = $ _____ cm

$A = $ _____ cm^2

9.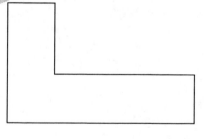

$P = $ _____ in

$A = $ _____ in^2

SIMILAR TRIANGLES

Two triangles are similar if the measurements of the three angles in both triangles are the same. If the three angles are the same, then their corresponding sides are proportional.

CORRESPONDING SIDES - The triangles below are similar. Therefore, the two shortest sides from each triangle, *c* and *f*, are corresponding. The two longest sides from each triangle, *a* and *d*, are corresponding. The two medium length sides, *b* and *e*, are corresponding.

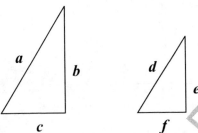

PROPORTIONAL - The corresponding sides of similar triangles are proportional to each other. This means if we know all the measurements of one triangle, and we only know one measurement of the other triangle, we can figure out the measurements of the other two sides with proportion problems. The two triangles below are similar.

Note: **To set up the proportion correctly, it is important to keep the measurements of each triangle on opposite sides of the equal sign.**

To find the short side:	To find the medium length side:

Step 1: Set up the proportion.

$$\frac{long\ side}{short\ side} \qquad \frac{12}{6} = \frac{16}{?}$$

Step 2: Solve the proportion.

$$16 \times 6 = 96$$
$$96 \div 12 = 8$$

Step 1: Set up the proportion.

$$\frac{long\ side}{medium} \qquad \frac{12}{9} = \frac{16}{??}$$

Step 2: Solve the proportion.

$$16 \times 9 = 144$$
$$144 \div 12 = 12$$

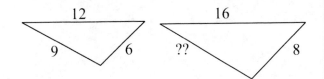

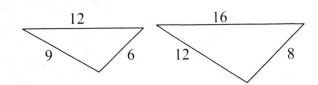

MORE SIMILAR TRIANGLES

Find the missing side from the following similar triangles.

1.
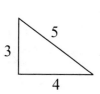
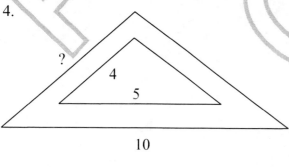

5.
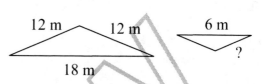
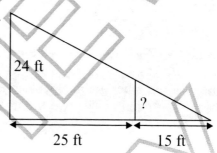

2.
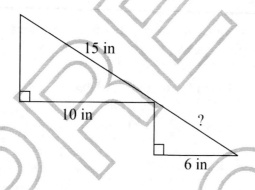

6.
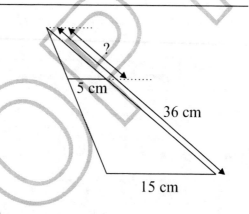

3.
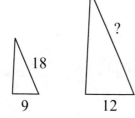

7.
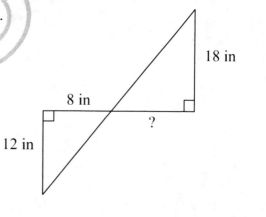

4.

8.

PYTHAGOREAN THEOREM

Pythagoras was a Greek mathematician and philosopher who lived around 600 B.C. He started a math club among Greek aristocrats called the Pythagoreans. Pythagoras formulated the **Pythagorean Theorem**, which states that in a **right triangle**, the sum of the squares of the legs of the triangle are equal to the square of the hypotenuse. Most often you will see this formula written as $a^2 + b^2 = c^2$. **This relationship is only true for right triangles.**

EXAMPLE: Find the length of side c.

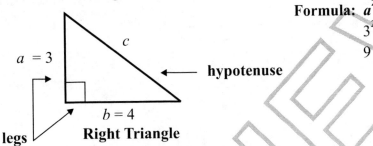

Formula: $a^2 + b^2 = c^2$

$3^2 + 4^2 = c^2$

$9 + 16 = c^2$

$25 = c^2$

$25 = c^2$

$\sqrt{25} = \sqrt{c^2}$

$5 = c$

Find the hypotenuse of the following triangles. Round the answers to two decimal places.

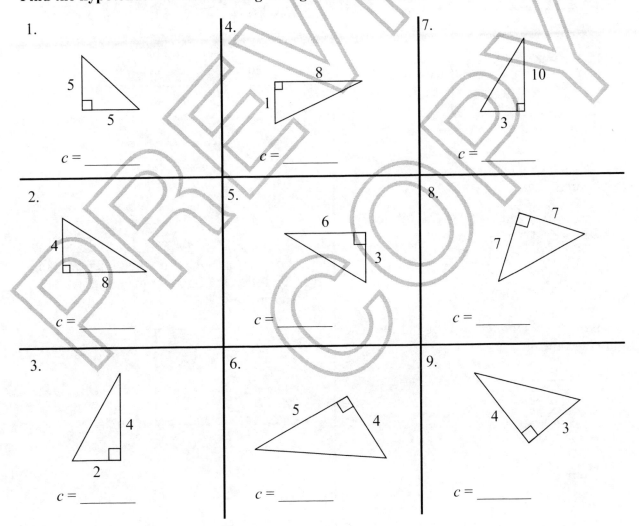

1.

5

5

$c = $ _____

4.

8

1

$c = $ _____

7.

10

3

$c = $ _____

2.

4

8

$c = $ _____

5.

6

3

$c = $ _____

8.

7

7

$c = $ _____

3.

4

2

$c = $ _____

6.

5

4

$c = $ _____

9.

4

3

$c = $ _____

CHAPTER 17 REVIEW

1. What is the length of line segment $\overline{WY}$?

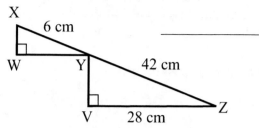

_____

2. Find the missing side. Round your answer to two decimal places.

_____

3. Find the area of the shaded region of the figure below.

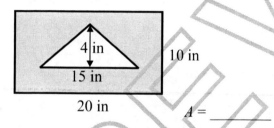

$A =$ _____

4. Calculate the perimeter of the following figure.

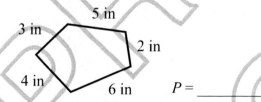

$P =$ _____

Calculate the perimeter and area of the following figures.

5.

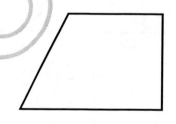

$P =$ _____

$A =$ _____

6.

7 in

4 in

$P =$ _____

$A =$ _____

Calculate the circumference and the area of the following circles.

7.

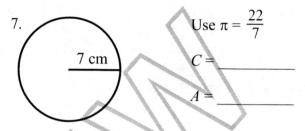

Use $\pi = \dfrac{22}{7}$

$C =$ _____

$A =$ _____

8.

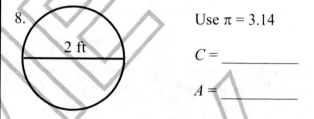

Use $\pi = 3.14$

$C =$ _____

$A =$ _____

9. Use $\pi = 3.14$ to find the area of the shaded part. Round your answer to the nearest whole number.

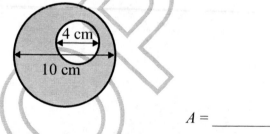

$A =$ _____

10. Use a ruler to measure the dimensions of the following figure in inches. Find the perimeter.

$P =$ _____

11. Use a ruler to measure the dimensions of the following figure in centimeters. Find the perimeter and area.

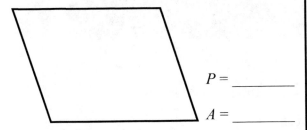

$P = $ _____

$A = $ _____

12. The shaded area below represents Grimes National Park. On the grid below, each square represents 10 square miles. Estimate the area of Grimes National Park.

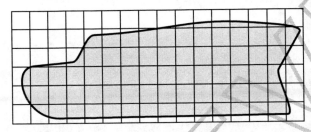

Area is about _____

13. The following two triangles are similar. Find the length of the missing side.

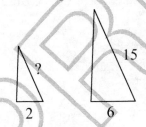

_____

Find the area of the following figures.

14.

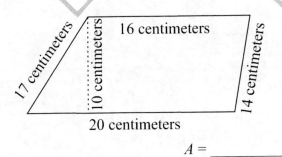

$A = $ _____

15.

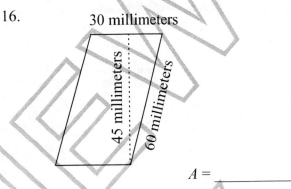

$A = $ _____

16.

$A = $ _____

17.

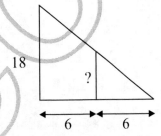

$A = $ _____

18. Find the missing side of the triangle below.

_____

19. What is the area of a square which measures 8 inches on each side?

$A = $ _____

Chapter 18 | Solid Geometry

In this chapter, you will learn about the following three-dimensional shapes.

SOLIDS

cube

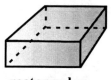
rectangular prism

cone

cylinder

sphere

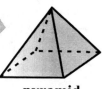
pyramid

UNDERSTANDING VOLUME

Volume - Measurement of volume is expressed in cubic units such as in^3, ft^3, m^3, cm^3, or mm^3. The volume of a solid is the number of cubic units that can be contained in the solid.

First, let's look at rectangular solids.

EXAMPLE: How many 1 cubic centimeter cubes will it take to fill up the figure below?

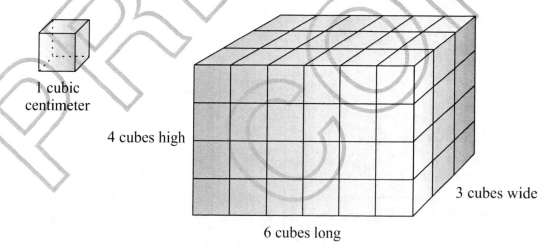

1 cubic centimeter

4 cubes high

3 cubes wide

6 cubes long

To find the volume, you need to multiply the length, times the width, times the height.

Volume of a rectangular solid = length $\times$ width $\times$ height ($V = l\,w\,h$).

$V = 6 \times 3 \times 4 = 72\ cm^3$

VOLUME OF RECTANGULAR PRISMS

You can calculate the volume (*V*) of a rectangular prism (box) by multiplying the length (*l*) by the width (*w*) by the height (*h*), as expressed in the formula $V = (lwh)$.

EXAMPLE: Find the volume of the box pictured on the right.

Step 1: Insert measurements from the figure into the formula.

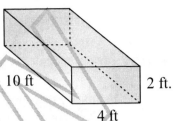

Step 2: Multiply to solve. $10 \times 4 \times 2 = 80$ ft^3

Note: **Volume is always expressed in cubic units such as in^3, ft^3, m^3, cm^3, or mm^3.**

Find the volume of the following rectangular prisms (boxes).

1.
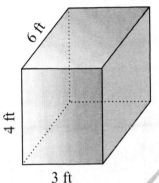

6 ft

4 ft

3 ft

V = _____

2.
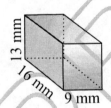

13 mm

16 mm

9 mm

V = _____

3.
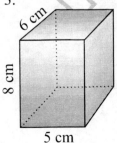

6 cm

8 cm

5 cm

V = _____

4.
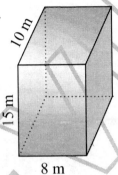

10 m

15 m

8 m

V = _____

5.
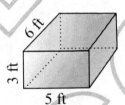

6 ft

3 ft

5 ft

V = _____

6.
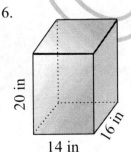

20 in

16 in

14 in

V = _____

7.
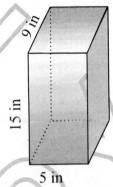

9 in

15 in

5 in

V = _____

8.
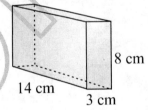

8 cm

14 cm

3 cm

V = _____

9.
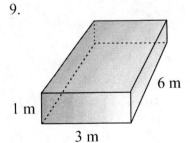

6 m

1 m

3 m

V = _____

VOLUME OF CUBES

A **cube** is a special kind of rectangular prism (box). Each side of a cube has the same measure. So, the formula for the volume of a cube is $V = s^3$ ($s \times s \times s$).

EXAMPLE: Find the volume of the cube pictured at the right.

Step 1: Insert measurements from the figure into the formula.

Step 2: Multiply to solve. $5 \times 5 \times 5 = 125 \text{ cm}^3$

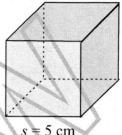

$s = 5$ cm

Note: Volume is always expressed in cubic units such as in^3, ft^3, m^3, cm^3, or mm^3.

Answer each of the following questions about cubes.

1. If a cube is 3 centimeters on each edge, what is the volume of the cube?

   _____

2. If the measure of the edge is doubled to 6 centimeters on each edge, what is the volume of the cube?

   _____

3. What if the edge of a 3 centimeter cube is tripled to become 9 centimeters on each edge? What will the volume be?

   _____

4. How many cubes with edges measuring 3 centimeters would you need to stack together to make a solid 12 centimeter cube?

   _____

5. What is the volume of a 2 centimeter cube?

   _____

6. Jerry built a 2 inch cube to hold his marble collection. He wants to build a cube with a volume 8 times larger. How much will each edge measure?

   _____

Find the volume of the following cubes.

7.
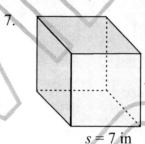

$s = 7$ in

$V =$ _____

8.
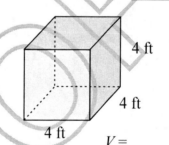

4 ft

4 ft

4 ft

$V =$ _____

9. 12 inches = 1 foot

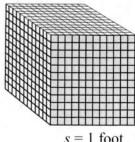

$s = 1$ foot

How many cubic inches are in a cubic foot? _____

VOLUME OF SPHERES, CONES, CYLINDERS, AND PYRAMIDS

To find the volume of a solid, insert the measurements given for the solid into the correct formula and solve. Remember, volumes are expressed in cubic units such as in^3, ft^3, m^3, cm^3, or mm^3.

Sphere
$$V = \frac{4}{3}\pi r^3$$

Cone
$$V = \frac{1}{3}\pi r^2 h$$

Cylinder
$$V = \pi r^2 h$$

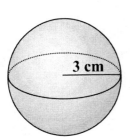

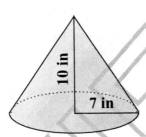

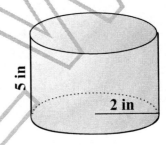

$V = \frac{4}{3}\pi r^3$ $\pi = 3.14$
$V = \frac{4}{3} \times 3.14 \times 27$
$V = 113.04 \ cm^3$

$V = \frac{1}{3}\pi r^2 h$ $\pi = 3.14$
$V = \frac{1}{3} \times 3.14 \times 49 \times 10$
$V = 512.87 \ in^3$

$V = \pi r^2 h$ $\pi = \frac{22}{7}$
$V = \frac{22}{7} \times 4 \times 5$
$V = 62\frac{6}{7} \ in^3$

Pyramids

$V = \frac{1}{3}Bh$ B = area of rectangular base

$V = \frac{1}{3}Bh$ B = area of triangular base

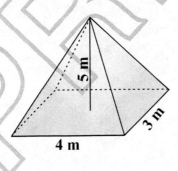

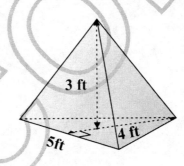

$V = \frac{1}{3}Bh$ $B = l \times w$
$V = \frac{1}{3} \times 4 \times 3 \times 5$
$V = 20 \ m^3$

$B = \frac{1}{2} \times 5 \times 4 = 10 \ ft^2$
$V = \frac{1}{3} \times 10 \times 3$
$V = 10 \ ft^3$

Find the volume of the following shapes. Use π = 3.14.

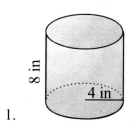

1. $V =$ _____

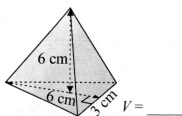

2. $V =$ _____

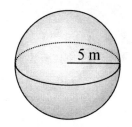

3. $V =$ _____

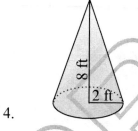

4. $V =$ _____

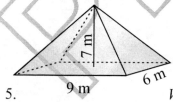

5. $V =$ _____

6. $V =$ _____

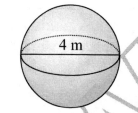

7. $V =$ _____

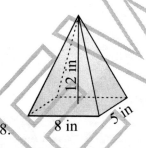

8. $V =$ _____

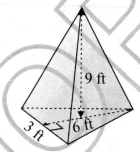

9. $V =$ _____

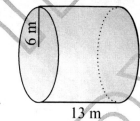

10. $V =$ _____

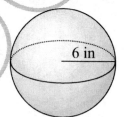

11. $V =$ _____

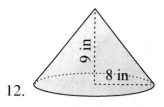

12. $V =$ _____

TWO-STEP VOLUME PROBLEMS

Some objects are made from two geometric figures, for example the tower below.

EXAMPLE: Find the maximum volume of the towel pictured at the right.

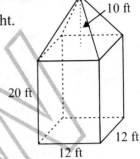

Step 1: Determine which formulas you will need. The tower is made from a pyramid and a rectangular prism, so you will need the formulas for the volume of these two figures.

Step 2: Find the volume of each part of the tower.
The bottom of the tower is a rectangular prism. $V = lwh$
$V = 12 \times 12 \times 20 = 2,880 \text{ ft}^3$

The top of the tower is a rectangular pyramid. $V = \frac{1}{3}Bh$
$V = \frac{1}{3} \times 12 \times 12 \times 10 = 480 \text{ ft}^3$

Step 3: Add the two volumes together. $2880 \text{ ft}^3 + 480 \text{ ft}^3 = 3,360 \text{ ft}^3$

Find the volume of the geometric figures below. *Hint:* If part of a solid has been removed, find the volume of the hole, and subtract it from the volume of the total object.

1.

2. Each side of the cubes in the figure below measures 3 inches.

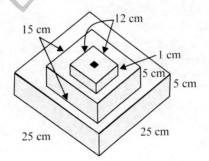

3. A rectangular hole passes through the middle of the figure below. The hole measures 1 cm on each side.

4. In the figure below, 3 cylinders are stacked on top of one another. The radii of the cylinders are 2 inches, 4 inches, and 6 inches. The height of each cylinder is 1 inch.

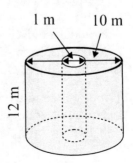

5.

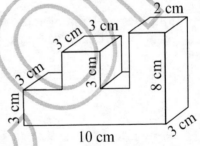

6. A hole, 1 meter in diameter, has been cut through the cylinder below.

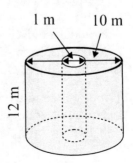

GEOMETRIC RELATIONSHIPS OF SOLIDS

In the previous chapter, you looked at geometric relationships between 2-dimensional figures. Now you will learn about the relationships between 3-dimensional figures. The formulas for finding the volumes of geometric solids are given below.

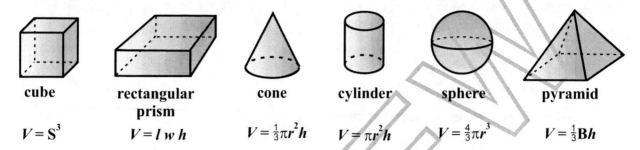

cube	rectangular prism	cone	cylinder	sphere	pyramid
$V = S^3$	$V = l\,w\,h$	$V = \frac{1}{3}\pi r^2 h$	$V = \pi r^2 h$	$V = \frac{4}{3}\pi r^3$	$V = \frac{1}{3}Bh$

By studying each formula and by comparing formulas between different solids, you can determine general relationships.

EXAMPLE 1: How would doubling the radius of a sphere affect the volume?

The volume of a sphere is $V = \frac{4}{3}\pi r^3$. Just by looking at the formula, can you see that by doubling the radius, the volume would increase by 8 times the original volume? So, a sphere with a radius of 2 would have a volume 8 times greater than a sphere with a radius of 1.

EXAMPLE 2: A cylinder and a cone have the same radius and the same height. What is the difference between their volumes?

Compare the formulas for the volume of a cone and the volume of a cylinder. They are identical except that the cone is multiplied by $\frac{1}{3}$. Therefore, the volume of a cone with the same height and radius as a cylinder would be one-third less. Or, the volume of a cylinder with the same height and radius as a cone would be three times greater.

EXAMPLE 3: If you double one dimension of a rectangular prism, how will the volume be affected? How about doubling two dimensions? How about doubling all three dimensions?

Do you see that doubling just one of the dimensions of a rectangular prism will also double the volume? Doubling two of the dimensions will cause the volume to increase 4 times the original volume. Doubling all three dimensions will cause the volume to increase 8 times the original volume.

EXAMPLE 4: A cylinder holds 100 cubic centimeters of water. If you triple the radius of the cylinder but keep the height the same, how much water would you need to fill the new cylinder?

Tripling the radius of a cylinder causes the volume to increase by 3^2, or 9 times the original volume. The volume of the new cylinder would hold 9×100 or 900 cubic centimeters of water.

Answer the following questions by comparing the volumes of two solids that share some of the same dimensions.

1. If you have a cylinder with a height of 8 inches and a radius of 4 inches, and you have a cone with the same height and radius, how many times greater is the volume of the cylinder than the volume of the cone?

2.

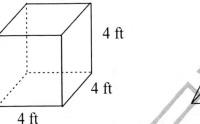

   In the two figures above, how many times larger is the volume of the cube than the volume of the pyramid?

3. How many times greater is the volume of a cylinder if you double the radius?

4. How many times greater is the volume of a cylinder if you double the height?

5. In a rectangular solid, how many times greater is the volume if you double the length?

6. In a rectangular solid, how many times greater is the volume if you double the length and the width?

7. In a rectangular solid, how many times greater is the volume if you double the length and the width and the height?

8. In the following two figures, how many cubes like Figure 1 will fit inside Figure 2?

   Figure 1 6 in, 6 in, 6 in   Figure 2 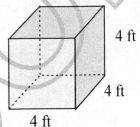 4 ft, 4 ft, 4 ft

9. A sphere has a radius of 1. If the radius is increased to 3, how many times greater will the volume be?

10. It takes 2 liters of water to fill cone A below. If the cone is stretched so the radius is doubled, but the height stays the same, how much water is needed to fill the new cone, B?

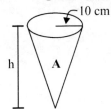 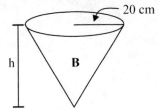

SURFACE AREA

The **surface area of a solid** is the total area of all the sides of a solid.

CUBE

There are six sides on a cube. To find the surface area of a cube, find the area of one side and multiply by 6.

Area of each side of the cube:
$3 \times 3 = 9 \text{ cm}^2$

Total surface area: $9 \times 6 = 54 \text{ cm}^2$

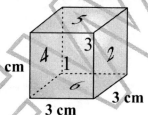

RECTANGULAR PRISM

There are 6 sides on a rectangular prism. To find the surface area, add the areas of the six rectangular sides.

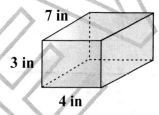

Top and Bottom

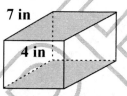

Area of top side:
$7 \text{ in} \times 4 \text{ in} = 28 \text{ in}^2$
Area of top and bottom:
$28 \text{ in} \times 2 \text{ in} = 56 \text{ in}^2$

Front and Back

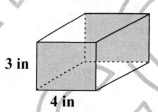

Area of front:
$3 \text{ in} \times 4 \text{ in} = 12 \text{ in}^2$
Area of front and back:
$12 \text{ in} \times 2 \text{ in} = 24 \text{ in}^2$

Left and Right

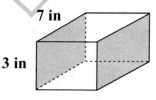

Area of left side:
$3 \text{ in} \times 7 \text{ in} = 21 \text{ in}^2$
Area of left and right:
$21 \text{ in} \times 2 \text{ in} = 42 \text{ in}^2$

Total surface area: $56 \text{ in}^2 + 24 \text{ in}^2 + 42 \text{ in}^2 = 122 \text{ in}^2$

Find the surface area of the following cubes and prisms.

1.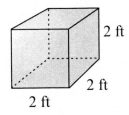

   2 ft
   2 ft
   2 ft

   SA = _____

2.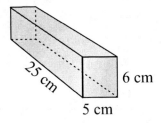

   25 cm
   6 cm
   5 cm

   SA = _____

3.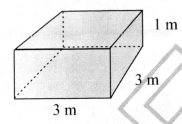

   1 m
   3 m
   3 m

   SA = _____

4.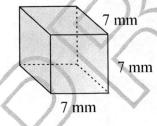

   7 mm
   7 mm
   7 mm

   SA = _____

5.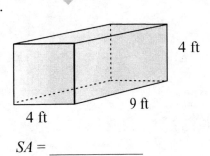

   4 ft
   9 ft
   4 ft

   SA = _____

6.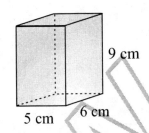

   9 cm
   5 cm   6 cm

   SA = _____

7.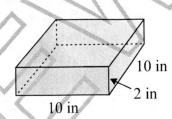

   10 in
   2 in
   10 in

   SA = _____

8.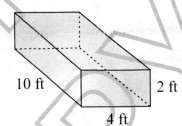

   10 ft   2 ft
   4 ft

   SA = _____

9.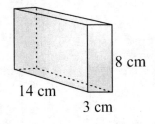

   5 m
   5 m
   5 m

   SA = _____

10.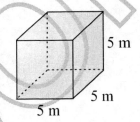

    8 cm
    14 cm
    3 cm

    SA = _____

PYRAMID

The pyramid below is made of a square base with 4 triangles on the sides.

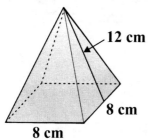

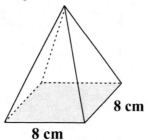

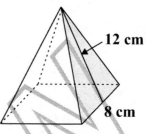

Area of square base:
$A = l \times w$
$A = 8 \times 8 = 64 \text{ cm}^2$

Area of sides:
Area of 1 side = $\frac{1}{2}bh$
$A = \frac{1}{2} \times 8 \times 12 = 48 \text{ cm}^2$
Area of 4 sides = $48 \times 4 = 192 \text{ cm}^2$

Total surface area: $64 + 192 = 256 \text{ cm}^2$

Find the surface area of the following pyramids.

1.

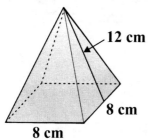

SA = _____

4.

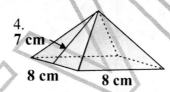

SA = _____

7.

SA = _____

2.

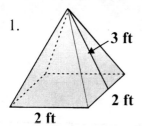

SA = _____

5.

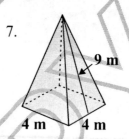

SA = _____

8.

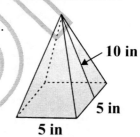

SA = _____

3.

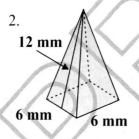

SA = _____

6.

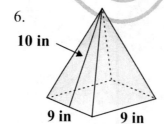

SA = _____

9.

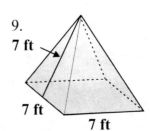

SA = _____

CYLINDER

If the side of a cylinder was slit from top to bottom and laid flat, its shape would be a rectangle. The length of the rectangle is the same as the circumference of the circle that is the base of the cylinder. The width of the rectangle is the height of the cylinder.

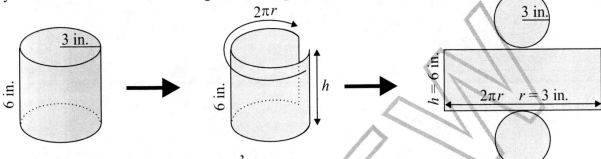

Total Surface Area of a Cylinder = $2\pi r^2 + 2\pi rh$

Area of top and bottom:
Area of a circle = πr^2
Area of top = $3.14 \times 3^2 = 28.26$ in.^{2}
Area of top and bottom = $2 \times 28.26 = 56.52$ in.^{2}

Area of side:
Area of rectangle = $l \times h$
$l = 2\pi r = 2 \times 3.14 \times 3 = 18.84$ in.
Area of rectangle = $18.84 \times 6 = 113.04$ in.^{2}

Total surface area = $56.52 + 113.04 = 169.56$ in^2

Find the surface area of the following cylinders. Use $\pi = 3.14$

1.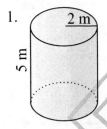
2 m
5 m

SA = _____

4.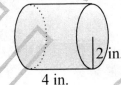
2 in.
4 in.

SA = _____

7.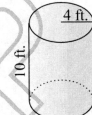
4 ft.
10 ft.

SA = _____

2.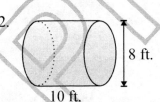
8 ft.
10 ft.

SA = _____

5.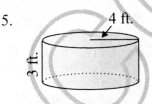
4 ft.
3 ft.

SA = _____

8.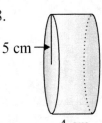
5 cm
4 cm

SA = _____

3.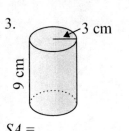
3 cm
9 cm

SA = _____

6.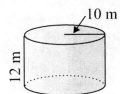
10 m
12 m

SA = _____

9.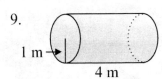
1 m
4 m

SA = _____

SPHERE

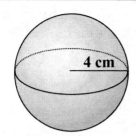

Surface area = $4\pi r^2$
Surface area = $4 \times 3.14 \times 4^2$
Surface area = 200.96 cm^2

Find the surface area of a sphere given the following measurements where *r* = radius and *d* = diameter. Use π = 3.14.

1. $r = 2$ in $SA =$ —————
2. $r = 6$ m $SA =$ —————
3. $r = \frac{3}{4}$ yd $SA =$ —————
4. $d = 8$ cm $SA =$ —————
5. $d = 50$ mm $SA =$ —————
6. $r = \frac{1}{4}$ ft $SA =$ —————

7. $d = 14$ cm $SA =$ —————
8. $r = \frac{1}{5}$ km $SA =$ —————
9. $d = 3$ in $SA =$ —————
10. $d = \frac{2}{3}$ ft $SA =$ —————
11. $r = 10$ mm $SA =$ —————
12. $d = 5$ yd $SA =$ —————

CONE

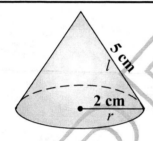

Total Surface Area : T = πr(r+s)

π = 3.14   r = radius of base   s = slant height
T = $3.14 \times 2(2 + 5)$
T = 6.28×7
T = 43.96 cm^2

Find the surface area of the following cones. Use π = 3.14.

1.

$SA =$ —————

3.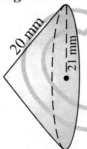

$SA =$ —————

5.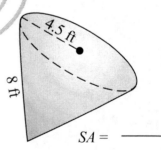

$SA =$ —————

2.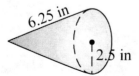

$SA =$ —————

4.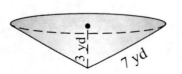

$SA =$ —————

6.

$SA =$ —————

260

NETS OF SOLID OBJECTS

Prisms

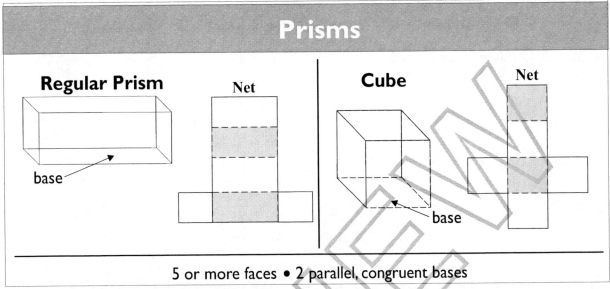

Regular Prism   Net

base

Cube   Net

base

5 or more faces • 2 parallel, congruent bases

Pyramid

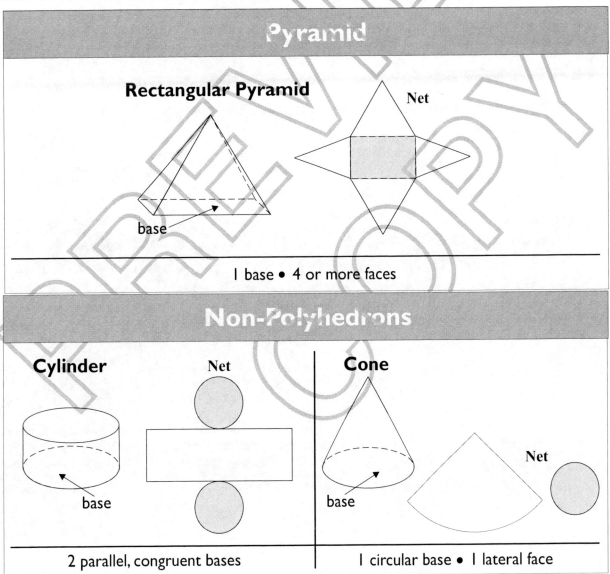

Rectangular Pyramid   Net

base

1 base • 4 or more faces

Non-Polyhedrons

Cylinder   Net

base

Cone   Net

base

2 parallel, congruent bases   1 circular base • 1 lateral face

USING NETS TO FIND SURFACE AREA

A **net** is a two dimensional representation of a three dimensional object. Nets clearly illustrate the plane figures that make up a solid.

EXAMPLE 1: Find the surface area of the figure shown below.

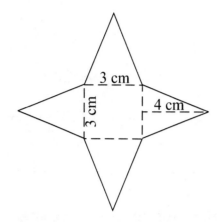

Step 1: Find the area of the 4 triangles.
$A = \frac{1}{2} bh$
$A = \frac{1}{2} \times 3 \times 4 = 6 \text{ cm}^2$
Area of all 4 triangles = $4 \times 6 = 24 \text{ cm}^2$

Step 2: Find the area of the base.
$A = lw$
$A = 3 \times 3 = 9 \text{ cm}^2$

Step 4: Find the sum of the areas of all the plane figures.
Surface Area = $24 \text{ cm}^2 + 9 \text{ cm}^2$
$SA = 33 \text{ cm}^2$

EXAMPLE 2: A net for a cone is shown below. Find the surface area.

Step 1: Find the area of the base.
$A = \pi r^2$
$A = 3.14 \times 6^2 = 3.14 \times 36 = 113.04 \text{ in}^2$

Step 2: Find the area of the cone section.
$A = \pi rl$
$A = 3.14 \times 6 \times 10$
$A = 188.40 \text{ in}^2$

Step 3: Find the sum of the areas of the base and the cone section.
Surface area = $113.04 \text{ in}^2 + 188.40 \text{ in}^2$
$SA = 301.44 \text{ in}^2$

The nets for various solids are given. Find the surface area of the objects. If needed, use π = 3.14.

1.

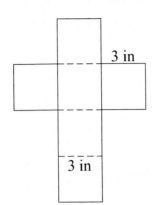

3 in

3 in

SA = _____

3.

11 cm

5 cm

SA = _____

2.

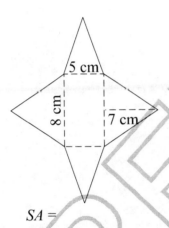

5 cm

8 cm

7 cm

SA = _____

4.

7 ft

15 ft

7 ft

SA = _____

Using a ruler, measure the dimensions of the following nets to the nearest tenth of a centimeter, and calculate the surface area of the object. If needed, use π = 3.14.

5.

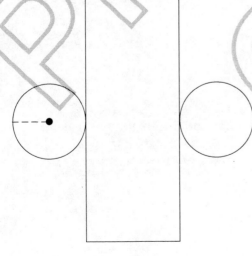

SA = _____

6.

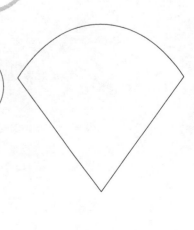

SA = _____

DIAGONALS OF 3D OBJECTS

In this section, we will find the length of the diagonal in cubes and rectangular prisms.

cube

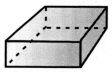
rectangular prism

Example: Find the length of the diagonal, $\overline{CF}$, in the rectangular prism shown below.

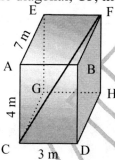

Step 1:   First, a planar cross-section must be made from segment $\overline{BF}$ to segment $\overline{CG}$, as shown in the diagram below. Segment $\overline{CF}$ is now the diagonal of a two dimensional plane and will be calculated using the Pythagorean Theorem.

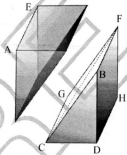 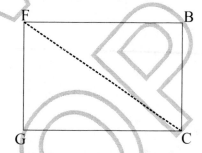

Step 2:   Next, find the length of the diagonal $\overline{CB}$ by using the Pythagorean theorem. The length of the planar cross-section is the hypotenuse of the right triangle $\triangle CBD$ and is calculated using the Pythagorean Theorem.

$$(\overline{CB})^2 = (\overline{CD})^2 + (\overline{BD})^2$$
$$(\overline{CB})^2 = (3)^2 + (4)^2$$
$$(\overline{CB})^2 = 9 + 16$$
$$\overline{CB} = \sqrt{25} = 5$$

Step 3:   Now that you have the length of $\overline{CB}$, you can find the length of $\overline{CF}$ using the Pythagorean theorem because $\overline{CB}$, $\overline{BF}$, and $\overline{CF}$ make a right triangle.

$$(\overline{CF})^2 = (\overline{CB})^2 + (\overline{BF})^2$$
$$(\overline{CF})^2 = (5)^2 + (7)^2$$
$$(\overline{CF})^2 = 25 + 49$$
$$\overline{CF} = \sqrt{74} = 8.6$$

Find the length of the diagonal for each solid.

1.

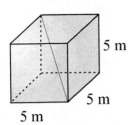

5 m
5 m
5 m

5.

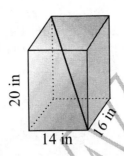

20 in
14 in
16 in

2.

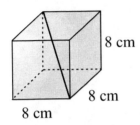

8 cm
8 cm
8 cm

6.

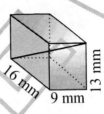

16 mm
9 mm
13 mm

3.

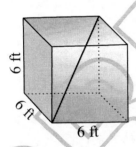

6 ft
6 ft
6 ft

7.

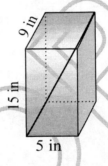

9 in
15 in
5 in

4.

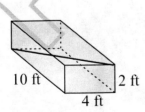

10 ft
4 ft
2 ft

8.

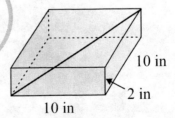

10 in
10 in
2 in

9. A lifeguard's pole must be able to reach every part of the kiddie pool while he remains seated in the lifeguard stand, 4' above the pool. The pool is 12' wide and 20' long. Find the length that the pole must be in order to reach the bottom depth of the pool (3') at the furthest point from the lifeguard stand.

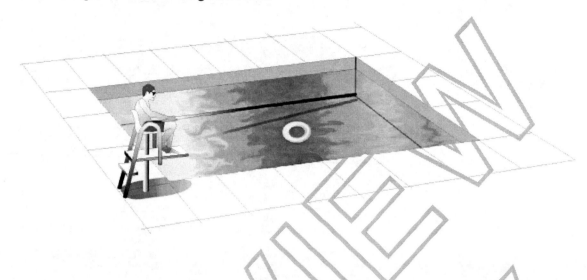

10. Tim and Kevin are trying to move a big sofa into their dorm room on the sixth floor of an old building. They hope they can use the elevator, which is only 5 feet wide and 4 feet deep, with a ceiling height of 7 feet. The sofa is 102" wide. Will it fit in the elevator?

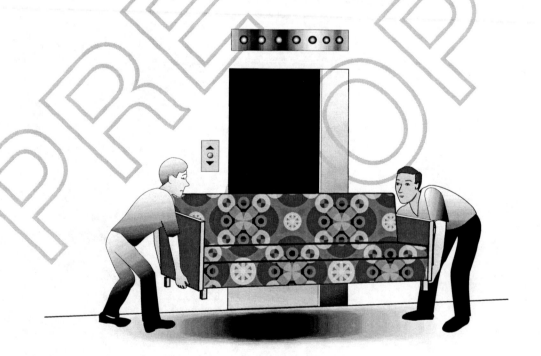

SOLID GEOMETRY WORD PROBLEMS

1. If an Egyptian pyramid has a square base that measures 500 yards by 500 yards, and the pyramid stands 300 yards tall, what would be the volume of the pyramid? Use the formula for volume of a pyramid, $V = \frac{1}{3}Bh$, where B is the area of the base.

   $V = $ _____

2. Robert is using a cylindrical barrel filled with water to flatten the sod in his yard. The circular ends have a radius of 1 foot. The barrel is 3 feet wide. How much water will the barrel hold? The formula for volume of a cylinder is $V = \pi r^2 h$. Use $\pi = 3.14$.

   $V = $ _____

3. If a basketball measures 24 centimeters in diameter, what volume of air will it hold? The formula for volume of a sphere is $V = \frac{4}{3}\pi r^3$. Use $\pi = 3.14$.

   $V = $ _____

4. What is the volume of a cone that is 2 inches in diameter and 5 inches tall? The formula for volume of a cone is $V = \frac{1}{3}\pi r^2 h$. Use $\pi = 3.14$.

   $V = $ _____

5. Kelly has a rectangular fish aquarium that measures 24 inches wide, 12 inches deep, and 18 inches tall. What is the maximum amount of water that the aquarium will hold?

   $V = $ _____

6. Jenny has a rectangular box that she wants to cover in decorative contact paper. The box is 10 cm long, 5 cm wide, and 5 cm high. How much paper will she need to cover all 6 sides?

   $SA = $ _____

7. Gasco needs to construct a cylindrical, steel gas tank that measures 6 feet in diameter and is 8 feet long. How many square feet of steel will be needed to construct the tank? Use the following formulas as needed: $A = l \times w$, $A = \pi r^2$, $C = 2\pi r$. Use $\pi = 3.14$.

   $SA = $ _____

8. Craig wants to build a miniature replica of San Francisco's Transamerica Pyramid out of glass. His replica will have a square base that measures 6 cm by 6 cm. The 4 triangular sides will be 6 cm wide and 60 cm tall. How many square centimeters of glass will he need to build his replica? Use the following formulas as needed: $A = l \times w$ and $A = \frac{1}{2}bh$.

   $SA = $ _____

9. Jeff built a wooden, cubic toy box for his son. Each side of the box measures 2 feet. How many square feet of wood did he use to build the toy box? How many cubic feet of toys will the box hold?

   $SA = $ _____

   $V = $ _____

FRONT, TOP, SIDE, AND CORNER VIEWS OF SOLID OBJECTS

Solid objects are 3-dimensional and therefore, are able to be viewed from several perspectives. You should be able to recognize the corner view of a solid given the front, top, and side views. Likewise, you should be able to draw and/or recognize the front, top, and side views of an object given its corner view.

EXAMPLE 1: Draw the front, top, and side views of the object shown below.

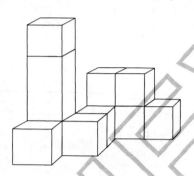

Solution:

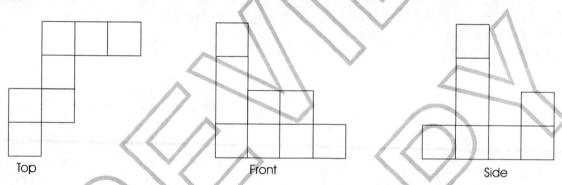

EXAMPLE 2: The top, front, and side views of an object are shown below. How many cubes would it take to build this object.

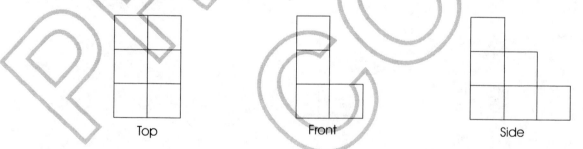

Solution: Draw the object first, and then count the number of cubes used to create the structure.

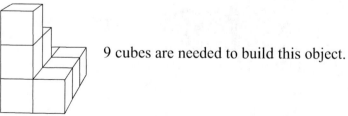

9 cubes are needed to build this object.

Refer to the object shown below to answer questions 1, 2, and 3.

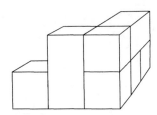

1. Which of the following is the top view of the solid?

A.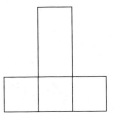

B.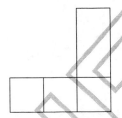

C.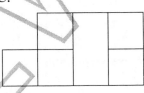

2. Which of the following is the side view of the solid?

A.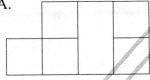

B.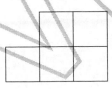

C.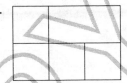

3. Which of the following is the front view of the solid?

A.

B.

C.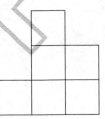

4. Given below are the front, top, and side views of a solid. Draw the object.

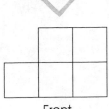 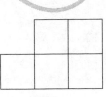 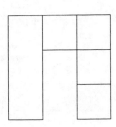

Front   Top   Side

Refer to the front, top, and side views of the object shown below to answer questions 5 and 6.

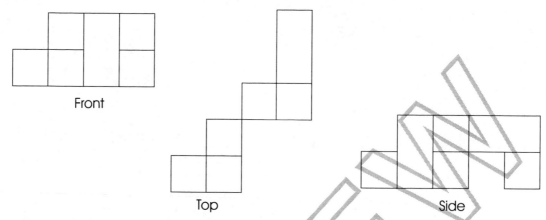

Front

Top

Side

5. How many rectangular boxes would be needed to build this object? _____

6. How many cubes would be needed to build this object? _____

7. Given the object shown below, draw the front, top, and side views.

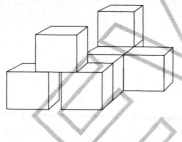

Front   Top   Side

8. The front, side, and top views of a solid are shown below. How many total blocks are needed to construct this object? _____

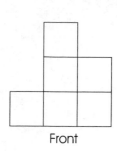

Front

Side

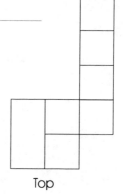

Top

CHAPTER 18 REVIEW

Find the volume and/or the surface area of the following solids.

1.

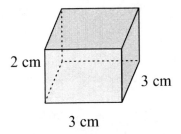

2 cm

3 cm

3 cm

V = _____

SA = _____

2.

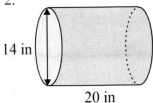

14 in

20 in

$V = \pi r^2 h$

$SA = 2\pi r^2 + 2\pi r h$

Use $\pi = \frac{22}{7}$

V = _____

SA = _____

3.

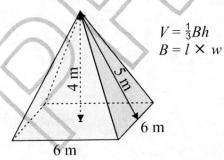

4 m

5 m

6 m

6 m

$V = \frac{1}{3}Bh$

$B = l \times w$

V = _____

SA = _____

4.

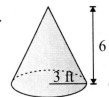

6 ft

3 ft

$V = \frac{1}{3}\pi r^2 h$

Use $\pi = 3.14$

V = _____

5.

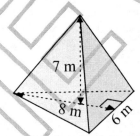

7 m

8 m

6 m

$V = \frac{1}{3}Bh$

B = area of the triangular base

V = _____

6.

7 in

$V = \frac{4}{3}\pi r^3$

Use $\pi = \frac{22}{7}$

V = _____

7.   The sandbox at the local elementary school is 60 inches wide and 100 inches long. The sand in the box is 6 inches deep. How many cubic inches of sand are in the sandbox?

_____

8. If you have cubes that are two inches on each edge, how many would fit in a cube that was 16 inches on each edge?

_____

9. If you double each edge of a cube, how many times larger is the volume?

_____

10. It takes 8 cubic inches of water to fill the cube below. If each side of the cube is doubled, how much water is needed to fill the new cube?

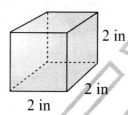

2 in
2 in
2 in

_____

11. If a ball is 4 inches in diameter, what is its surface area? Use π = 3.14

_____

12. A grain silo is in the shape of a cylinder. If the silo has an inside diameter of 10 feet and a height of 35 feet, what is the maximum volume inside the silo?

Use $\pi = \frac{22}{7}$

_____

13. A closed cardboard box is 30 centimeters long, 10 centimeters wide, and 20 centimeters high. What is the total surface area of the box?

14. Siena wants to build a wooden toy box with a lid. The dimensions of the toy box are 3 feet long, 4 feet wide, and 2 feet tall. How many square feet of wood will she need to construct the box?

_____

15. How many 1-inch cubes will fit inside a larger 1 foot cube? (Figures are not drawn to scale.)

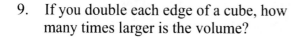

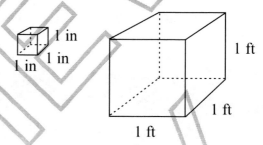

1 in
1 in
1 in

1 ft
1 ft
1 ft

_____

16. The cylinder below has a volume of 240 cubic inches. The cone below has the same radius and the same height as the cylinder. What is the volume of the cone?

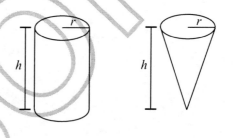

r
h

r
h

_____

17. Estimate the volume of the base and dome below.

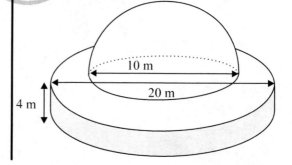

10 m
20 m
4 m

18. Find the volume of the figure below.

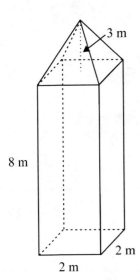

3 m

8 m

2 m

2 m

_____

19. Find the volume of the figure below.
    Each side of each cube measures 4 feet.

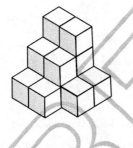

_____

Chapter 19 | Measurement

When measuring length, distance, volume, mass, weight, or temperature, most of the world uses the **metric system**. The United States, however, commonly uses the **English system** of measurement. You should be familiar with both the metric system and the English system of measure.

CUSTOMARY MEASURE

Customary measure in the United States is based on the English system. The following chart gives common customary units of measure as well as the standard units for time.

English System of Measure

Measure	Abbreviations	Appropriate Instrument
Time: 1 week = 7 days 1 day = 24 hours 1 hour = 60 minutes 1 minute = 60 seconds	week = week hour = hr or h minutes = min seconds = sec	calendar clock clock clock
Length: 1 mile = 5,280 feet 1 yard = 3 feet 1 foot = 12 inches	mile = mi yard = yd foot = ft inch = in	odometer yard stick, tape line ruler, yard stick
Volume: 1 gallon = 4 quarts 1 quart = 2 pints 1 pint = 2 cups 1 cup = 8 ounces	gallon = gal quart = qt pint = pt ounce = oz	quart or gallon container quart container cup, pint, or quart container cup
Weight: 1 pound = 16 ounces	pound = lb ounce = oz	scale or balance
Temperature: Fahrenheit Celsius	^{o}F ^{o}C	thermometer thermometer
Area: 1 acre = 4,840 square yards 1 sq yard = 9 sq ft	acre = acre	tape measure

APPROXIMATE ENGLISH MEASURE

Match the item on the left with its approximate (not exact) measure on the right. You may use some answers more than once.

_____ 1. The height of an average woman is about _____ .

_____ 2. An average candy bar weighs about _____ .

_____ 3. An average donut is about _____ across (in diameter).

_____ 4. A piece of notebook paper is about _____ long.

_____ 5. A tennis ball is about _____ across (in diameter).

_____ 6. The average basketball is about _____ across.

_____ 7. The average month is about _____ .

_____ 8. How long is the average lunch table?

_____ 9. About how much does a computer disk weigh?

_____ 10. What is the average height of a table?

A. 1 yard

B. 2 yards

C. $5\frac{1}{2}$ feet

D. 4 weeks

E. $2\frac{1}{2}$ inches

F. 2 ounces

G. 1 foot

CONVERTING UNITS USING DIMENSIONAL ANALYSIS

An easy way to solve math problems involving the conversion of units is by using **dimensional analysis**. Don't let this fancy term intimidate you. Dimensional analysis is really a simple, step-by-step way to convert units of measure by using **conversion factors**. A conversion factor is a fraction always equal to 1. The easiest way to understand dimensional analysis and conversion factors is to look at an example.

EXAMPLE 1:   How many inches are in 3 feet?

Step 1:   First, what is a conversion factor for feet to inches? You know that there are 12 inches to 1 foot. Therefore, the conversion factors are simply the following fractions:

$$\frac{1 \text{ ft}}{12 \text{ in}} \quad \text{or} \quad \frac{12 \text{ in}}{1 \text{ ft}} \longleftarrow \textbf{conversion factors}$$

Step 2:   Multiply the number with the old unit by the conversion factor to get the number with the new unit. Choose the conversion factor that will put the old unit on the bottom of the fraction so that it will cancel the other old unit.

$$3 \text{ ft} \times \frac{12 \text{ in}}{1 \text{ ft}} = 36 \text{ in}$$

Notice that the unit of "feet" cancel each other, and you are left with inches.

WRONG: If you had chosen the wrong conversion factor, this is what you would have:

$$3 \text{ ft} \times \frac{1 \text{ ft}}{12 \text{ in}} = \frac{3 \text{ ft} \times \text{ ft}}{12 \text{ in}} \text{ or } \frac{1 \text{ ft}^2}{4 \text{ in}}$$

Notice, feet multiplied by feet are feet squared. None of the units cancel, so you know right away that this is wrong.

EXAMPLE 2:   4 oz. equal how many pounds?

Step 1:   You should know that there are 16 ounces in 1 pound, so your conversion factors are

$$\frac{16 \text{ oz}}{1 \text{ lb}} \quad \text{or} \quad \frac{1 \text{ lb}}{16 \text{ oz}}$$

Step 2: $4 \cancel{\text{ oz}} \times \dfrac{1 \text{ lb}}{16 \cancel{\text{ oz}}} = \dfrac{4 \text{ lb}}{16} \quad \text{or} \quad \dfrac{1}{4} \text{ lb}$

Again, choose the conversion factor that will cause the old unit to cancel. In this case, multiplying by the correct conversion factor means dividing 4 by 16 to get the correct units of pounds.

The two examples just given are really simple, and you may think it would be just as easy to do the math in your head. That may be true for very simple unit conversions. However, when converting more than one unit, unit conversions become more complicated. It is harder to figure out in your head whether you should divide or multiply. Get in the habit now of using dimensional analysis to solve easy problems, and more complex problems will seem very easy. Look at two more examples:

EXAMPLE 3:   John drives on the interstate 60 miles per hour. How many feet per second is he traveling?

$$\frac{60 \cancel{\text{ miles}}}{\cancel{\text{hour}}} \times \frac{1 \cancel{\text{ hour}}}{60 \cancel{\text{ min}}} \times \frac{1 \cancel{\text{ min}}}{60 \text{ sec}} \times \frac{5280 \text{ ft}}{1 \cancel{\text{ mile}}} = \frac{60 \times 5280}{60 \times 60} = 88 \text{ ft/sec}$$

convert hours to minutes   convert minutes to seconds   convert miles to feet   units in feet per second

EXAMPLE 4:   Karen is a cashier at a grocery store. She can scan 20 items per minute. How many items can she scan in 15 seconds?

$$\frac{20 \text{ items}}{\cancel{\text{minute}}} \times \frac{1 \cancel{\text{ minute}}}{60 \cancel{\text{ seconds}}} \times 15 \cancel{\text{ seconds}} = \frac{20 \times 15}{60} = 5 \text{ items}$$

276

Using dimensional analysis and the table on previous pages, make the following conversions.

1. 15 yards to feet   _____
2. 3 pounds to ounces   _____
3. 4 feet to inches   _____
4. 2 yards to inches   _____
5. 2 gallons to quarts   _____
6. 6 pints to cups   _____
7. 1 gallon to cups   _____
8. 3 inches to feet   _____
9. $1\frac{1}{2}$ feet to inches   _____
10. 42 inches to feet   _____

11. Jared can work 54 math problems in 1 hour. How many problems can he work in 10 minutes?

12. A machine makes 360 gadgets every 24 hours. How many gadgets does it make every 20 minutes?

13. Perry's car can travel an average of 22 miles on 1 gallon of gas. How many miles can his car travel on 1 quart of gas?

14. A fast food restaurant serves an average of 40 customers per hour. On average, how many minutes does each customer wait for his or her order? (Hint: find minutes per customer, not customers per minute.)

15. Leah travels 22 feet per second on her bicycle. How many miles per hour does she travel?

THE METRIC SYSTEM

The metric system uses units based on multiples of ten. The basic units of measure in the metric system are the **meter**, the **liter**, and the **gram**. Metric prefixes tell what multiple of ten the basic unit is multiplied by. Below is a chart of metric prefixes and their values. The ones rarely used are shaded.

Prefix	kilo (k)	hecto (h)	deka (da)	unit (m, L, g)	deci (d)	centi (c)	milli (m)
Meaning	1000	100	10	1	0.1	0.01	0.001

To help you remember the order of the metric prefixes, use the following sentence:

Kings **H**ave **D**ances **U**ntil **D**ragons **C**hange **M**usic.

Or, make up your own sentence to help you memorize these prefixes.

UNDERSTANDING METERS

The basic unit of **length** in the metric system is the **meter**. Meter is abbreviated "m".

Metric Unit	Abbreviation	Memory Tip	Equivalents
1 millimeter	mm	Thickness of a dime	10 mm = 1 cm
1 centimeter	cm	Width of the tip of the little finger	100 cm = 1 m
1 meter	m	Distance from the nose to the tip of fingers (a little longer than a yard)	1000 m = 1 km
1 kilometer	km	A little more than half a mile	

UNDERSTANDING LITERS

The basic unit of **liquid volume** in the metric system is the **liter**. Liter is abbreviated "L".

The liter is the volume of a cube measuring 10 cm on each side. A milliliter is the volume of a cube measuring 1 cm on each side. A capital L is used to signify liter, so it is not confused with the number 1.

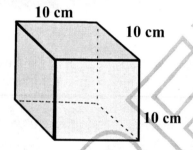

Volume = 1000 cm^3 = 1 Liter
(a little more than a quart)

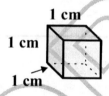

Volume = 1 cm^3 = 1 mL
(an eyedropper holds 1 mL)

UNDERSTANDING GRAMS

The basic unit of **mass** in the metric system is the **gram**. Gram is abbreviated "g".

A **gram** is the **mass** of **one cubic centimeter** of **water** at 4^0 C.

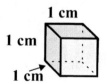

A large paper clip has a mass of about 1 gram (1g).
A nickel has a mass of 5 grams (5 g).
1000 grams = 1 kilogram (kg) = a little over 2 pounds

1 milligram (mg) = 0.001 gram. This is an extremely small amount and is used in medicine.

An aspirin tablet has a mass of 300 mg.

ESTIMATING METRIC MEASUREMENTS

Choose the best estimates.

1. The height of an average man
    A. 18 cm   B. 1.8 m   C. 6 km   D. 36 mm

2. The volume of a coffee cup
    A. 300 mL   B. 20 L   C. 5 L   D. 1 kL

3. The width of this book
    A. 215 mm   B. 75 cm   C. 2 m   D. 1.5 km

4. The mass of an average man
    A. 5 mg   B. 15 cg   C. 25 g   D. 90 kg

5. The mass of a dime
    A. 3 g   B. 30 g   C. 10 cg   D. 1 kg

6. The length of a basketball court
    A. 1000 mm   B. 250 cm   C. 28 m   D. 2 km

Choose the best units of measure.

7. The distance from Baton Rouge to Indianapolis
    A. millimeter   B. centimeter   C. meter   D. kilometer

8. The length of a house key
    A. millimeter   B. centimeter   C. meter   D. kilometer

9. The thickness of a nickel
    A. millimeter   B. centimeter   C. meter   D. kilometer

10. The width of a classroom
    A. millimeter   B. centimeter   C. meter   D. kilometer

11. The length of a piece of chalk
    A. millimeter   B. centimeter   C. meter   D. kilometer

12. The height of a pine tree
    A. millimeter   B. centimeter   C. meter   D. kilometer

CONVERTING UNITS IN THE METRIC SYSTEM

In general, converting between units within the metric system is easier than converting within the English system because the units are in multiples of ten. Conversions between the same unit of measure means simply moving the decimal point.

EXAMPLE 1:   Convert 34.5 meters to centimeters.

List or visualize the metric prefixes to figure out how many places to move the decimal point.

k   h   da   u   .   d   c   m

Start at "u" (for unit) and count over to the "c" (for centi). The decimal point needs to move to the right 2 places to convert the unit measure of meters to centimeters.

So, 3 4 . 5 0 meters is 3450 centimeters. Notice you have to add a zero.

EXAMPLE 2:   Convert 250.1 millimeters to dekameters.

k   h   da   u   d   c   m   .

Start at the "m" (for milli) and count over to "da" (for deka). The decimal point needs to move to the left 4 places to convert millimeters to dekameters.

So, 0 2 5 0 . 1 millimeters is .02501 dekameters. Again, add a zero.

Practice converting the following metric measurements.

1.	35 mg to g	_____	11.	0.06 daL to dL	_____
2.	6 km to m	_____	12.	0.417 kg to cg	_____
3.	21.5 mL to L	_____	13.	18.2 cL to L	_____
4.	4.9 mm to cm	_____	14.	81.2 dm to cm	_____
5.	5.35 kL to mL	_____	15.	72.3 cm to mm	_____
6.	32.1 mg to kg	_____	16.	0.003 kL to L	_____
7.	156.4 m to km	_____	17.	5.06 g to mg	_____
8.	25 mg to cg	_____	18.	1.058 mL to cL	_____
9.	17.5 L to mL	_____	19.	43 hm to km	_____
10.	4.2 g to kg	_____	20.	2.057 m to cm	_____

MORE METRIC CONVERSIONS

Sometimes metric conversions can be a little more complicated. In these cases, you can use dimensional analysis.

EXAMPLE 1: How many cubic centimeters are in a liter?
First, what conversions do you know? You should know that one cubic centimeter is equal to one milliliter. You should also know that there are 1000 milliliters in one liter.

Multiply 1 liter by conversion factors until you get cubic centimeters.

$$1 \cancel{L} \times \frac{1000 \cancel{mL}}{1 \cancel{L}} \times \frac{1 \text{ cm}^3}{1 \cancel{mL}} = 1000 \text{ cm}^3$$

EXAMPLE 2: How many cubic millimeters are in a cubic meter?

Again, use what you know. You know that there are 1000 millimeters in 1 meter, so the conversion would look like the following:

$$1 \cancel{m^3} \times \frac{1000 \text{ mm}}{1 \cancel{m}} \times \frac{1000 \text{ mm}}{1 \cancel{m}} \times \frac{1000 \text{ mm}}{1 \cancel{m}} =$$

$$1000 \text{ mm} \times 1000 \text{ mm} \times 1000 \text{ mm} = 1{,}000{,}000{,}000 \text{ mm}^3$$

Remember that m^3 is $m \times m \times m$, and $mm \times mm \times mm$ is mm^3.

Practice using dimensional analysis to make the following metric or English conversions.

1. How many milliliters are in three liters?

2. How many cubic inches are in one cubic foot?

3. Holly's aspirin has a mass of 582 milligrams. What is the mass of the aspirin in grams?

4. Raymond determines that 20 drops from his eyedropper equals exactly 1 milliliter. How many drops would it take to fill a liter container?

5. A container holds 250 mL. How many liters does it hold?

6. Runners prepare to compete in the 300 meter dash. How many centimeters long is this race?

7. A hiker moves from one campsite to the next and travels 12.2 km in one day. How many millimeters did the hiker travel?

8. Sakura buys a 2 liter bottle of soda. Her drinking cup holds 400 mL. How many drinking cups can she fill with the bottle of soda? (Note: You are finding a number that does not have units, so all units must cancel.)

DISTANCE

EXAMPLE: Jessie traveled for 7 hours at an average rate of 58 miles per hour. How far did she travel?

Solution: Multiply the number of hours by the average rate of speed.

7 hours $\times$ 58 miles/hour $=$ 406 miles

---

1. Myra traveled for 9 hours at an average rate of 45 miles per hour. How far did she travel?

2. A tour bus drove 4 hours, averaging 58 miles per hour. How many miles did it travel?

3. Tina drove for 7 hours at an average speed of 53 miles per hour. How far did she travel?

4. Dustin raced for 3 hours, averaging 176 miles per hour. How many miles did he race?

5. Kris drove 5 hours and averaged 49 miles per hour. How far did she travel?

6. Oliver drove at an average of 93 miles per hour for 3 hours. How far did he travel?

7. A commercial airplane traveled 514 miles per hour for 2 hours. How far did it fly?

8. A train traveled at 125 miles per hour for 4 hours. How many miles did it travel?

9. Carl drove a constant 65 miles an hour for 3 hours. How many miles did he drive?

10. Jasmine drove for 5 hours, averaging 40 miles per hour. How many miles did she drive?

RATE

EXAMPLE: Laurie traveled 312 miles in 6 hours. What was her average rate of speed?

Solution: Divide the number of miles by the number of hours.

$$\frac{312 \text{ miles}}{6 \text{ hours}} = 52 \text{ miles/hour}$$

Laurie's average rate of speed was 52 miles per hour (or 52 mph).

Find the average rate of speed in each problem below.

1. A race car went 500 miles in 4 hours. What was its average rate of speed?

2. Carrie drove 124 miles in 2 hours. What was her average speed?

3. After 7 hours of driving, Chad had gone 364 miles. What was his average speed?

4. Anna drove 360 miles in 8 hours. What was her average speed?

5. After 3 hours of driving, Paul had gone 183 miles. What was his average speed?

6. Nicole ran 25 miles in 5 hours. What was her average speed?

7. A train traveled 492 miles in 6 hours. What was its average rate of speed?

8. A commercial jet traveled 1,572 miles in 3 hours. What was its average speed?

9. Jillian drove 195 miles in 3 hours. What was her average speed?

RATIO PROBLEMS

In some word problems, you may be asked to express answers as a **ratio**. Ratios can look like fractions, they can be written with a colon, or they can be written in word form with "to" between the numbers. Numbers must be written in the order they are requested. In the following problem, 8 cups of sugar are mentioned before 6 cups of strawberries. But in the question part of the problem, you are asked for the ratio of STRAWBERRIES to SUGAR. The amount of strawberries IS THE FIRST WORD MENTIONED, so it must be the **top** number of the fraction. The amount of sugar, THE SECOND WORD MENTIONED, must be the **bottom** number of the fraction.

EXAMPLE: The recipe for jam requires 8 cups of sugar for every 6 cups of strawberries. What is the ratio of strawberries to sugar in this recipe?

First number requested $\dfrac{6}{8}$ **cups strawberries**
Second number requested   **cups sugar**

Answers may be reduced to lowest terms. $\dfrac{6}{8} = \dfrac{3}{4}$

This ratio is also correctly expressed as 3:4 or 3 to 4.

Practice writing ratios for the following word problems and reduce to lowest terms. DO NOT CHANGE ANSWERS TO MIXED NUMBERS. Ratios should be left in fraction form.

1. Out of the 248 seniors, 112 are boys. What is the ratio of boys to the total number of seniors?

   _____

2. It takes 7 cups of flour to make 2 loaves of bread. What is the ratio of cups of flour to loaves of bread?

   _____

3. A skyscraper that stands 620 feet tall casts a shadow that is 125 feet long. What is the ratio of the shadow to the height of the skyscraper?

   _____

4. The newborn weighs 8 pounds and is 22 inches long. What is the ratio of weight to length?

   _____

5. Jack paid $6.00 for 10 pounds of apples. What is the ratio of the price of apples to the pounds of apples?

   _____

6. Twenty boxes of paper weigh 520 pounds. What is the ratio of boxes to pounds?

   _____

SOLVING PROPORTIONS

Two **ratios (fractions)** that are **equal** to each other are called **proportions**. For example, $\frac{1}{4} = \frac{2}{8}$. Read the following example to see how to find a number missing from a proportion.

EXAMPLE: $\frac{5}{15} = \frac{8}{x}$

Step 1: To find x, you first multiply the two numbers that are diagonal to each other. **15 × 8 = 120**

$$\frac{5}{\boxed{15}} = \frac{\boxed{8}}{x}$$

Step 2 Then, divide the product (120) by the other number in the proportion (5). **120 ÷ 5 = 24**

Therefore, $\frac{5}{15} = \frac{8}{24}$ $x = 24$

Practice finding the number missing from the following proportions. First, multiply the two numbers that are diagonal from each other. Then divide by the other number.

1. $\frac{2}{5} = \frac{6}{x}$

2. $\frac{9}{3} = \frac{x}{5}$

3. $\frac{x}{12} = \frac{3}{4}$

4. $\frac{7}{x} = \frac{3}{9}$

5. $\frac{12}{x} = \frac{2}{5}$

6. $\frac{12}{x} = \frac{4}{3}$

7. $\frac{27}{3} = \frac{x}{2}$

8. $\frac{1}{x} = \frac{3}{12}$

9. $\frac{15}{2} = \frac{x}{4}$

10. $\frac{7}{14} = \frac{x}{6}$

11. $\frac{5}{6} = \frac{10}{x}$

12. $\frac{4}{x} = \frac{3}{6}$

13. $\frac{x}{5} = \frac{9}{15}$

14. $\frac{9}{18} = \frac{x}{2}$

15. $\frac{5}{7} = \frac{35}{x}$

16. $\frac{x}{2} = \frac{8}{4}$

17. $\frac{15}{20} = \frac{x}{8}$

18. $\frac{x}{40} = \frac{5}{100}$

19. $\frac{4}{7} = \frac{x}{28}$

20. $\frac{7}{6} = \frac{42}{x}$

21. $\frac{x}{8} = \frac{1}{4}$

RATIO AND PROPORTION WORD PROBLEMS

You can use ratios and proportions to solve problems.

EXAMPLE: A stick one meter long is held perpendicular to the ground and casts a shadow 0.4 meters long. At the same time, an electrical tower casts a shadow 112 meters long. Use ratio and proportion to find the height of the tower.

Step 1: Set up a proportion using the numbers in the problem. Put the shadow lengths on one side of the equation, and put the heights on the other side. The 1 meter height is paired with the 0.4 meter length, so let them both be top numbers. Let the unknown height be x.

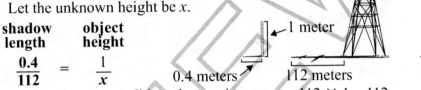

$$\frac{\text{shadow}}{\text{length}} \qquad \frac{\text{object}}{\text{height}}$$

$$\frac{0.4}{112} = \frac{1}{x}$$

Step 2: Solve the proportion as you did on the previous page. $112 \times 1 = 112$
$112 \div 0.4 = 280$ **Answer:** The tower height is 280 meters.

Use ratio and proportion to solve the following problems.

1. Rudolph can mow a lawn that measures 1000 square feet in 2 hours. At that rate, how long would it take him to mow a lawn 3500 square feet?

   _____

2. Faye wants to know how tall her school building is. On a sunny day, she measures the shadow of the building to be 6 feet. At the same time, she measures the shadow cast by a 5 foot statue to be 2 feet. How tall is her school building?

   _____

3. Out of every 5 students surveyed, 2 listen to country music. At that rate, how many students in a school of 800 listen to country music?

   _____

4. Bailey, a Labrador Retriever, had a litter of 8 puppies. Four of the puppies were black. At that rate, how many would be black in a litter of 10 puppies?

   _____

5. According to the instructions on a bag of fertilizer, 5 pounds of fertilizer are needed for every 100 square feet of lawn. How many square feet will a 25 pound bag cover?

   _____

6. A race car can travel 2 laps in 5 minutes. How long will it take the race car to complete 100 laps at that rate?

   _____

7. If it takes 7 cups of flour to make 4 loaves of bread, how many loaves of bread can you make from 35 cups of flour?

   _____

8. If 3 pounds of jelly beans cost $6.30, how much would 2 pounds cost?

   _____

9. For the first 4 home football games, the concession stand sold 600 hotdogs. If that ratio stays constant, how many hotdogs will sell for all 10 home games?

   _____

MAPS AND SCALE DRAWINGS

EXAMPLE 1: On a map drawn to scale, 5 cm represents 30 kilometers. A line segment connecting two cities is 7 cm long. What distance does this line segment represent?

Step 1: Set up a proportion using the numbers in the problem. Keep centimeters on one side of the equation and kilometers on the other. The 5 cm is paired with the 30 kilometers, so let them both be top numbers. Let the unknown distance be x.

$$\overset{\textbf{cm}}{\frac{5}{7}} = \overset{\textbf{km}}{\frac{30}{x}}$$

Step 2: Solve the proportion as you have previously. $7 \times 30 = 210$
$210 \div 5 = 42$ **Answer:** 7 cm represents 42 km.

Sometimes the answer to a scale drawing problem will be a fraction or mixed number.

EXAMPLE 2: On a scale drawing, 2 inches represents 30 feet. How many inches long is a line segment that represents 5 feet?

Step 1: Set up the proportion as you did above.

$$\overset{\textbf{inches}}{\frac{2}{x}} = \overset{\textbf{feet}}{\frac{30}{5}}$$

Step 2: **First, multiply the two numbers that are diagonal from each other. Then, divide by the other number.**

$2 \times 5 = 10$ $10 \div 30$ is less than 1 so express the answer as a fraction and reduce.

$10 \div 30 = \frac{10}{30} = \frac{1}{3}$ inch   **Answer:** $\frac{1}{3}$ of an inch represents 5 feet.

Set up proportions for each of the following problems and solve.

1. If 2 inches represent 50 miles on a scale drawing, how long would a line segment be that represents 25 miles? _____

2. On a scale drawing, 2 cm represent 15 km. A line segment on the drawing is 3 cm long. What distance does this line segment represent? _____

3. On a map drawn to scale, 5 cm represent 250 km. How many kilometers are represented by a line 6 cm long? _____

4. If 2 inches represent 80 miles on a scale drawing, how long would a line segment be that represents 280 miles? _____

5. On a map drawn to scale, 5 cm represent 200 km. How long would a line segment be that represents 260 km? _____

6. On a scale drawing of a house plan, one inch represents 5 feet. How many feet wide is the bathroom if the width on the drawing is 3 inches? _____

286

USING A SCALE TO FIND DISTANCES

By using a **map scale**, you can determine the distance between two places in the real world. The **map scale** shows distances in both miles and kilometers. You will need your ruler to do these exercises. On the scale below, you will notice that 1 inch = 800 miles. To find the distance between Calgary and Ottawa, measure with a ruler between the two cities. You will find it measures about $2\frac{1}{2}$ inches. From the scale, you know 1 inch = 800 miles. Use multiplication to find the distance in miles. $2.5 \times 800 = 2,000$. The cities are about 2,000 miles apart.

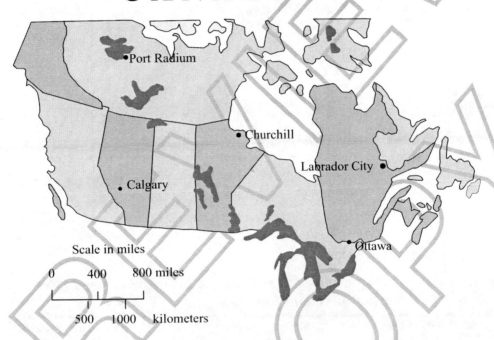

Find these distances in miles.

1. Calgary to Churchill _____

2. Churchill to Ottawa _____

3. Port Radium to Churchill _____

4. Port Radium to Ottawa _____

5. Labrador City to Ottawa _____

6. Calgary to Labrador City _____

Find these distances in kilometers.

7. Churchill to Labrador City _____

8. Ottawa to Port Radium _____

9. Port Radium to Calgary _____

10. Churchill to Ottawa _____

11. Calgary to Churchill _____

12. Calgary to Ottawa _____

USING A SCALE ON A BLUEPRINT

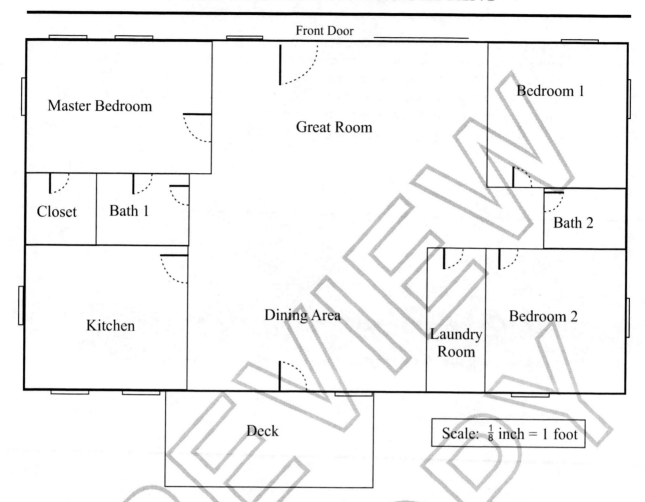

Front Door

Master Bedroom

Great Room

Bedroom 1

Closet

Bath 1

Bath 2

Kitchen

Dining Area

Laundry Room

Bedroom 2

Deck

Scale: $\frac{1}{8}$ inch = 1 foot

Use a ruler to find the measurements of the rooms on the blueprint above. Convert to feet using the scale. The first problem is done for you.

	long wall		short wall	
	ruler measurement	room measurement	ruler measurement	room measurement
1. Kitchen	$1\frac{3}{4}$ inch	14 feet	$1\frac{1}{2}$ inch	12 feet
2. Deck				
3. Closet				
4. Bedroom 1				
5. Bedroom 2				
6. Master Bedroom				
7. Bath 1				
8. Bath 2				

DIMENSIONAL ANALYSIS

A **dimension** is a property that can be measured such as length, time, mass, or temperature; or it is calculated by multiplying or dividing other dimensions. Some examples include length/time (velocity), length^3 (volume), or mass/length^3 (density).

A measured quantity can be expressed in any appropriate dimension. The equivalence between two expressions of a given quantity may be written as a ratio:

$$\frac{16 \text{ ounces}}{1 \text{ pound}} \quad \text{or} \quad \frac{2000 \text{ pounds}}{1 \text{ ton}}$$

Ratios of equivalent values expressed in different units like these are known as **conversion factors**. To convert given quantities in one set of units to their equivalent values in another set of units, we set up **dimensional equations**. Writing these with vertical and horizontal bars and carrying along units often helps avoid mistakes in these types of equations.

EXAMPLE 1: **How many inches are in four yards?**

Step 1: First we need to develop our conversion factors. We are given a value in yards and we want to find the value in inches. We know that there are 3 feet in 1 yard, and that there are 12 inches in 1 foot, so we will set up a dimensional equation using these conversion factors. We will then be able to cross out the units top and bottom to make sure that we are left with the units that we want. Then we will multiply the numbers across the top and divide by the numbers on the bottom to get the answer.

$$\frac{4 \text{ yards}}{} \bigg| \frac{3 \text{ feet}}{1 \text{ yard}} \bigg| \frac{12 \text{ inches}}{1 \text{ foot}} = 144 \text{ inches} \qquad \frac{(4) \times (3) \times (12)}{(1) \times (1)} = 144$$

EXAMPLE 2: **How many seconds are there in March?**

$$\frac{60 \text{ seconds}}{1 \text{ minute}} \bigg| \frac{60 \text{ minutes}}{1 \text{ hour}} \bigg| \frac{24 \text{ hours}}{1 \text{ day}} \bigg| \frac{31 \text{ days}}{\text{March}} = \frac{2{,}678{,}400 \text{ seconds}}{\text{March}}$$

We are often asked to convert the units of area or volume, which are square or cubic terms. It is important to note that the numerical coefficient **must** also be squared or cubed when converting units that are squared or cubed.

We are often asked to convert the units of area or volume, which are square or cubic terms. It is important to note that the numerical coefficient **must** also be squared or cubed when converting units that are squared or cubed.

EXAMPLE 3: The volume of a barrel is 10 ft³. How many in³ of water will a barrel hold?

$$\frac{10 \text{ ft}^3}{} \left| \frac{(12)^3 \text{ in}^3}{1 \text{ ft}^3} \right. = \frac{(10) \times (12) \times (12) \times (12)}{(1)} = 17,280 \text{ in}^3$$

EXAMPLE 4: How many kilograms per cubic meter (kg/m³) are there in 3 grams per cubic centimeter(g/cm³)?

$$\frac{3 \text{ grams}}{\text{cm}^3} \left| \frac{1 \text{ kg}}{1000 \text{ grams}} \right| \frac{(100)^3 \text{ cm}^3}{1 \text{ m}^3} = 3000 \frac{\text{kg}}{\text{m}^3}$$

This is an example of converting units of **density**, which is a measure in mass per unit volume.

Another useful scientific term which often needs converting is **velocity,** measured in length per unit time.

EXAMPLE 5: What is the highway speed limit of 65 miles per hour in feet per second?

$$\frac{65 \text{ miles}}{1 \text{ hour}} \left| \frac{5,280 \text{ feet}}{1 \text{ mile}} \right| \frac{1 \text{ hour}}{60 \text{ minutes}} \left| \frac{1 \text{ minute}}{60 \text{ seconds}} \right. = 95.3 \frac{\text{feet}}{\text{second}}$$

Problems:

1. Jared can work 54 math problems in one hour. How many problems can he write in 10 minutes?
2. Leah rides 22 feet per second on her bicycle. How many miles per hour does she ride?
3. A jar of honey has a density of 14 kg per m³. What is its density in g/mm³?
4. If a pitcher will hold 2 ft³ of lemonade, how many in³ of lemonade will it hold?
5. Juan's car gets an average of 24 miles per gallon of gas. How far can Juan go on 1 quart of gas?
6. From question 5, how many gallons of gas will it take Juan to travel 528 miles?
7. How many cubic centimeters (cm³) of water are in 1 m³?
8. How many cubic feet are in a hole that is 3 feet deep, 4 feet wide, and 6 feet long?
9. How many yards are in 3 miles?
10. John Smoltz throws his fastball 99 miles per hour. If he starts 60 feet, 6 inches from home plate, how many seconds does it take the ball to get to the plate?

290

CONVERTING UNITS OF TIME

	Time	
1 week	= 7	days
1 day	= 24	hours
1 hour	= 60	minutes
1 minute	= 60	seconds

Abbreviations	
year	= yr
week	= wk
hour	= hr or h
minutes	= min
seconds	= sec

EXAMPLE: Simplify: 2 days 34 hr 75 min

Step 1   75 minutes is more than 1 hour. There are 60 minutes in an hour, so divide 75 by 60.

$$\begin{array}{r} 1\ hr \\ 60\overline{)\ 75} \\ -60 \\ \hline 15\ min \end{array}$$

$$\begin{array}{r} 2\ days\ \ 34\ hr\ \ 75\ min \\ +\ \ \ \ \ \ \ \ 1\ hr\ \ 15\ min \\ \hline 2\ days\ \ 35\ hr\ \ 15\ min \end{array}$$

Step 2   35 hours is more than 1 day. There are 24 hours in a day, so divide 35 hours by 24.

$$\begin{array}{r} 1\ day \\ 24\overline{)\ 35} \\ -24 \\ \hline 11\ hr. \end{array}$$

$$\begin{array}{r} 2\ days\ \ 35\ hr.\ \ 15\ min. \\ +\ 1\ day\ \ 11\ hr. \\ \hline 3\ days\ \ 11\ hr\ \ 15\ min. \end{array}$$

Simplify the following:

1.   5 years 18 months

2.   4 hours 84 minutes

3.   1 minute 76 seconds

4.   3 weeks 8 days

5.   1 week 10 days

6.   3 minutes 80 seconds

7.   5 hours 75 minutes

8.   5 days 30 hours 78 min

9.   5 wk 8 days 36 hr

10.   2 hr 55 min 86 sec

11.   12 hr 86 min 87 sec

12.   7 years 13 months

1. Out of the 100 coins, 45 are in mint condition. What is the ratio of mint condition coins to the total number of coins?

———————

2. The ratio of boys to girls in the ninth grade is 6:5. If there are 135 girls in the class, how many boys are there?

———————

3. Twenty out of the total 235 seniors graduated with honors. What is the ratio of seniors graduating with honors to the total number of seniors?

———————

4. Aunt Bess uses 3 cups of oatmeal to bake 6 dozen oatmeal cookies. How many cups of oatmeal would she need to bake 15 dozen cookies?

———————

5. On a map, 2 centimeters represents 150 kilometers. If a line between two cities measures 5 centimeters, how many kilometers apart are they?

———————

6. Shondra used six ounces of chocolate chips to make two dozen cookies. At that rate, how many ounces of chocolate chips would she need to make seven dozen cookies?

———————

7. When Rick measures the shadow of a yard stick, it is 5 inches. At the same time, the shadow of the tree he would like to chop down is 45 inches. How tall is the tree in yards?

———————

Solve the following proportions:

8. $\frac{8}{x} = \frac{1}{2}$

9. $\frac{2}{5} = \frac{x}{10}$

10. $\frac{x}{6} = \frac{3}{9}$

11. $\frac{4}{9} = \frac{8}{x}$

12. On a scale drawing of a house floor plan, 1 inch represents 2 feet. The length of the kitchen measures 5 inches on the floor plan. How many feet does that represent?

———————

13. If 4 inches represent 8 feet on a scale drawing, how many feet does 6 inches represent?

———————

14. On a scale drawing, 3 centimeters represent 100 miles. If a line segment between two points measured 5 centimeters, how many miles would it represent?

———————

15. On a map scale, 2 centimeters represent 5 kilometers. If two towns on the map are 20 kilometers apart, how long would the line segment be between the two towns on the map?

———————

16. If 3 inches represent 10 feet on a scale drawing, how long will a line segment be that represents 15 feet?

———————

Fill in the blanks below with the appropriate unit of measurement.

17. A box of assorted chocolates might weigh about 1 _____ (English).

18. A compact disc is about 7 _____ (English) across.

19. In Europe, gasoline is sold in _____ (metric).

20. A vitamin C tablet has a mass of 500 _____ (metric) .

Fill in the blanks below with the appropriate English or metric conversions.

21. Two gallons equals _____ cups.

22. 4.2 L equals _____ mL.

23. $3\frac{1}{2}$ yards equals _____ inches.

24. 6,800 m equals _____ kilometers.

25. 36 oz. equals _____ pounds.

26. 730 mg equals _____ kg.

Use dimensional analysis to make the following conversions.

27. Convert 30 miles per hour to feet per second.

28. An outlet pipe releases 450 gallons of water per hour. How many cups of water are released per second?

29. If Janet can type 36 words per minute, how many words can she type in 10 seconds?

30. A bicyclist travels 22 feet per second. How many miles per hour is the bicyclist traveling?

31. How many square inches are in a floor tile that measures 2 square feet?

32. A water bottle holds 500 cm^3. How many cubic inches does the water bottle hold? Use the conversion factor of 1 inch = 2.5 cm.

33. An airplane travels at 540 miles per hour. How many feet per second does it travel?

34. Jose's bathroom measures 10 feet by 10 feet. How many square centimeters does his bathroom measure? Use the conversion factor of 1 foot = 30 centimeters.

35. Paulo's faucet drips 2 cubic inches per day. How many cubic centimeters does it drip? Use the conversion factor of 1 inch = 2.5 cm.

36. Yulisa has 681 grams of candy. How many pounds of candy does she have if 1 pound = 454 grams?

Chapter 20 | Data Analysis and Probability

MEAN

Statistics is a branch of mathematics. Using statistics, mathematicians organize data (numbers) into forms that are easily understood. In statistics, the **mean** is the same as the **average**. To find the **mean** of a list of numbers, first, add together all the numbers in the list, and then divide by the number of items in the list.

EXAMPLE: Find the mean of 38, 72, 110, 548.

Step 1: First add $38 + 72 + 110 + 548 = $ **768**

Step 2: There are 4 numbers in the list, so divide the total by 4.
The mean is **192**.

$$4\overline{)768} \quad = 192$$

Practice finding the mean (average). Round to the nearest tenth if necessary.

1. Dinners served:

   489 561 522 450

   Mean = _____

2. Prices paid for shirts:

   $4.89 $9.97 $5.90 $8.64

   Mean = _____

3. Piglets born:

   23 19 15 21 22

   Mean = _____

4. Student absences:

   6 5 13 8 9 12 7

   Mean = _____

5. Paychecks received:

   $89.56 $99.99 $56.54

   Mean = _____

6. Choir attendance:

   56 45 97 66 70

   Mean = _____

7. Long distance calls:

   33 14 24 21 19

   Mean = _____

8. Train boxcars:

   56 55 48 61 51

   Mean = _____

9. Cookies eaten:

   5 6 8 9 2 4 3

   Mean = _____

Find the mean (average) of the following word problems.

10. Val's science grades were 95, 87, 65, 94, 78, and 97. What was her average? _____

11. Ann runs a business from her home. The number of orders for the last 7 business days were 17, 24, 13, 8, 11, 15, and 9. What was the average number of orders per day? _____

12. Melissa tracked the number of phone calls she had per day: 8, 2, 5, 4, 7, 3, 6, 1. What was the average number of calls she received? _____

7.6.3

FINDING DATA MISSING FROM THE MEAN

EXAMPLE: Mara knew she had an 88 average in her biology class, but she lost one of her papers. The three papers she could find had scores of 98%, 84%, and 90%. What was the score on her fourth paper?

Step 1: Figure the total score on four papers with an 88% average. $.88 \times 4 = 3.52$

Step 2: Add together the scores from the three papers you have. $.98 + .84 + .90 = 2.72$

Step 3: Subtract the scores you know from the total score. $3.52 - 2.72 = .80$ She had 80% on her fourth paper.

---

1. Gabriel earned 87% on his first geography test. He wants to keep a 92% average. What does he need to get on his next test to bring his average up?

2. Rian earned $68.00 on Monday. How much money must he earn on Tuesday to have an average of $80 earned for the two days?

3. Haley, Chuck, Dana, and Chris entered a contest to see who could bake the most chocolate chip cookies in an hour. They baked an average of 75 cookies. Haley baked 55, Chuck baked 70, and Dana baked 90. How many did Chris bake?

4. Four wrestlers made a pact to lose some weight before the competition. They lost an average of 7 pounds each, over the course of 3 weeks. Carlos lost 6 pounds, Steve lost 5 pounds, and Greg lost 9 pounds. How many pounds did Wes lose?

5. Three boxes are ready for shipment. The boxes average 26 pounds each. The first box weighs 30 pounds; the second weighs 25 pounds. How much does the third box weigh?

MEDIAN

In a list of numbers ordered from lowest to highest, the **median** is the middle number. To find the **median,** first arrange the numbers in numerical order. If there is an odd number of items in the list, the **median** is the middle number. If there is an even number of items in the list, the **median** is the **average of the two middle numbers.**

EXAMPLE 1: Find the median of 42, 35, 45, 37, and 41.

Step 1: Arrange the numbers in numerical order: 35 37 (41) 42 45.

Step 2: Find the middle number. **The median is 41.**

EXAMPLE 2: Find the median of 14, 53, 42, 6, 14, and 46.

Step 1: Arrange the numbers in numerical order: 6 14 (14 42) 46 53.

Step 2: Find the average of the 2 middle numbers.
$(14 + 42) \div 2 = 28$. **The median is 28.**

Circle the median in each list of numbers.

1. 35, 55, 40, 30, and 45
2. 7, 2, 3, 6, 5, 1, and 8
3. 65, 42, 60, 46, and 90
4. 15, 16, 19, 25, and 20
5. 75, 98, 87, 65, 82, 88, and 100
6. 33, 42, 50, 22, and 19
7. 401, 758, and 254
8. 41, 23, 14, 21, and 19
9. 5, 8, 3, 10, 13, 1, and 8

10.	11.	12.
19	9	45
14	3	32
12	10	66
15	17	55
18	6	61

7.6.3

MODE

In statistics, the **mode** is the number that occurs most frequently in a list of numbers.

EXAMPLE: Exam grades for a Math class were as follows:
70 88 92 85 99 85 70 85 99 100 88 70 99 88 88 99 88 92 85 88.

Step 1: Count the number of times each number occurs in the list.

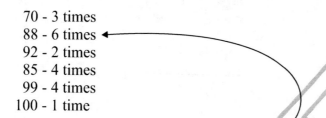

      70 - 3 times
      88 - 6 times
      92 - 2 times
      85 - 4 times
      99 - 4 times
     100 - 1 time

Step 2: Find the number that occurs most often.
The mode is 88 because it is listed 6 times. No other number is listed as often.

Find the mode in each of the following lists of numbers.

1.	2.	3.	4.	5.	6.	7.
88	54	21	56	64	5	12
15	42	16	67	22	4	41
88	44	15	67	22	9	45
17	56	78	19	15	8	32
18	44	21	56	14	4	16
88	44	16	67	14	7	12
17	56	21	20	22	4	12

mode ___   mode ___   mode ___   mode ___   mode ___   mode ___   mode ___

8. 48, 32, 56, 32, 56, 48, 56   **mode** _____   17. 22, 45, 48, 12, 22, 41, 22   **mode** _____

9. 12, 16, 54, 78, 16, 25, 20   **mode** _____   18. 62, 44, 78, 62, 54, 44, 62   **mode** _____

10. 5, 4, 8, 3, 4, 2, 7, 8, 4, 2   **mode** _____   19. 54, 22, 54, 78, 22, 78, 22   **mode** _____

11. 11, 9, 7, 11, 7, 5, 7, 7, 5   **mode** _____   20. 14, 17, 33, 21, 33, 17, 33   **mode** _____

12. 84, 22, 79, 22, 87, 22, 22   **mode** _____   21. 65, 51, 8, 21, 8, 65, 70, 8   **mode** _____

13. 95, 87, 65, 94, 78, 95   **mode** _____   22. 17, 24, 13, 8, 11, 8, 15, 9   **mode** _____

14. 8, 2, 5, 4, 7, 2, 3, 6, 1   **mode** _____   23. 51, 45, 84, 51, 65, 74, 51   **mode** _____

15. 89, 7, 11, 89, 17, 56   **mode** _____   24. 8, 74, 65, 15, 9, 10, 74   **mode** _____

16. 15, 48, 52, 41, 8, 48   **mode** _____   25. 62, 54, 2, 7, 89, 2, 7, 54, 2   **mode** _____

7.6.3

STEM-AND-LEAF PLOTS

A **stem-and-leaf plot** is a way to organize and analyze statistical data. To make a stem-and-leaf plot, first draw a vertical line.

Final Math Averages
85 92 87 62 75 84 96 52
45 77 98 75 71 79 85 82
87 74 76 68 93 77 65 84
79 65 77 82 86 84 92 60
99 75 88 74 79 80 63 84
87 90 75 81 73 69 73 75
31 86 89 65 69 75 79 76

Stem	Leaves
3	1
4	5
5	2
6	0,2,3,5,5,5,8,9,9
7	1,3,3,3,4,4,5,5,5,5,5,5,6,6,7,7,7,9,9,9
8	0,1,2,2,4,4,4,4,5,5,6,6,7,7,7,8,9
9	0,2,2,3,6,8,9

On the left side of the line, list all the numbers that are in the tens place from the set of data. Next, list each number in the ones place on the right side of the line in ascending order. It is easy to see at a glance that most of the students scored in the 70's or 80's with a majority having averages in the 70's. It is also easy to see that the maximum average is 99, and the lowest average is 31. Stem-and-leaf plots are a way to organize data making it easy to read.

Make a stem-and leaf-plot from the data below, and then answer the questions that follow.

1. **Speeds on Turner Road**

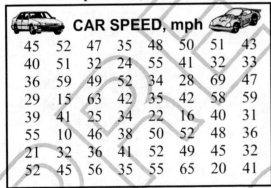

CAR SPEED, mph
45 52 47 35 48 50 51 43
40 51 32 24 55 41 32 33
36 59 49 52 34 28 69 47
29 15 63 42 35 42 58 59
39 41 25 34 22 16 40 31
55 10 46 38 50 52 48 36
21 32 36 41 52 49 45 32
52 45 56 35 55 65 20 41

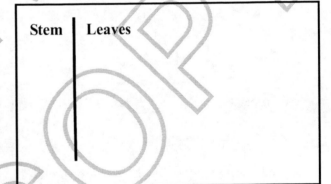

Stem	Leaves

2. What was the fastest speed recorded?

3. What was the slowest speed recorded?

4. Which speed was most often recorded?

5. If the speed limit is 45 miles per hour, how many were speeding?

6. If the speed limit is 45 miles per hour, how many were at least 20 mph over or under the speed limit?

7.6.1

MORE STEM-AND-LEAF PLOTS

Two sets of data can be displayed on the same stem-and-leaf plot.

EXAMPLE: The following is an example of a back-to-back stem-and-leaf plot.

Bryan's Math scores {60,65,72,78,85,90}
Bryan's English scores {78,88,89,89,92,95,100}

Math		English
	6	
5,0	6	
2,8	7	8
5	8	8,9,9
0	9	2,5
	10	0

2|7 means 72   8|9 means 89

Read the stem-and-leaf plot below and answer the questions that follow.

3rd grade Boys' Weights		3rd Grade Girls' Weights
8,7,5,3,2	4	0,2, 4, 7
6, 4, 1, 0	5	1,8,8,8, 9
5	6	0 6, 6, 8, 8
0	9	8

4|5 means 54   6|8 means 68

1. What is the median for the girls' weights?

2. What is the median for the boys' weights?

3. What is the mode for the girls' weights?

4. What is the weight of the lightest boy?

5. What is the weight of the heaviest boy?

6. What is the weight of the heaviest girl?

7. Create a stem-and-leaf plot for the data given below.

Automobile Speeds on I-85

60	65	80	75	92	81	63
65	67	75	78	79	77	69
62	57	64	65	68	71	69
71	73	56	69	69	70	74

Automobile Speeds on I-75

72	56	62	65	63	60	58
55	57	70	69	59	53	61
58	61	63	67	57	63	67
56	58	59	62	64	63	69

8. What is the median speed for I-75?

9. What is the median speed for I-85?

10. What is the mode speed for I-75?

11. What is the mode speed for I-85?

12. What was the fastest speed on either interstate?

7.6.1

QUARTILES AND EXTREMES

In statistics, large sets of data are separated into four equal parts. These parts are called **quartiles**. The **median** separates the data into two halves. Then, the median of the upper half is the **upper quartile**, and the median of the lower half is the **lower quartile**. The difference between the upper and lower quartile is the **interquartile range.**

The **extremes** are the highest and lowest values in a set of data. The lowest value is called the **lower extreme**, and the highest value is called the **upper extreme**.

EXAMPLE 1: The following set of data shows the high temperatures (in degrees Fahrenheit) in cities across the United States on a particular autumn day. Find the median, the upper quartile, the lower quartile, the upper extreme, and the lower extreme of the data.

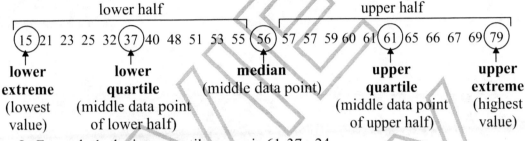

In Example 1, the interquartile range is 61-37= 24.

EXAMPLE 2: The following set of data shows the fastest race car qualifying speeds in miles per hour. Find the median, the upper quartile, the lower quartile, the upper extreme, and the lower extreme of the data.

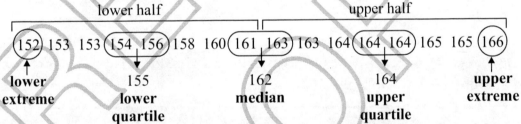

Note: When you have an even number of data points, the median is the average of the two middle points. The lower middle number is then included in the lower half of the data, and the upper middle number is included in the upper half.

Find the median, the upper quartile, the lower quartile, the upper extreme, and the lower extreme of each set of data given below.

1. 0 0 1 1 1 2 2 3 3 4 5

2. 15 16 18 20 22 22 23

3. 62 75 77 80 81 85 87 91 94

4. 74 74 76 76 77 78

5. 3 3 3 5 5 6 6 7 7 7 8 8

6. 190 191 192 192 194 195 196

7. 6 7 9 9 10 10 11 13 15

8. 21 22 24 25 27 28 32 35

8.6.3

BOX-AND-WHISKER PLOTS

Box-and-whisker plots are used to summarize data as well as to display data. A box-and-whisker plot summarizes data using the median, upper and lower quartiles, and the lower and upper extreme values. Consider the data below–a list of employees' ages at the Acme Lumber Company:

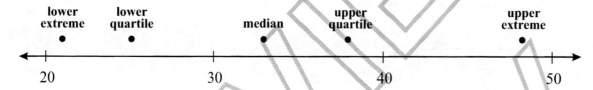

Step 1: Find the median, upper quartile, lower quartile, upper extreme, and lower extreme just like you did on the previous page.

Step 2: Plot the 5 data points found in step 1 above on a number line as shown below.

Step 3: Draw a box around the quartile values, and draw a vertical line through the median value. Draw whiskers from each quartile to the extreme value data points.

This box-and-whisker displays five types of information: lower extreme, lower quartile, median, upper quartile, and upper extreme.

Draw a box-and-whisker plot for the following sets of data.

1.
10 12 12 15 16 17 19 21 22 22 25 27 31 35 36 37 38 38 41 43 45 50 51 56 57 58 59

2.
5 5 6 7 9 9 10 11 12 15 15 16 17 18 19 19 20 22 24 26 27 27 30 31 31 35 37

8.6.4

SCATTER PLOTS

A **scatter plot** is a graph of ordered pairs involving two sets of data. These plots are used to detect whether two sets of data, or variables, are truly related.

In the example to the right, two variables, income and education, are being compared to see if they are related or not. Twenty people were interviewed, ages 25 and older, and the results were recorded on the chart.

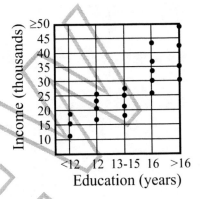

Imagine drawing a line on the scatter plot where half the points are above the line and half the points are below it. In the plot on the right, you will notice that this line slants upward and to the right. This line direction means there is a **positive** relationship between education and income. In general, for every increase in education, there is a corresponding increase in income.

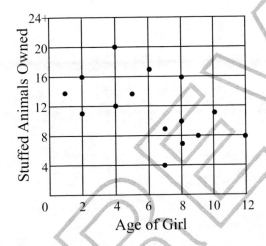

Now, examine the scatter plot on the left. In this case, 15 girls ages 2-12 were interviewed and asked, "How many stuffed animals do you currently have?" If you draw an imaginary line through the middle points, you will notice that the line slants downward and to the right. This plot demonstrates a **negative** relationship between the age of girls and their stuffed animal ownership. In general, as the girls' ages increase, the number of stuffed animals owned decreases.

Finally, look at the scatter plot shown on the right. In this plot, Rita wanted to see the relationship between the temperature in the classroom and the grades she received on tests she took at that temperature. As you look to your right, you will notice that the points are distributed all over the graph. Because this plot is not in a pattern, there is no way to draw a line through the middle of the points. This type of point pattern indicates there is **no** relationship between Rita's grades on tests and the classroom temperature.

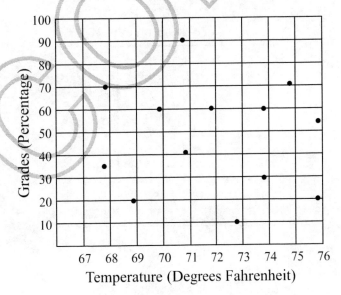

8.6.5

301

Examine each of the scatter plots below. On the line below each plot, write whether the relationship shown between the two variables is "positive", "negative", or "no relationship".

1.

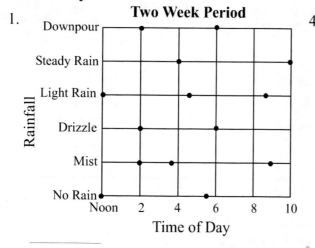

_____

2.

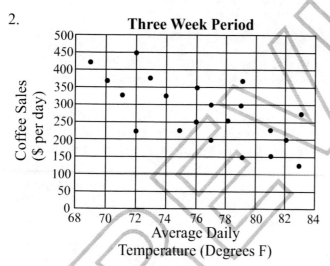

_____

3

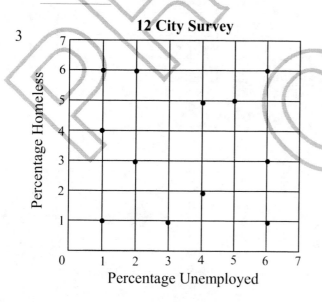

_____

4.

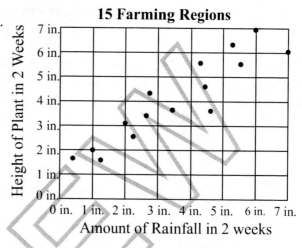

_____

5.

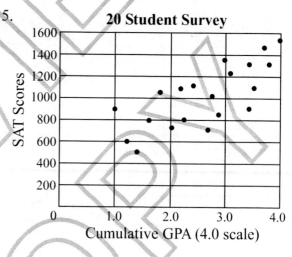

_____

6.

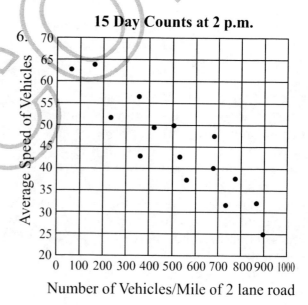

_____

8.6.5

THE LINE OF BEST FIT

At this point, you now understand how to plot points on a Cartesian plane. You also understand how to find the data trend on a Cartesian plane. These skills are necessary to accomplish the next task, determining the line of best fit.

In order to find a line of best fit, you must first draw a scatterplot of all data points. Once this is accomplished, draw an oval around all of the points plotted. Draw a line through the points in such a way that the line separates half the points from one another. You may now use this line to answer questions.

Example: The following data set contains the heights of children between 5 and 13 years old. Make a scatter plot and draw the line of best fit to represent the trend. Using the graph, determine the height for a 14 year-old child.

Age 5: 4'6", 4'4", 4' 5"   Age 8: 4'8", 4'6", 4'7"   Age 11: 5'0", 4'10"
Age 6: 4'7", 4'5", 4'6"   Age 9: 4'9", 4'7", 4'10"   Age 12: 5'1", 4'11", 5'0", 5'3"
Age 7: 4'9", 4'7", 4'6", 4'8"   Age 10: 4'9", 4'8", 4' 10"   Age 13: 5'3", 5'2", 5'0", 5'1"

In this example, the data points lay in a positive sloping direction. To determine the line of best fit, all data points were circled, then a line of best fit was drawn. Half of the points lay below, half above the line of best fit drawn bisecting the narrow length of the oval.

To find the height of a 14-year old, simply continue the line of best fit forward. In this case, the height is 62 inches.

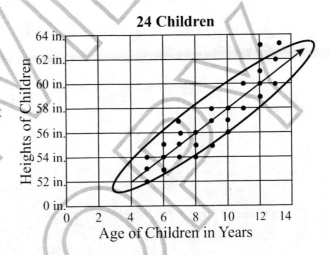

Plot the data sets below, then draw the line of best fit. Next, use the line to estimate the value of the next measurement.

1. Selected Values of the Sleekster Brand Light Compact Vehicles: New Vehicle: $13,000.
    1 year old: $12,000, $11,000, $12,500   3 year old: $8,500, $8,000, $9,000
    2 year old: $9,000, $10,500, $9,500   4 year old: $7,500, $6,500, $6,000
    5 year old: ?

2. The relationship between string length and kite height for the following kites:
    (L=500ft., H=400ft.) (L=250ft, H=150ft.) (L=100ft., H=75 ft.) (L=500ft., H=350ft.)
    (L=250ft., H=200ft.) (L=100ft., H=50ft.) (L=600ft., H=?)

3. Relationship between Household Incomes(HI) and Household Property Values (HPV):
    (HI=$30,000, HPV=$100,000) (HI=$45,000, HPV=$120,000) (HI=$60,000,
    HPV=$135,000) (HI=$50,000, HPV=115,000) (HI=$35,000, HPV=105,000) (HI=65,000,
    HPV=155,000) (HI=$90,000, HPV = ?)

8.6.5

303

MAKING PREDICTIONS

Use what you know about number patterns to answer the following questions.

Corn plants grow as tall as they will get in about 20 weeks. Study the chart of the rate of corn plant growth below, and answer the questions that follow.

Corn Growth	
Beginning Week	**Height (inches)**
2	9
7	39
11	63
14	??

1. If the growth pattern continues, how high will the corn plant be beginning week 14?

2. If the growth pattern was constant (at the same rate from week to week), how high was the corn in the beginning of the 8th week?

Pierre Lebont is painting masks today to be sold during Mardi-Gras. He paints large masks that take longer to dry in the beginning of the day and smaller and smaller masks as the day progresses.

Time	# Masks Completed per Hour
Hour 1	3
Hour 3	5
Hour 6	8

3. How many masks did Pierre paint during his second hour of work?

4. How many masks will Pierre have painted by the end of an 8 hour day?

Charles Boudreaux is picking up crabs from the traps he laid near the Gulf of Mexico. As he travels east, he notices that the crab catches are getting larger.

Buoy	Traps	Number of Crabs Caught
West	Trap 1	12
↓	Trap 2	unrecorded
	Trap 3	30
	Trap 4	unrecorded
East	Trap 5	48

5. How many crabs would he likely pull up in the sixth trap?

6. If he has 6 traps, how many is he likely to pull up altogether?

7.6.2

MISLEADING STATISTICS

As you read magazines and newspapers, you will see many charts and graphs which present statistical data. This data will present you with how measurements change over time or how one measurement corresponds to another measurement. However, some charts and graphs are presented to make changes in data appear greater than they actually are. The people presenting the data create these distortions to make exaggerated claims.

There is one method to arrange the data in ways which can exaggerate statistical measurements. A statistician can create a graph in which the number line does not begin with zero.

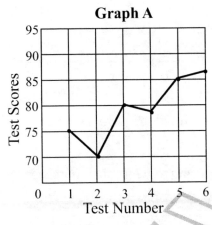

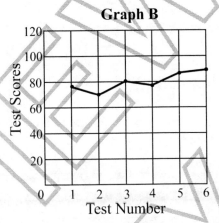

In the two graphs above, notice how each graph displays the same data. However, the way the data is displayed in graph A appears more striking than the data display in graph B. Graph A's data presentation is more striking because the test score numbers do not begin at zero.

Another form of misleading information is through the use of the wrong statistical measure to determine what is the middle. For instance, the mean, or average, of many data measurements allows **outliers** (data measurements which lie well outside the normal range) to have a large effect. Examine the measurements in the chart below.

Address	Household Income	Address	Household Income
341 Spring Drive	$19,000	346 Spring Drive	$30,000
342 Spring Drive	$17,000	347 Spring Drive	$32,000
343 Spring Drive	$26,000	348 Spring Drive	$1,870,000
344 Spring Drive	$22,000	349 Spring Drive	$31,000
345 Spring Drive	$25,000	350 Spring Drive	$28,000

Average (Mean) Household Income: $210,000
Median Household Income: $27,000

In this example, the outlier, located at 348 Spring Drive, inflates the average household income on this street to the extent that it is over eight times the median income for the area.

Read the following charts and graphs, and then answer the questions below.

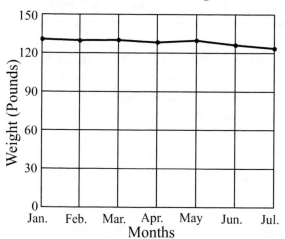

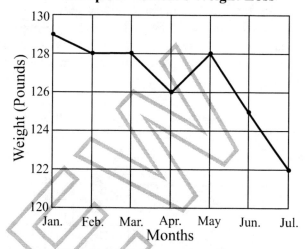

1. Which graph above presents misleading statistical information? Why is the graph misleading?

Twenty teenagers were asked how many electronic and computer games they purchased per year. The following table shows the results.

Number of Games	0	1	2	3	4	5	58
Number of Teenagers	4	2	5	3	4	1	1

2. Find the mean of the data.
3. Find the median of the data.
4. Find the mode of the data.
5. Which measurement is most misleading?
6. Which measurement would depict the data most accurately?
7. Is the *mean* of a set of data affected by outliers? Justify your answer with the example above.

Examine the two bar graphs below.

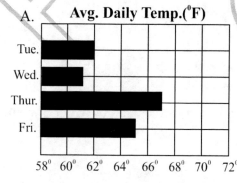

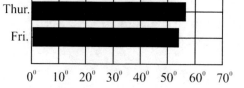

8. Which graph is misleading? Why?

7.6.3
7.6.4

READING TABLES

A **table** is a concise way to organize large quantities of information using rows and columns. **Read each table carefully, and then answer the questions that follow.**

Some employers use a tax table like the one below to figure how much Federal Income Tax should be withheld from a single person paid weekly. The number of withholding allowances claimed is also commonly referred to as the number of deductions claimed.

Federal Income Tax Withholding Table				
SINGLE Persons – WEEKLY Payroll Period				
If the wages are –		And the number of withholding allowances claimed is –			
At least	But less than	0	1	2	3
		The amount of income tax to be withheld is –			
$250	260	31	23	16	9
$260	270	32	25	17	10
$270	280	34	26	19	12
$280	290	35	28	20	13
$290	300	37	29	22	15

1. David is single, claims 2 withholding allowances, and earned $275 last week. How much Federal Income Tax was withheld?

2. Cecily claims 0 deductions, and she earned $297 last week. How much Federal Income Tax was withheld?

3. Sherri claims 3 deductions and earned $268 last week. How much Federal Income Tax was withheld from her check?

4. Mitch is single and claims 1 allowance. Last week, he earned $291. How much was withheld from his check for Federal Income Tax?

5. Ginger earned $275 this week and claims 0 deductions. How much Federal Income Tax will be withheld from her check?

BAR GRAPHS

Bar graphs can be either vertical or horizontal. There may be just one bar or more than one bar for each interval. Sometimes each bar is divided into two or more parts. In this section, you will work with a variety of bar graphs. Be sure to read all titles, keys, and labels to completely understand all the data that is presented. **Answer the questions about each graph below.**

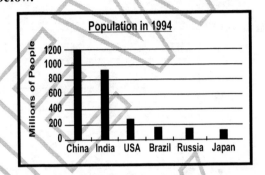

1. Which country has over 1 billion people?

2. How many countries have fewer than 200,000,000 people?

3. How many more people does India have than Japan?

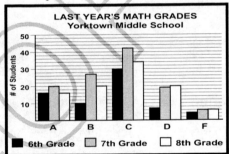

4. How many of last year's 6th graders made C's in math?

5. How many more math students made B's in the 7th grade than in the 8th Grade?

6. Which letter grade is the mode of the data?

7. How many 8th graders took math last year?

8. How many students made A's in math last year?

LINE GRAPHS

Line graphs often show how data change over time. Study the line graph below charting temperature changes for a day in Sandy, Nevada. Then, answer the questions that follow.

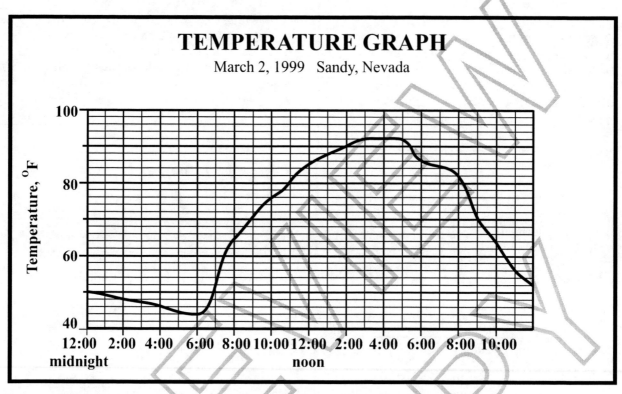

TEMPERATURE GRAPH
March 2, 1999   Sandy, Nevada

Study the graph, and then answer the questions below.

1. When was the coolest time of the day? _____

2. When was the hottest time of the day? _____

3. How much did the temperature rise between 6:00 a.m. and 2:00 p.m.? _____

4. How much did the temperature drop between 6:00 p.m. and 11:00 p.m.? _____

5. What is the difference in temperature between 8:00 a.m. and 8:00 p.m.? _____

6. Between which two hour time period was the greatest increase in temperature? _____

7. Between which hours of the day did the temperature continually increase? _____

8. Between which two hours of the day did the temperature change the least? _____

9. How much did the temperature decrease from 2:00 a.m. to 6:00 a.m.? _____

10. During which two times of day was the temperature 70°F? _____

8.6.4
7.6.1

MULTIPLE LINE GRAPHS

Multiple line graphs are a way to present a large quantity of data in a small space. It would often take several paragraphs to explain in words the same information that one graph could do.

On the graph below, there are three lines. You will need to read the **key** to understand the meaning of each.

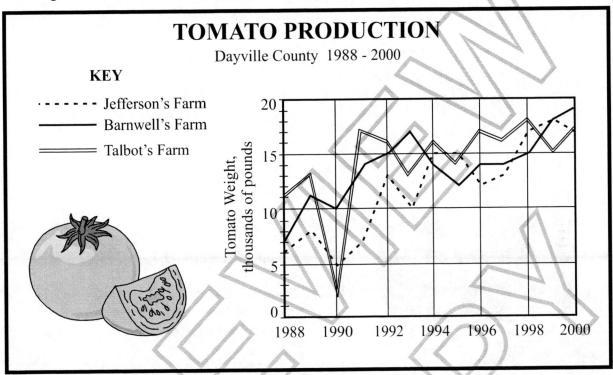

TOMATO PRODUCTION
Dayville County 1988 - 2000

KEY
- - - - - Jefferson's Farm
——— Barnwell's Farm
===== Talbot's Farm

Study the graph, and then answer the questions below.

1. In what year did Barnwell's Farm produce 8,000 pounds of tomatoes more than Talbot's Farm?

2. In which year did Dayville County produce the most pounds of tomatoes?

3. In 1993, how many more pounds of tomatoes did Barnwell's Farm produce than Talbot's Farm?

4. How many pounds of tomatoes did Dayville County's three farms produce in 1992?

5. In which year did Dayville County produce the fewest pounds of tomatoes?

6. Which farm had the most dramatic increase in production from one year to the next?

7. How many more pounds of tomatoes did Jefferson's Farm produce in 1992 than in 1988?

8. Which farm produced the most pounds of tomatoes in 1995?

8.6.4
7.6.1

CIRCLE GRAPHS

Circle graphs represent data expressed in percentages of a total. The parts in a circle graph should always add up to 100%. Circle graphs are sometimes called **pie graphs** or **pie charts**.

To figure the value of a percent in a circle graph, multiply the percent by the total. Use the circle graphs below to answer the questions. The first question is worked for you as an example.

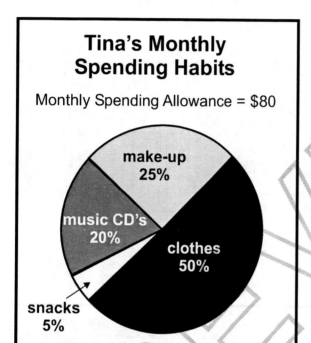

1. How much did Tina spend each month on music CD's?

$$\$80 \times 0.20 = \$16.00$$

                                         $16.00

2. How much did Tina spend each month on make-up?

               _____

3. How much did Tina spend each month on clothes?

               _____

4. How much did Tina spend each month on snacks?

               _____

Fill in the following chart.

Favorite Activity	Number of Students
5. watching TV	
6. talking on the phone	
7. playing video games	
8. surfing the Internet	
9. playing sports	
10. reading	

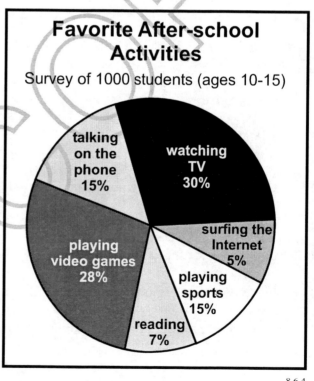

8.6.4
7.6.1

PROBABILITY

Probability is the chance something will happen. Probability is most often expressed as a fraction, a decimal, a percent, or can also be written out in words.

EXAMPLE 1:

Billy had 3 red marbles, 5 white marbles, and 4 blue marbles on the floor. His cat came along and batted one marble under the chair. What is the **probability** it was a red marble?

Step 1: The number of red marbles will be the top number of the fraction.

Step 2: The total number of marbles is the bottom number of the fraction.

$$\frac{3}{12} = \frac{1}{4}$$

Expressed as a decimal, $\frac{1}{4} = .25$, as a percent, $\frac{1}{4} = 25\%$, and written out in words $\frac{1}{4}$ is one out of four.

EXAMPLE 2: Determine the probability that the pointer will stop on a shaded wedge or the number 1.

Step 1: Count the number of possible wedges that the spinner can stop on to satisfy the above problem. There are 5 wedges that satisfy it (4 shaded wedges and one number 1). The top number of the fraction is 5.

Step 2: Count the total number of wedges, 7. The bottom number of the fraction is 7.
The answer is $\frac{5}{7}$ or **five out of seven.**

EXAMPLE 3: Refer to the spinner above. If the pointer stops on the number 7, what is the probability that it will **not** stop on 7 on the next Spin?

Step 1: Ignore the information that the pointer stopped on the number 7 on the previous spin. The probability of the next spin does not depend on the outcome of the previous spin. Simply find the probability that the spinner will **not** stop on 7. Remember, if P is the probability of An event occurring, 1− P is the probability of an event **not** occurring. In this example, the probability of the spinner landing on 7, is $\frac{1}{7}$.

Step 2: The probability is that the spinner will not stop on 7 is $1 - \frac{1}{7}$ which equals $\frac{6}{7}$. The answer is $\frac{6}{7}$ or **six out of seven.**

8.6.6

Find the probability of the following problems. Express as a <u>PERCENT</u>.

1. A computer chose a random number between 1 and 50. What is the probability of you guessing the same number that the computer chose in 1 try?

2. There are 24 candy-coated chocolate pieces in a bag. Eight have defects in the coating that can be seen only with close inspection. What is the probability of pulling out a defective piece without looking?

3. Seven sisters have to choose which day each will wash the dishes. They put equal size pieces of paper each labeled with a day of the week in a hat. What is the probability that the first sister who draws will choose a weekend day?

4. For his garden, Clay has a mixture of 12 white corn seeds, 24 yellow corn seeds, and 16 bi-color corn seeds. If he reaches for a seed without looking, what is the probability that Clay will plant a bi-color corn seed first?

5. Mom just got a new department store credit card in the mail. What is the probability that the last digit is an odd number?

6. Alex has a paper bag of cookies that includes 8 chocolate chip, 4 peanut butter, 6 butterscotch chip, and 12 ginger. Without looking, his friend John reaches in the bag for a cookie. What is the probability that the cookie is peanut butter?

7. An umpire at a little league baseball game has 14 balls in his pockets. Five of the balls are brand A, 6 are brand B, and 3 are brand C. What is the probability that the next ball he throws to the pitcher is a brand C ball?

MORE PROBABILITY

EXAMPLE: You have a cube with one number, 1, 2, 3, 4, 5 and 6 painted on each face of the cube. What is the probability that if you throw the cube 3 times, you will get the number 2 each time?

If you roll the cube once, you have a 1 in 6 chance of getting the number 2. If you roll the cube a second time, you again have a 1 in 6 chance of getting the number 2. If you roll the cube a third time, you again have a 1 in 6 chance of getting the number 2. The probability of rolling the number 2 three times in a row is:

$$\frac{1}{6} \times \frac{1}{6} \times \frac{1}{6} = \frac{1}{216}$$

Find the probability that each of the following events will occur.

1. There are 10 balls in a box, each with a different digit on it: 0, 1, 2, 3, 4, 5, 6, 7, 8, & 9. A ball is chosen at random and then put back in the box.
    A. What is the probability that if you picked out a number ball 3 times, you would get the number 7 each time?
    B. What is the probability you would pick a ball with 5, then 9, and then 3?
    C. What is the probability that if you picked out a ball four times, you would always get an odd number?

2. A couple has 4 children ages 9, 6, 4, and 1. What is the probability that they are all girls?

3. There are 26 letters in the alphabet allowing a different letter to be on each of 26 cards. The cards are shuffled. After each card is chosen at random, it is put back in the stack of cards, and the cards are shuffled again.
    A. What is the probability that you pick 3 cards, one at a time, and you would draw first a "y", then an "e", and then an "s"?
    B. What is the probability that you would draw 4 cards and get the letter "z" each time?
    C. What is the probability that you would draw twice and get a letter that is in the word "random" both times?

Express the following as a <u>FRACTION</u>.

1. Prithi has two boxes. Box 1 contains 3 red, 2 silver, 4 gold, and 2 blue combs. She also has a second box containing 1 black and 1 clear brush. What is the probability that Prithi selected a red brush from box 1 and a black brush from box 2?

2. Terrell cast his line into a pond containing 7 catfish, 8 bream, 3 trout, and 6 northern pike. He immediately caught a bream. What are the chances that Terrell will catch a second bream when he casts his line?

3. Gloria Quintero entered a contest in which the person who draws his or her initials out of a box containing all 26 letters of the alphabet wins the grand prize. Gloria reaches in and draws a "G", keeps it, then draws another letter. What is the probability that Gloria will next draw a "Q"?

4. Steve Marduke had two spinners in front of him. The first one was numbered 1-6, and the second was numbered 1-3. If Steve Spins each spinner once, what is the probability that the first spinner will show an odd number and the second spinner will show a "1"?

5. Carrie McCallister flipped a coin twice and got heads both times. What is the probability that Carrie will get tails the third time she flips the coin?

6. Vince Macaluso is pulling two socks out of a washing machine in the dark. The washing machine contains three tan, one white, and two black socks. If Vince reaches in and pulls the socks out one at a time, what is the probability that Vince will pull out two tan socks in his first two tries?

TREE DIAGRAMS

Drawing a **tree diagram** is another method of determining the probability of an event occuring.

EXAMPLE: If you toss two six-sided numbered cubes, what is the probability you will get two cubes that add up to 9? One way to determine the probability is to make a tree diagram.

Cube 1	Cube 2	Cube 1 plus Cube 2
1	1	2
	2	3
	3	4
	4	5
	5	6
	6	7
2	1	3
	2	4
	3	5
	4	6
	5	7
	6	4
3	1	8
	2	5
	3	6
	4	7
	5	8
	6	⑨
4	1	5
	2	6
	3	7
	4	8
	5	⑨
	6	10
5	1	6
	2	7
	3	8
	4	⑨
	5	10
	6	11
6	1	7
	2	8
	3	⑨
	4	10
	5	11
	6	12

Alternative method

Write down all of the numbers in both cubes which would add up to 9.

Cube 1	Cube 2
4	5
5	4
6	3
3	6

Numerator = 4 combinations

For Denominator: Multiply the number of sides on one cube times the number of sides on the other cube.

$6 \times 6 = 36$

Numerator:
Denominator: $\dfrac{4}{36} = \dfrac{1}{9}$

There are 36 possible ways the cubes could land. Out of those 36 ways, the two cubes add up to 9 only 4 times. The probability you will get two cubes that add up to 9 is $\dfrac{4}{36}$ or $\dfrac{1}{9}$.

7.6.7

Read each of the problems below. Then, answer the questions.

1. Jake has a spinner. The spinner is divided into eight equal regions numbered 1-8. In two spins, what is the probability that the numbers added together would equal 12?

2. Charlie and Libby each spin one spinner one time. The spinner is divided into 5 equal regions numbered 1-5. What is the probability that these two spins added together would equal 7?

3. Gail spins a spinner twice. The spinner is divided into 9 equal regions numbered 1-9. In two spins, what is the probability that the difference between the two numbers would equal 4?

4. Diedra throws two 10-sided dice. What is the probability that the difference between the two numbers would equal 7?

5. Cameron throws two six-sided dice. What is the probability that the difference between the two numbers would equal 3?

6. Tesla spins one spinner twice. The spinner is divided into 11 equal regions numbered 1-11. What is the probability that the two numbers added together would equal 11?

7. Samantha decides to roll two five-sided dice. What is the probability that the two numbers added together would equal 4?

8. Mary Ellen spins a spinner twice. The spinner is divided into 7 equal regions numbered 1-7. What is the probability that the product of the two numbers would equal 10?

9. Conner decides to roll two six-sided dice. What is the probability that the product of the two numbers equals 4?

10. Tabitha spins one spinner twice. The spinner is divided into 9 equal regions numbered 1-9. What is the probability that the sum of the two numbers would equal 10?

11. Darnell decides to roll two 15-sided dice. What is the probability that the difference between the two numbers would be 13?

12. Inez spins one spinner twice. The spinner is divided into 12 equal regions numbered 1-12. What is the probability that the sum of the two numbers will be equal to 10?

13. Gina spins one spinner twice. The spinner is divided into 8 equal regions numbered 1-8. What is the probability that the two numbers added together would equal 9?

14. Celia rolls two six-sided dice. What is the probability that the difference between the two numbers will be 2?

15. Brett spins one spinner twice. The spinner is divided into 4 equal regions numbered 1-4. What is the probability that the difference between the two numbers will be 3?

DISJOINT EVENTS

Disjoint events are two or more events that cannot take place at the same time. They are usually within the same space, but occur independently of one another.

EXAMPLE: Determine if the following two events are disjoint.
In a 26 card alphabet deck and only one card is drawn at a time, drawing a P or drawing a H

Answer: Yes, these two events are disjoint because if one card is drawn at a time, then you cannot draw a P and a H at the same time.

Determine if the following events are disjoint. If they are disjoint, write D, and if they are not disjoint write ND.

1. eating a green piece of candy or eating a red piece of candy

2. rolling a six or rolling a two using a cube with 1, 2, 3, 4, 5, or 6 on each side

3. team A is winning against team B, and team C is winning against team D.

4. pulling two nickels or pulling 10¢ out of a bag of coins

5. team A is winning against team B or team C is winning against team B.

6. eating a dinner of steak and potatoes or eating a potato

7. going swimming outside or getting a suntan

8. mowing the lawn or mopping the kitchen

PROBABILITY OF DISJOINT EVENTS

To find the probability that one or the other of two disjoint events will occur, you must find the probability of each event happening, then add the two probabilities together. (Hint: two disjoint events are almost always separated by the word "or.")

EXAMPLE: Using two six-sided cubes numbered 1, 2, 3, 4, 5, or 6 on each side, find the probability of rolling a sum of seven or a sum of twelve.

Step 1: First, write out all the different combinations that can occur.

Number rolled on the second cube

		1	2	3	4	5	6
Number rolled on the first cube	1	(1, 1)	(1, 2)	(1, 3)	(1, 4)	(1, 5)	(1, 6)
	2	(2, 1)	(2, 2)	(2, 3)	(2, 4)	(2, 5)	(2, 6)
	3	(3, 1)	(3, 2)	(3, 3)	(3, 4)	(3, 5)	(3, 6)
	4	(4, 1)	(4, 2)	(4, 3)	(4, 4)	(4, 5)	(4, 6)
	5	(5, 1)	(5, 2)	(5, 3)	(5, 4)	(5, 5)	(5, 6)
	6	(6, 1)	(6, 2)	(6, 3)	(6, 4)	(6, 5)	(6, 6)

Step 2:   By looking at the chart you can find the total number of combinations and the number of combinations for the sum of seven and twelve. There are thirty-six different combinations when rolling the cubes. Therefore, this will be the bottom number of the fractions of the probabilities. Now, you must find the number of combinations that make the sum of seven and twelve.

sum of seven: (1, 6), (2, 5), (3, 4), (4, 3), (5, 2), (6, 1)   six different ways
sum of twelve: (6, 6)   one different way

Step 3:   Find the probabilities of each event, then add them together.
The probability of rolling the sum of seven is $\frac{6}{36}$ (six ways out of thirty-six different possibilities), and the probability of rolling the sum of twelve is $\frac{1}{36}$.

$$\frac{6}{36} + \frac{1}{36} = \frac{7}{36}$$

Answer: The probability of either rolling a sum of seven or a sum of twelve is $\frac{7}{36}$.

Find the probability of the disjoint events. If the two events are not disjoint, write Not Possible.

1. Using a six-sided cube numbered 1, 2, 3, 4, 5, or, 6 on each side, find the probability that you will roll a sum of eight or a sum of six.

2. If you place four Alphabet card decks with 26 letters in each deck face down, what is the probability of either drawing a F or drawing a W?

3. In a bag of candy, there are five cherry pieces of candy, six strawberry pieces of candy, four lemon pieces of candy, and six grape pieces of candy. What is the probability of picking a strawberry piece of candy or a lemon piece of candy?

4. There are fifty pieces of paper numbered one through fifty. If all fifty pieces of paper are face down on the floor, what is the probability of picking up an even number or a multiple of nine?

5. An umpire at a little league baseball game has 35 balls in the bucket next to him. Ten of the balls are brand A, eight are brand B, thirteen are brand C, and four are brand D. What is the probability that the next ball he throws to the pitcher is either brand A or Brand C?

6. Terrell cast his line into a pond containing 9 catfish, 6 bream, 5 trout, and 6 northern pike. He immediately caught a bream, then released it back into the water. What are the chances that Terrell will catch a either a trout or a catfish the next time he casts his line?

CHAPTER 20 REVIEW

1. There are 50 students in the school orchestra in the following sections:

   25 string section
   15 woodwind
   5 percussion
   5 brass

   One student will be chosen at random to present the orchestra director with an award. What is the probability the student will be from the woodwind section?

2. Fluffy's cat treat box contains 6 chicken-flavored treats, 5 beef-flavored treats, and 7 fish-flavored treats. If Fluffy's owner reaches in the box without looking and chooses one treat, what is the probability that Fluffy will get a chicken-flavored treat?

3. The spinner on the right stopped on the number 5 on the first spin. What is the probability that it will not stop on the number 5 on the second spin?

4. Three cakes are sliced into 20 pieces each. Each cake contains 1 gold ring. What is the probability that one person who eats one piece of cake from each of the 3 cakes will find 3 gold rings?

5. Brianna tossed a coin 4 times. What is the probability she got all tails?

6. Sherri turned the spinner on the right 3 times. What is the probability that the pointer always landed on a shaded number?

Read and answer the following.

There are 9 slips of paper in a hat, each with a number from 1 to 9. The numbers correspond to a group of students who must answer a question when the number for their group is drawn. Each time a number is drawn, the number is put back in the hat.

7. What is the probability that the number 6 will be drawn twice in a row?

8. What is the probability that the first 5 numbers drawn will be odd numbers?

9. What is the probability that the second, third, and fourth numbers drawn will be even numbers?

10. What is the probability that the first five times a number is drawn it will be the number 5?

11. What is the probability that the first five numbers drawn will be 1, 2, 3, 4, 5 in that order?

12. If you toss two six-sided dice, what is the probability they will add up to 7? (Make a tree diagram.)

13. Make a tree diagram to show the probability of a couple with 3 children having a boy and two girls.

Find the mean, median, and mode for each of the following sets of data, and fill in the table below.

❶ Miles Run by Track Team Members

Jeff	24
Eric	20
Craig	19
Simon	20
Elijah	25
Rich	19
Marcus	20

❸ Hardware Store Payroll June Week 2

Erica	$280
Dane	$206
Sam	$240
Nancy	$404
Elsie	$210
Gail	$305
David	$280

❷ 1992 SUMMER OLYMPIC GAMES
Gold Medals Won

Unified Team	45	Hungary	11
United States	37	South Korea	12
Germany	33	France	8
China	16	Australia	7
Cuba	14	Japan	3
Spain	13		

Data Set Number	Mean	Median	Mode
❶			
❷			
❸			

4. Jenica bowled three games and scored an average of 116 points per game. She scored 105 on her first game and 128 on her second game. What did she score on her third game?

5. Concession stand sales for each game in the season were $320, $540, $230, $450, $280, and $580. What was the mean sales per game?

6. Cedrick D'Amitrano works Friday and Saturday delivering pizza. He delivers 8 pizzas on Friday. How many pizzas must he deliver on Saturday to average 11 pizzas per day?

7. Long cooked three Vietnamese dinners that weighed a total of 40 ounces. What was the average weight for each dinner?

Over the past 2 years, Coach Strive has kept a record of how many points his basketball team, the Bearcats, has scored in each game:

29 32 35 35 36 38 39 40 40
41 42 43 44 44 45 47 49 50
52 53 62

8. Create a stem-and-leaf plot for the data.

Stem	Leaves

9. Create a box-and-whisker plot for the Data.

10. What is the median?
11. What is the upper quartile?
12. What is the lower quartile?
13. What is the upper extreme?
14. What is the lower extreme?
15. What is the interquartile range?

The Riveras and the Rogers families are meeting for a Fourth of July family reunion in the same park. The ages of the Rivera family members are 48, 79, 20, 2, 14, 84, 32, 61, 48, 92, 87, 54, 41, 27, 18, 21, 36, 44, 27, 66, 27, 16, 54, 48, 48, 6, and 4. The ages of the Roger's family members are 26, 84, 14, 7, 30, 50, 55, 41, 29, 33, 1, 15, 48, 16, and 20. Plot each of their ages on the stem-and-leaf plot at the top of the next column and answer the questions that follow.

318

16.

Nine cooks were asked "When you use a thermometer what is the actual temperature inside your oven when it is set at 350°F?" The responses are in the chart below. Answer the questions that follow.

Temperature(°F)	104	347	348	349	350	351	352
# of Cooks	1	1	1	2	1	2	1

17. In the data above, which age is the mode of the data for the Rivera family?

18. Which age is the median in the Rogers family?

19. What ages are the two oldest Riveras?

20. What age are the two youngest Riveras?

21. Which family has the older median age?

24. Find the mean of the data above.
25. Find the median of the data above.
26. Which measurement is misleading? Why?

Examine each set of information below. If you drew a scatter plot for the information in each of these tables, write whether the relationship would be "positive", "negative", or "no relationship".

For each plot, write whether the Relationship shown between the two variables is "positive", "negative", or "No relationship".

22.

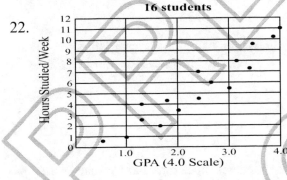

Hair Color	Weight
Brown	200
Black	168
Blonde	179
Auburn	189
White	128
Blonde	100
Black	139
Red	201
Dark Blond	175

23.

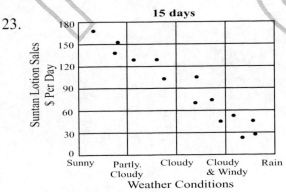

27. If the hair colors were ordered from dark to light, what is the relationship?

Hours of Daylight by Latitude

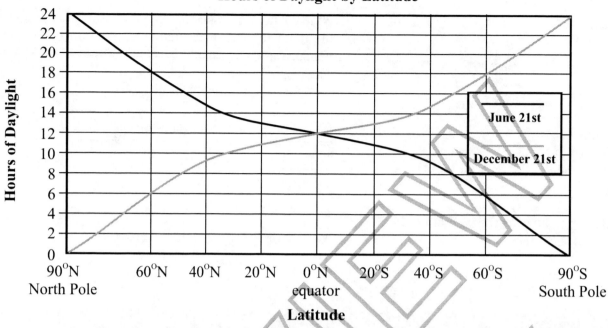

28. On June 21st, how many hours of daylight are available at the equator? _____

29. How many more hours of daylight does a person at 60°N latitude have
on June 21st than a person at 60°S latitude? _____

30. Where would a person experience an entire 24 hours of daylight on
December 21st? _____

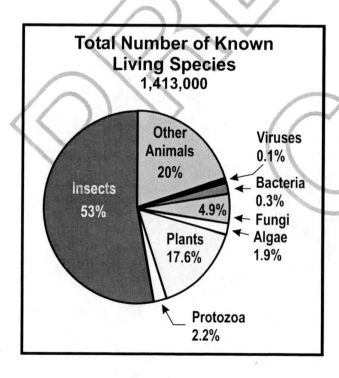

31. Which category given in the pie chart represents the greatest number of known living species?

_____

32. Of the total 1,413,000 known living species, how many are viruses?

_____

33. Of the total 1,413,000 known living species, how many are insects?

_____

320   Copyright © American Book Company

Permutations and Combinations

PERMUTATIONS

A **permutation** is an arrangement of items in a specific order. If a problem asks how many ways you can arrange 6 books on a bookshelf, it is asking you how many permutations there are for 6 items.

EXAMPLE 1: Ron has 4 items: a model airplane, a trophy, an autographed football, and a toy sports car. How many ways can he arrange the 4 items on a shelf?

Solution: Ron has 4 choices for the first item on the shelf. He then has 3 choices left for the second item. After choosing the second item, he has 2 choices left for the third item and only one choice for the last item. The diagram below shows the permutations for arranging the 4 items on a shelf if he chooses to put the trophy first.

1st item   **2nd item**   **3rd item**   **4th item**

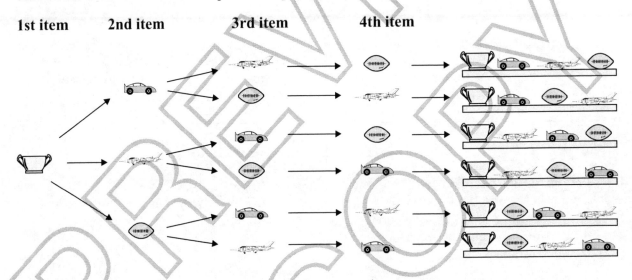

Count the number of permutations if Ron chooses the trophy as the first item. There are 6 permutations. Next, you could construct a tree diagram of permutations choosing the model car first. That tree diagram would also have 6 permutations. Then, you could construct a tree diagram choosing the airplane first. Finally, you could construct a pyramid choosing the football first. You would then have a total of 4 tree diagrams, each having 6 permutations. The total number of permutations is 6 × 4 = 24. There are 24 ways to arrange the 4 items on a bookshelf.

You probably don't want to draw tree diagrams for every permutation problem. What if you want to know the permutations for arranging 30 objects? Fortunately, mathematicians have come up with a formula for calculating permutations.

For the problem above, Ron has 4 items to arrange. Therefore, multiply 4 × 3 × 2 × 1 = 24. Another way of expressing this calculation is 4!, stated as 4 factorial. 4! = 4 × 3 × 2 × 1.

EXAMPLE 2: How many ways can you line up 6 students?

Solution: The number of permutations for 6 students = 6! = 6 × 5 × 4 × 3 × 2 × 1 = 720. There are 6 choices for the first position, 5 for the second, 4 for the third, 3 for the fourth, 2 for the fifth, and 1 for the sixth.

Work the following permutation problems.

1. How many ways can you arrange five books on a bookshelf?

2. Myra has six novels to arrange on a bookshelf. How many ways can she arrange the novels?

3. Seven sprinters signed up for the 100 meter dash. How many ways can the seven sprinters line up on the start line?

4. Keri wants an ice-cream cone with one scoop of chocolate, one scoop of vanilla, and one scoop of strawberry. How many ways can the scoops be arranged on the cone?

5. How many ways can you arrange the letters A, B, C, and D?

6. At Sam's party, the DJ has four song requests. In how many different orders can he play the four songs?

7. Yvette has five comic books. How many different ways can she stack the comic books?

8. Sandra's couch can hold three people. How many ways can she and her two friends sit on the couch?

9. How many ways can you arrange the numbers 2, 3, 5?

10. At a busy family restaurant, four tables open up at the same time. How many different ways can the hostess seat the next four families waiting to be seated?

MORE PERMUTATIONS

EXAMPLE: If there are 6 students, how many ways can you line up any 4 of them?

Solution: Multiply $6 \times 5 \times 4 \times 3 = 360$. There are 6 choices for the first position in line, 5 for the second position, 4 for the third position, and 3 for the last position. There are 360 ways to line up 4 of the 6 students.

Find the number of permutations for each of the problems below.

1. How many ways can you arrange four out of eight books on a shelf?

   _____

2. How many 3 digit numbers can be made using the numbers 2, 3, 5, 8, and 9?

   _____

3. How many ways can you line up four students out of a class of twenty?

   _____

4. Kim worked in the linen department of a store. Eight new colors of towels came in. Her job was to line up the new towels on a long shelf. How many ways could she arrange the eight colors?

   _____

5. Terry's CD player holds 5 CD's. Terry owns 12 CD's. How many different ways can he arrange his CD's in the CD player?

   _____

6. Erik has eleven shirts he wears to school. How many ways can he choose a different shirt to wear on Monday, Tuesday, Wednesday, Thursday, and Friday?

   _____

7. Deb has a box of twelve markers. The art teacher told her to choose three markers and line them up on her desk. How many ways can she line up three markers from the twelve?

   _____

8. Jeff went into an ice cream store serving 32 flavors of ice cream. He wanted a cone with two different flavors. How many ways could he order two scoops of ice cream, one on top of the other?

   _____

9. In how many ways can you arrange any three letters from the 26 letters in the alphabet?

   _____

COMBINATIONS

In a **permutation**, objects are arranged in a particular order. In a **combination**, the order does not matter. In a **permutation**, if someone picked two letters of the alphabet, **k, m** and **m, k**; they would be considered 2 different permutations. In a **combination, k, m** and **m, k** would be the same combination. A different order does not make a new combination.

EXAMPLE: How many combinations of three letters from the set {a, b, c, d, e} are there?

Step 1: Find the **permutation** of 3 out of 5 objects.

Step 2: Divide by the permutation of the **number of objects** to be chosen from the total (3). This step eliminates the duplicates in finding the permutations.

$$\frac{5 \times 4 \times 3}{3 \times 2 \times 1} = 10$$

Step 3: Cancel common factors and simplify.

Find the number of combinations for each problem below.

1. How many combinations of 4 numbers can be made from the set of numbers {2, 4, 6, 7, 8, 9}?

2. Johnston Middle School wants to choose 3 students at random from the 7th grade to take an opinion poll. There are 124 seventh graders in the school. How many different groups of 3 students could be chosen? (Use a calculator for this one.)

3. How many combinations of 3 students can be made from a class of 20?

4. Fashion Ware catalog has a sweater that comes in 8 colors. How many combinations of 2 different colors does a shopper have to choose from?

5. Angelo's Pizza offers 10 different pizza toppings. How many different combinations can be made of pizzas with four toppings?

6. How many different combinations of 5 flavors of jelly beans can you make from a store that sells 25 different flavors of jelly beans?

7. The track team is running the relay race in a competition this Saturday. There are 14 members of the track team. The relay race requires 4 runners. How many combinations of 4 runners can be formed from the track team?

8. Kerri got to pick 2 prizes from a grab bag containing 12 prizes. How many combinations of 2 prizes are possible?

MORE COMBINATIONS

Another kind of combination involves selection from several categories.

EXAMPLE: At Joe's Deli, you can choose from 4 kinds of bread, 5 meats, and 3 cheeses when you order a sandwich. How many different sandwiches can be made with Joe's choices for breads, meats, and cheeses if you choose 1 kind of bread, 1 meat, and 1 cheese for each sandwich?

JOE'S SANDWICHES

Breads	**Meats**	**Cheeses**
White	Roast Beef	Swiss
Pumpernickel	Corned Beef	American
Light rye	Pastrami	Mozzarella
Whole wheat	Roast Chicken	
	Roast Turkey	

Solution: Multiply the number of choices in each category. There are 4 breads, 5 meats, and 3 cheeses, so $4 \times 5 \times 3 = 60$. There are 60 combinations of sandwiches.

Find the number of combinations that can be made in each of the problems below.

1. Angie has 4 pairs of shorts, 6 shirts, and 2 pairs of tennis shoes. How many different outfit combinations can be made with Angie's clothes?

    _____

2. Raymond has 7 baseball caps, 2 jackets, 10 pairs of jeans, and 2 pairs of sneakers. How many combinations of the 4 items can he make?

    _____

3. Claire has 6 kinds of lipstick, 4 eye shadows, 2 kinds of lip liner, and 2 mascaras. How many combinations can she use to make up her face?

    _____

4. Clarence's dad is ordering a new truck. He has a choice of 5 exterior colors, 3 interior colors, 2 kinds of seats, and 3 sound systems. How many combinations does he have to pick from?

    _____

5. A fast food restaurant has 8 kinds of sandwiches, 3 kinds of french fries, and 5 kinds of soft drinks. How many combinations of meals could you order if you ordered a sandwich, fries, and a drink?

    _____

6. In summer camp, Tyrone can choose from 4 outdoor activities, 3 indoor activities, and 3 water sports. He has to choose one of each. How many combinations of activities can he choose?

    _____

7. Jackie won a contest at school and gets to choose one pencil and one pen from the school store and an ice-cream from the lunch room. There are 5 colors of pencils, 3 colors of pens, and 4 kinds of ice-cream. How many combinations of prize packages can she choose?

    _____

CHAPTER 21 REVIEW

Answer the following permutation and combination problems.

1. Daniel has 7 trophies he has won playing soccer. How many different ways can he arrange them in a row on his bookshelf?

2. Missy has 12 colors of nail polish. She wears 1 color each day, 7 different colors a week. How many combinations of 7 colors can she make before she has to repeat the same 7 colors in a week?

3. Eileen has a collection of 12 antique hats. She plans to donate 5 of the hats to a museum. How many combinations of hats are possible for her donation?

4. Julia has 5 porcelain dolls. How many ways can she arrange 3 of the dolls on a display shelf?

5. Ms. Randal has 10 students. Every day she randomly draws the names of 2 students out of a bag to turn in their homework for a test grade. How many combinations of 2 students can she draw?

6. In the lunch line, students can choose 1 out of 3 meats, 1 out of 4 vegetables, 1 out of 3 desserts, and 1 out of 5 drinks. How many lunch combinations are there?

7. Andrea has 7 teddy bears in a row on a shelf in her room. How many ways can she arrange the bears in a row on her shelf?

8. Adrianna has 4 hats, 8 shirts, and 9 pairs of pants. Choosing one of each, how many different clothes combinations can she make?

9. The buffet line offers 5 kinds of meat, 3 different salads, a choice of 4 desserts, and 5 different drinks. If you chose one food from each category, from how many combinations would you have to choose?

10. How many pairs of students can Mrs. Smith choose to go to the library if she has 20 students in her class?

Chapter 22 | Problem-Solving

MISSING INFORMATION

Problems can only be solved if you are given enough information. Sometimes you are not given enough information to solve a problem.

EXAMPLE: Chuck has worked on his job for 1 year now. At the end of a year, his employer gave him a 12% raise. How much does Chuck make now?

To solve this problem, you need to know how much Chuck made when he began his job one year ago.

Each problem below does not give enough information for you to solve it. Beneath each Problem, describe the information you would need to solve the problem.

1. Fourteen percent of the coated chocolate candies in Nate's bag were yellow. At that rate, how many of the candies were yellow?

2. Patrick is putting up a fence around all four sides of his back yard. The fence costs $2.25 per foot, and his yard is 150 feet wide. How much will the fence cost?

3. Staci worked 5 days last week. She made $6.75 per hour before taxes. What was her total earnings before taxes were taken out?

4. Which is a better buy: 4 oz bar of soap for 88¢ or a bath bar for $1.20?

5. Randy bought a used car for $4,568 plus sales tax. What was the total cost of the car?

6. The Hall family ate at a restaurant, and each of their dinners cost $5.95. They left a 15% tip. What was the total amount of the tip?

7. If a kudzu plant grows 3 feet per day, in what month will it be 90 feet long?

8. Bethany traveled by car to her sister's house in Raleigh. She traveled at an average speed of 52 miles per hour. She arrived at 4:00 p.m. How far did she travel?

9. Terrence earns $7.50 per hour plus 5% commission on total sales over $500 per day. Today he sold $6,500 worth of merchandise. How much did he earn for the day?

10. Michelle works at a department store and gets an employee's discount on all of her purchases. She wants to buy a sweater that sells for $38.00. How much will the sweater cost after her discount?

11. John filled his car with 10 gallons of gas and paid for the gas with a $20 bill. How much change did he get back?

12. Olivia budgets $5.00 per work day for lunch. How much does she budget for lunch each month?

13. Joey worked 40 hours and was paid $356.00. His friend Pete worked 38 hours. Who was paid more per hour?

14. A train trip from Columbia to Boston took $18\frac{1}{4}$ hours. How many miles apart are the two cities?

15. Caleb spent 35% of his check on rent, 10% on groceries, and 18% on utilities. How much money did he have left from his check?

16. The Lyons family spent $54.00 per day plus tax on lodging during their vacation. How much tax did they pay for lodging per day?

17. Richard bought cologne at a 30% off sale. How much did he save buying the cologne on sale?

18. The bottling machine works 7 days a week and fills 1,000 bottles per hour. How many bottles did it fill last week?

19. Tyler, who works strictly on commission, brought in $25,000 worth of sales in the last 10 days. How much was his commission?

EXACT INFORMATION

Most word problems supply exact information and ask for exact answers. The following problems are the same as those on the previous two pages with the missing information given. Find the exact solution.

1. Fourteen percent of the coated chocolate candies in Nate's bag were yellow. If there were 50 pieces in the bag, how many of the candies were yellow?

2. Patrick is putting up a fence around all four sides of his back yard. The fence costs $2.25 per foot. His yard is 150 feet wide and 200 feet long. How much will the fence cost?

3. Staci worked 5 days last week, 8 hours each day. She made $6.75 per hour before taxes. How much did she make last week before taxes were taken out?

4. Which is a better buy: 4 oz bar of soap for 88¢? or a 6 oz bath bar for $1.20?

5. Randy bought a used car for $4,568 plus 6% sales tax. What was the total cost of the car?

6. The Hall family ate at a restaurant, and each of the 4 dinners cost $5.95. They left a 15% tip. What was the total amount of the tip?

7. If a kudzu plant grows 3 feet per day, in what month will it be 90 feet long if it took root in the middle of May?

8. Bethany traveled from South Carolina by car to her sister's house in Raleigh, North Carolina. She traveled at an average speed of 52 miles per hour. She left at 10:00 a.m. and arrived at 4:00 p.m. How far did she travel?

9. Terrence earns $7.50 per hour plus 5% commission on total sales over $500 per day. Today he sold $6,500 worth of merchandise and worked 7 hours. How much did he earn for the day?

10. Michelle works at a department store and gets a 20% employee's discount on all of her purchases. She wants to buy a sweater that sells for $38.00. How much will the sweater cost after her discount?

11. John filled his car with 10 gallons of gas priced at $1.24 per gallon. He paid for the gas with a $20 bill. How much change did he get back?

12. Olivia budgets $5.00 per work day for lunch. How much does she budget for lunches if she works 21 days this month?

13. Joey worked 40 hours and was paid $356.00. His friend Pete worked 38 hours at $8.70 per hour. Who was paid more per hour?

14. A train trip from Columbia, SC to Boston, MA took $18\frac{1}{4}$ hours. How many miles apart are the two cities if the train travels at an average speed of 50 miles per hour?

15. Caleb spent 35% of his check on rent, 10% on groceries, and 18% on utilities. How much money did he have left from his $260 check?

16. The Lyons family spent $54.00 per day plus 10% tax on lodging during their vacation. How much tax did they pay per day?

17. Richard bought cologne at a 30% off sale. The cologne was regularly priced at $44. How much did he save buying the cologne on sale?

18. The bottling machine works 7 days a week, 14 hours per day and fills 1,000 bottles per hour. How many bottles did it fill last week?

19. Tyler, who works strictly on commission, brought in $25,000 worth of sales in the last 10 days. He earns 15% commission on his sales. How much was his commission?

329

EXTRA INFORMATION

In each of the following problems, there is extra information given. **Look closely at the question, and use only the information you need to answer it.**

EXAMPLE: Gary was making $6.50 per hour. His boss gave him a $0.52 per hour raise. Gary works 40 hours per week. What percent raise did Gary receive?

Solution: To figure the percent of Gary's raise, you do **not** need to know how many hours per week Gary works. That is extra information not needed to answer the question. To figure the percent increase, simply divide the change in pay, 52¢, by the original wages, $6.50. $0.52 \div 6.50 = 0.08$

Gary received an 8% raise.

In the following questions, determine what information is needed from the problem to answer the question and solve.

1.   Leah wants a new sound system that is on sale for 15% off the regular price of $420. She has already saved $325 toward the cost. What is the dollar amount of the discount?

2.   Praveen bought a shirt for $34.80 and socks for $11.25. He gets $10.00 per week for his allowance. He paid $2.76 sales tax. What was his change from three $20 bills?

3.   Marty worked 38 hours this week, and he earned $8.40 per hour. His taxes and insurance deductions amount to 34% of his gross pay. What is his total gross pay?

4.   Tamika went shopping and spent $4.80 for lunch. She wants to buy a sweater that is on sale for $\frac{1}{4}$ off the regular $56.00 price. How much will she save?

5.   Nick drove an average of 52 miles per hour for 7 hours. His car gets 32 miles per gallon. How far did he travel?

6.   The odometer on Melody's car read 45,920 at the beginning of her trip and 46,460 at the end of her trip. Her speed averaged 54 miles per hour, and she used 20 gallons of gasoline. How many miles per gallon did she average?

7.   Eighty percent of the eighth graders attended the end of the year class picnic. There are 160 eighth graders and 54% of them ride the bus to school each day. How many students went to the class picnic?

8.   Matt has $5.00 to spend on snacks. Tastee Potato Chips cost $2.57 for a one pound bag at the grocery store. T-Mart sells the same bag of chips for $1.98. How much can he save if he buys the chips at T-Mart?

9.   Elaina wanted to make 10 cakes for the band bake sale. She needed $1\frac{3}{4}$ cups of flour and $2\frac{1}{4}$ cups of sugar for each cake. How many cups of flour did she need in all?

ESTIMATED SOLUTIONS

Some problems require an estimated solution. In order to have enough product to complete the job, you often need to buy more than you actually need because of the sizes the product come in.

In the following problems, be sure to round up your answer to the next whole number to find the correct solution.

1. Endicott Publishing received an order for 550 books. Each shipping box holds 30 books. How many boxes do the packers need to ship the order?

2. Elena's 250 chickens laid 314 eggs in the last 2 days. How many egg cartons holding one dozen eggs would be needed to hold all the eggs?

3. Antoinetta's Italian restaurant uses $1\frac{1}{4}$ quarts of olive oil every day. The restaurant is open 7 days a week. For the month of September, how many gallons should they order to have enough?

4. Eastmont High School is taking 316 students and 22 chaperones on a field trip. Each bus holds 44 persons. How many buses will the school need?

5. Fran volunteered to hem 11 choir robes that came in too long. Each robe is 7 feet around at the bottom. Hemming tape comes three yards to a pack. How many packs will Fran need to buy to go around all the robes?

6. Tonya is making matching vests for the children's choir. Each vest has 5 buttons on it, and there are 23 children in the choir. The button she picked comes 6 buttons to a card. How many cards of buttons does she need?

7. Tiffany is making the bread for the banquet. She needs to make 6 batches with $2\frac{1}{4}$ lb of flour in each batch. How many 10 lb bags of flour will she need to buy?

8. The homeless shelter is distributing 250 sandwiches per day to hungry guests. It takes one foot of plastic wrap to wrap each sandwich. There are 150 feet of plastic wrap per box. How many boxes will Mary need to buy to have enough plastic wrap for the week?

9. An advertising company has 15 different kinds of one-page flyers. The company needs 75 copies of each kind of flyer. How many reams of paper will the company need to produce the flyers? One ream equals 500 sheets of paper.

TWO-STEP PROBLEMS

Some problems require two steps to solve.

Read each of the following problems carefully and solve.

1. For a family picnic, Renee bought 10 pounds of hamburger meat and used $\frac{1}{4}$ pound of meat to make each hamburger patty. Renee's family ate 32 hamburgers. How many pounds of hamburger meat did she have left?

2. Vic sold 45 raffle tickets. His brother sold twice as many. How many tickets did they sell together?

3. Erin earns $2,200 per month. Her deductions amount to 28% of her paycheck. How much does she take home each month?

4. Matheson Middle School Band is selling t-shirts to raise money for new uniforms. They need to raise $1260. They are selling t-shirts for $12 each. There is a $6 profit for each shirt sold. So far, they have sold 85 t-shirts. How many more t-shirts do they need to sell to raise the $1260?

5. Alphonso was earning $1,860 per month and then got a 12% raise. How much will he make per month now?

6. Barbara and Jeff ate out for dinner. The total came to $15.00. They left a 15% tip. How much was the tip and the meal together?

7. Hillary is bicycling across Montana taking an 845 mile course. The first week, she covered 320 miles. The second week she traveled another 350 miles. How many more miles does she have to travel to complete the course?

8. Jason budgets 30% of his $1,100 income each month for food. How much money does he have to spend for everything else?

9. After Madison makes a 12% down payment on a $2,000 motorcycle, how much will she still owe?

10. Randy bought a pair of shoes for $51, a tie for $18, and a new belt for $23. If the sales tax is 8%, how much sales tax did he pay?

PATTERN PROBLEMS

Some problems follow a pattern. You must read these problems carefully and recognize the pattern.

EXAMPLE: Jason wants to swim in the ocean from the shore to the end of a pier 38 feet out. He must swim against the tide. For every 10 feet he swims forward, the tide takes him back 3 feet. How many total feet will he swim to reach the end of the pier?

Step 1: Draw a diagram or create a table to help you visualize the problem.

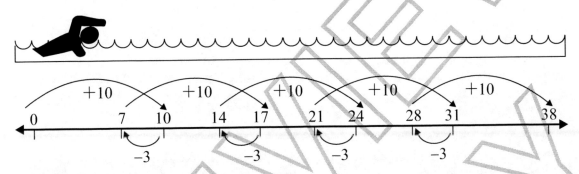

Step 2: Determine how many feet he swam by going forward 10 feet at a time. Look at the diagram above. There are 5 forward arrows that each represent 10 feet to get Jason to the 38 feet mark.
$10 \times 5 = 50$ feet

Read each of the following questions carefully. Draw a diagram or create a table to help you visualize the pattern. Then, answer the question.

1. In a strong man contest, the contestants must pull a car up an incline 13 feet long. Joe is the first contestant. With every tug, Joe pulls the car up the incline 3 feet. Before he can tug again, the car rolls down 1 foot. How many tugs will it take Joe to get the car up the incline?

2. Greta goes on a jungle safari and steps in quicksand. She sinks 25 inches. Her partner loops a rope around a tree limb directly above her so she can pull herself out. For every 5 inches she is able to pull up, she slips back down 1 inch. How many total inches must she climb up to completely free herself from the 25 inches of quicksand?

3. Randal is starting a job as a commissioned salesman. He establishes 5 clients the first week of his job. His goal is to double his clientele every week. If he can accomplish his goal, how many weeks will it take him to establish 320 clients?

4. A five-year old decides to run the opposite way on the moving sidewalk at the airport. For every 7 feet he runs forward, the sidewalk moves him 4 feet backward. How many total feet will he run to get to the end of a sidewalk that is 19 feet long?

5. A certain bacteria culture doubles its population every minute. Every 3 minutes, the entire culture decreases by half. If a culture starts with a population of 1000, what would be the population of the bacteria at 6 minutes? **Hint: The population equals 1000 at 0 minutes.**

6. Rita's mom needs to bake 5 dozen chocolate chip cookies for a bake sale. For every dozen cookies she bakes, her children eat two. How many dozen cookies will she have to bake to have 5 dozen for the bake sale?

7. A retail store plans on building 3 new stores each year. They expect that 1 store will go out of business every $2\frac{1}{2}$ years. How many years would it take to establish 26 stores?

8. Jasmine and her family sold Girl Scout cookies this spring at the mall. After every seven boxes she sold, her family bought a box to celebrate. Jasmine sold a total of 64 boxes. How many of those boxes did her family buy?

9. A pitcher plant has leaves modified as pitchers for trapping and digesting insects. An insect is lured to the edge of a 7 inch leaf and slips down 3 inches. For every 3 inches it slips down, it manages to climb up 2 inches. It then starts slipping back down 3 inches towards the digestive juices at the bottom of the leaf. How many total inches will the insect slip down until it reaches the bottom?

PATTERNS IN EXCHANGES

Some problems involve people exchanging things. Carefully read the example below.

EXAMPLE: Skip, Joe, Pete, and Don are new neighbors. The four neighbors meet to exchange phone numbers in case of an emergency. Each neighbor writes down his number for each of the other neighbors. What is the total number of telephone numbers written down?

Method 1: Make a list or draw a diagram to figure out the total number of phone numbers written. Skip writes his number once for Joe, once for Pete, and once for Don. Joe does the same for Skip, Pete, and Don. Pete copies his number for Skip, Joe, and Don. Finally, Don writes his number for Skip, Joe, and Pete. If you count the total number of phone numbers exchanged, you should get 12.

Method 2: Each neighbor knows his own number, so each writes his number for the other 3 neighbors. So 4 neighbors write their numbers down 3 times. $4 \times 3 = 12$.

In this type of problem where people are exchanging a unique item, you can use the following formula:

Total exchanges = (the number of people) $\times$ (the number of people $-$ 1)

Read the following questions carefully, and then solve each by using one of the methods above. Please note that the formula given in method 2 may not be appropriate to solve every problem.

---

1. At summer camp, 5 friends exchanged home addresses so that they could all keep in touch by writing. What is the total number of addresses exchanged between the 5 friends?

2. In Mrs. Kirk's homeroom class, all 20 students exchanged their yearbooks to be signed by each of their peers. How many signatures were written?

3. Three co-workers exchanged e-mail addresses with one another. How many e-mail addresses were exchanged?

4. A creative writing class held a contest for best written short story. Four friends in the class decided to write two short stories each. They then exchanged their written stories with each other to proof-read before turning them in. How many stories did each friend receive to read?

5. A class of 15 students had a party. Each student brought a different kind of candy to share with the other students. Each student started with an empty bag. The students then passed their bags around to each of the other students to receive some of the candy each brought in. How many times must each student pass his or her bag to sample the candy from all the other students? How many total passes were made among all the students?

MORE PATTERNS IN EXCHANGES

How does the number of exchanges between people change if the exchange is something like a handshake? Look at the example below where two people exchange one thing, like a handshake.

EXAMPLE: In a doubles tennis match, a team of two players competes against a team of two other players. After a match, it is customary for each of the players to shake hands at the net including a handshake between teammates. How many total handshakes are exchanged?

Long Method: Make a list or draw a diagram to figure out the total number of handshakes. If you count the handshakes represented by the diagram below, you should get 6.

Short Method: Use the same formula as you did on the previous page, but since two people are exchanging one thing, divide your answer by 2. $4 \times 3 = 12$ and $12 \div 2 = 6$.

When people are exchanging a mutual item, you can use the following formula:

$$\text{Total exchanges} = \frac{\text{(the number of people)} \times \text{(the number of people} - 1)}{2}$$

Read the following questions carefully, and then solve each by using one of the methods above. Please note that the formula given in the short method above may not be appropriate to solve every problem.

1. The last day of summer camp, 5 girls who had become good friends gave each other a hug. How many hugs were exchanged?

2. Before the coin toss at the football game, three captains from the Bruins team shook hands with each of the four captains from the Warriors team. How many handshakes were exchanged?

3. Seven basketball teams in a conference play each other twice in a season. How many total games are played in a season?

4. The teacher gave each of her 10 students a pegboard with 9 pegs. Each student was asked to connect each of the pegs on his or her pegboard with a rubber band. How many rubber bands did each student use to connect all the pegs? How many rubber bands did the entire class use?

5. Six different peaceful tribes built roads between each of their villages for easy trade. How many roads did they create if a different road connects each village?

NUMBER PATTERNS

In each of the examples below, there is a sequence of data that follows a pattern. You must find the pattern that holds true for each number in the data. Once you determine the pattern, you can write out an equation that fits the data and figure out any other number in the sequence.

	Sequence	Pattern	Next Number in Sequence	20th Number in the Sequence
EXAMPLE 1:	3, 4, 5, 6, 7	$n + 2$	$n + 2 = 8$	$20 + 2 = 22$

In number patterns, the sequence is the output. The input can be the set of whole numbers starting with 1. However, you must determine the "rule" or pattern. Look at the table below.

input	sequence
1 →	3
2 →	4
3 →	5
4 →	6
5 →	7

What pattern or "rule" can you come up with that gives you the first number in the sequence, 3, when you input 1? $n + 2$ will work because when $n = 1$, the first number in the sequence = 3. Does this pattern hold true for the rest of the numbers in the sequence? Yes, it does. When $n = 2$, the second number in the sequence = 4. When $n = 3$, the third number in the sequence = 5, and so on. Therefore, $n + 2$ is the pattern. Even without knowing the algebraic form of the pattern, you could figure out that 8 is the next number in the sequence. The equation describing this pattern would be $n + 2$. To find the 20th number in the pattern, use $n = 20$ to get 22.

	Sequence	Pattern	Next Number in Sequence	20th Number in the Sequence
EXAMPLE 2:	1, 4, 9, 16, 25	n^2	$n^2 = 36$	400
EXAMPLE 3:	−2, −4, −6, −8, −10	$-2n$	$-2n = -12$	−40

Find the pattern and the next number in each of the sequences below.

	Sequence	Pattern	Next Number in Sequence	20th number in the sequence
1.	−2, −1, 0, 1, 2	_____	_____	_____
2.	5, 6, 7, 8, 9	_____	_____	_____
3.	3, 7, 11, 15, 19	_____	_____	_____
4.	−3, −6, −9, −12, −15	_____	_____	_____
5.	3, 5, 7, 9, 11	_____	_____	_____
6.	2, 4, 8, 16, 32	_____	_____	_____
7.	1, 8, 27, 64, 125	_____	_____	_____
8.	0, −1, −2, −3, −4	_____	_____	_____
9.	2, 5, 10, 17, 26	_____	_____	_____
10.	4, 6, 8, 10, 12	_____	_____	_____

337

USING DIAGRAMS TO SOLVE PROBLEMS

Problems that require logical reasoning cannot always be solved with a set formula. Sometimes, drawing diagrams can help you see the solution.

EXAMPLE: Yvette, Barbara, Patty, and Nicole agreed to meet at the movie theater around 7:00 p.m. Nicole arrived before Yvette. Barbara arrived after Yvette. Patty arrived before Barbara but after Yvette. What is the order of their arrival?

Nicole	Yvette	Patty	Barbara
1st	2nd	3rd	4th

Arrange the names in a diagram so that the order agrees with the problem.

Use a diagram to answer each of the following questions.

1. Javy, Thomas, Pat, and Keith raced their bikes across the playground. Keith beat Thomas but lost to Pat and Javy. Pat beat Javy. Who won the race?

    _____

2. Jeff, Greg, Pedro, Lisa, Macy, and Kay eat lunch together at a round table. Kay wants to sit beside Pedro, Pedro wants to sit next to Lisa, Greg wants to sit next to Macy, and Jeff wants to sit beside Kay. Macy would rather not sit beside Lisa. Which two people should sit on each side of Jeff?

    _____

    _____

3. Three teams play a round-robin tournament where each team plays every other team. Team A beat Team C. Team B beat Team A. Team B beat Team C. Which team is the best?

    _____

4. Caleb, Thomas, Ginger, Alex, and Janice are in the lunch line. Thomas is behind Alex. Caleb is in front of Alex but behind Ginger and Janice. Janice is between Ginger and Caleb. Who is third in line?

    _____

5. Ray, Fleta, Paula, Joan, and Henry hold hands to make a circle. Joan is between Ray and Paula. Fleta is holding Ray's other hand. Paula is also holding Henry's hand. Who must be holding Henry's other hand?

    _____

6. The Bears, the Cavaliers, the Knights, and the Lions all competed in a track meet. One team from each school ran the 400 meter relay race. The Bears beat the Knights but lost to the Cavaliers. The Lions beat the Cavaliers. Who finished first, second, third, and fourth?

    1st _____

    2nd _____

    3rd _____

    4th _____

TRIAL AND ERROR PROBLEMS

Sometimes problems can only be solved by trial and error. You have to guess at a solution, and then check to see if it will satisfy the problem. If it does not, you must guess again until you get the right answer.

Solve the following problems by trial and error. Make a chart of your attempts so that you don't repeat the same attempt twice.

1. Becca had 5 coins consisting of one or more quarters, dimes, and nickels that totaled 75¢. How many quarters, dimes, and nickels did she have?

   quarters _____

   dimes _____

   nickels _____

2. Ryan needs to buy 42 cans of soda for a party at his house. He can get a six pack for $1.80, a box of 12 for $3.00, or a case of 24 for $4.90. What is the least amount of money Ryan must spend to purchase the 42 cans of soda?

   _____

3. Jana had 10 building blocks that were numbered 1 to 10. She took three of the blocks and added up the three numbers to get 27. Which three blocks did she pick?

   _____

4. Hank had 10 coins. He had 3 quarters, 3 dimes, and 4 nickels. He bought a candy bar for 75¢. How many different ways could he spend his coins to pay for the candy bar?

   _____

5. Refer to question 4. If Hank used 6 coins to pay for the candy bar, how many of his quarters did he spend?

6. The junior varsity basketball team needs to order 38 pairs of socks for the season. The coach can order 1 pair for 2.45, 6 pairs for $12.95, or 10 pairs for $20.95. What is the least amount of money he will need to spend to purchase exactly 38 pairs of socks?

   _____

7. Tyler has 5 quarters, 10 dimes, and 15 nickels in change. He wants to buy a notebook for $2.35 using the change that he has. If he wants to use as many of the coins as possible, how many quarters will he spend?

   _____

8. Kevin is packing up his room to move to another city. He has the following items left to pack.

   comic book collection ...   7 pounds
   track trophy ...   3 pounds
   coin collection ...   13 pounds
   soccer ball ...   1 pound
   model car ...   6 pounds

   If Kevin has a large box that will hold 25 pounds, what items should he pack in it to get the most weight without going over the box's weight limit?

INDUCTIVE REASONING AND PATTERNS

Humans have always observed what happened in the past and used these observations to predict what would happen in the future. This is called **inductive reasoning**. Although mathematics is referred to as the "deductive science," it benefits from inductive reasoning. We observe patterns in the mathematical behavior of a phenomenon, and then find a rule or formula for describing and predicting its future mathematical behavior. There are lots of different kinds of predictions that may be of interest.

EXAMPLE 1: Nancy is watching her nephew, Drew, arrange his marbles in rows on the kitchen floor. The figure below shows the progression of his arrangement.

Row 1
Row 2
Row 3
Row 4

QUESTION 1: Assuming this pattern continues, how many marbles would Drew place in a fifth row?

ANSWER 1: It appears that Drew doubles the number of marbles in each successive row. In the 4th row he had 8 marbles, so in the 5th row we can predict 16 marbles.

QUESTION 2: How many marbles will Drew place in the nth row?

ANSWER 2: To find a rule for the number of marbles in the nth row, we look at the pattern suggested by the table below.

Which row	1st	2nd	3rd	4th	5th
Number of marbles	1	2	4	8	16

Observing closely, you will notice that the nth row contains 2^{n-1} marbles.

QUESTION 3: Suppose Nancy tells you that Drew now has 6 rows of marbles on the floor. What is the total number of marbles in his arrangement?

ANSWER 3: Again, organizing the data in a table could be helpful.

Number of rows	1	2	3	4	5
Total number of marbles	1	3	7	15	31

With careful observation, one will notice that the total number of marbles is always 1 less than a power of 2; indeed, for n rows there are $2^{n}-1$ marbles total.

340

QUESTION 4: If Drew has 500 marbles, what is the maximum number of *complete* rows he can form?

ANSWER 4: With 8 complete rows, Drew will use $2^{8-1} = 255$ marbles, and to form 9 complete rows he would need $2^{9-1} = 511$ marbles; thus, the answer is 8 complete rows.

EXAMPLE 2: Manuel drops a golf ball from the roof of his high school while Carla videos the motion of the ball. Later, the video is analyzed, and the results are recorded concerning the height of each bounce of the ball.

QUESTION 1: What height do you predict for the fifth bounce?

Initial height	1st bounce	2nd ounce	3rd bounce	4th bounce
30 ft	18 ft	10.8 ft	6.48 ft	3.888 ft

ANSWER 1: To answer this question, we need to be able to relate the height of each bounce to the bounce immediately preceding it. Perhaps the best way to do this is with **ratios** as follows:

$$\frac{\text{Height of 1st bounce}}{\text{Initial bounce}} = 0.6 \qquad \frac{\text{Height of 2nd bounce}}{\text{Height of 1st bounce}} = 0.6 \ldots \frac{\text{Height of 4th bounce}}{\text{Height of 3rd bounce}} = 0.6$$

Since the ratio of the height of each bounce to the bounce before it appears constant, we have some basis for making predictions.

Using this, we can reason that the fifth bounce will be equal to 0.6 of the fourth bounce.

Thus, we predict the fifth bounce to have a height of **0.6 × 3.888 = 2.3328 ft.**

QUESTION 2: Which bounce will be the last one with a height of one foot or greater?

ANSWER 2: For this question, keep looking at predicted bounce heights until a bounce less than 1 foot is reached.

The sixth bounce is predicted to be 1.39968 ft.
The seventh bounce is predicted to be 0.839808 ft.

Thus, the last bounce with a height greater than 1 foot is predicted to be the sixth one.

Read each of the following questions carefully. Use inductive reasoning to answer each question. You may wish to make a table or a diagram to help you visualize the pattern in some of the problems. Show your work.

George is stacking his coins as shown below.

1. How many coins do you predict he will place in the fourth stack?

2. How many coins in an *n*th row?

3. If George has exactly 6 "complete" stacks, how many coins does he have?

4. If George has 2,000 coins, how many complete stacks can he form?

Bob and Alice have designed and created a Web site for their high school. The first week they had 5 visitors to the site; during the second week, they had 10 visitors; and during the third week, they had 20 visitors.

5. If current trends continue, how many visitors can they expect in the fifth week?

6. How many in the *n*th week?

7. How many weeks will it be before they get more than 500 visitors in a single week?

8. In 1979 (the first year of classes), there were 500 students at Brookstone High. In 1989, there were 1000 students. In 1999, there were 2000 students. How many students would you predict at Brookstone in 2009 if this pattern continues (and no new schools are built)?

9. The number of new drivers' licenses issued in the city of Boomtown, USA was 512 in 1992, 768 in 1994, 1,152 in 1996, and 1,728 in 1998. Estimate the number of new drivers' licenses that will be issued in 2000.

10. The average combined (math and verbal) SAT score for seniors at Brookstone High was 1,000 in 1996, 1,100 in 1997, 1,210 in 1998, and 1331 in 1999. Predict the combined SAT score for Brookstone seniors in 2000.

Juan wants to be a medical researcher, inspired in part by the story of how penicillin was discovered as a mold growing on a laboratory dish. One morning, Juan observes a mold on one of his lab dishes. Each morning thereafter, he observes and records the pattern of growth. The mold appeared to cover about 1/32 of the dish surface on the first day, 1/16 on the second day, and 1/8 on the third day.

11. If this rate of growth continues, on which day can Juan expect the entire dish to be covered with mold?

12. Suppose that whenever the original dish gets covered with mold Juan transfers half of the mold to another dish. How long will it be before *both* dishes are covered again?

13.   Every year on the last day of school, the Brookstone High cafeteria serves the principal's favorite dish–Broccoli Surprise. In 1988, 1024 students chose to eat Broccoli Surprise on the last day of school, 512 students in 1992, and 256 students in 1996. Predict how many will choose Broccoli Surprise on the last day of school in 2000.

Part of testing a new drug is determining the rate at which it will break down (*decay*) in the blood. The decay results for a certain antibiotic after a 1000 milligram injection are given in the table below.

12:00 PM	1:00 PM	2:00 PM
1000 mg	800 mg	640 mg

14.   Predict the number of milligrams that will be in the patient's bloodstream at 3:00 PM.

15.   At which hour can the measurer expect to record a result of less than 300 mg?

16.   Marie has a daylily in her mother's garden. Every Saturday morning in the spring, she measures and records its height in the table below. What height do you predict for Marie's daylily on April 29? (Hint: Look at the *change* in height each week when looking for the pattern.)

April 1	April 8	April 15	April 22
12 in	18 in	21 in	22.5 in

17.   Bob puts a glass of water in the freezer and records the temperature every 15 minutes. The results are displayed in the table below. If this pattern of cooling continues, what will be the temperature at 2:15 PM? (Hint: Again, look at the *changes* in temperature in order to see the pattern.)

1:00 PM	1:15 PM	1:30 PM	1:45 PM
92°F	60°F	44°F	36°F

Suppose you cut your hand on a rusty nail that deposits 25 bacteria cells into the wound. Suppose also that each bacterium splits into two bacteria every 15 minutes.

18.   How many bacteria will there be after two hours?

19.   How many 15-minute intervals will pass before there are over a million bacteria?

20.   Elias performed a psychology experiment at his school. He found that when someone is asked to pass information along to someone else, only about 70% of the original information is actually passed to the recipient. Suppose Elias gives the information to Brian, Brian passes it along to George, and George passes it to Montel. Using Elias's results from past experiments, what percentage of the original information does Montel actually receive?

FINDING A RULE FOR PATTERNS

EXAMPLE: Mr. Applegate wants to put desks together in his math class so that students can work in groups. The diagram below shows how he wishes to do it.

With 1 table he can seat 4 students, with 2 tables he can seat 6, with 3 tables 8, and with 4 tables 10.

QUESTION 1: How many students can he seat with 5 tables?

ANSWER 1: With 5 tables, he could seat 5 students along the sides of the tables and 1 student on each end; thus, a total of 12 students could be seated.

QUESTION 2: Write a rule that Mr. Applegate could use to tell how many students could be seated at n tables. Explain how you got the rule.

ANSWER 2: For n tables, there would be n students along each of 2 sides and 2 students on the ends (1 on each end); thus, a total of $2n + 2$ students could be seated at n tables.

EXAMPLE 2: When he isn't playing football for the Brookstone Bears, Tim designs Web pages. A car dealership paid Tim $500 to start a site with photos of its cars. The dealer also agreed to pay Tim $50 for each customer who buys a car first viewed on the Web site.

QUESTION 1: Write and explain a rule that tells how much the dealership will pay Tim for the sale of n cars from his Web site.

ANSWER 1: Tim's payment will be the initial $500 plus $50 for each sale. Translated into mathematical language, if Tim sells n cars, he will be paid a total of $500 + 50n$ dollars.

QUESTION 2: How many cars have to be sold from his site in order for Tim to get $1,000 from the dealership?

ANSWER 2: He earned $500 just by establishing the site, so he only needs to earn an additional $500, which at $50 per car requires the sale of only 10 cars. (Note: Another way to solve this problem is to use the rule found in the first question. In that case, you simply solve the equation $500 + 50n = 1000$ for the variable n.)

EXAMPLE 3: Eric is baking muffins to raise money for the homecoming dance. He makes 18 muffins with each batch of batter, but he must give one muffin each to his brother, his sister, and his dog, and himself (of course!) each time a batch is finished baking.

QUESTION 1: Write a rule for the number of muffins Eric produces for the fund raiser with n batches.

ANSWER 1: He bakes 18 with each batch, but only 14 are available for the fund raiser. Thus, with n batches he will produce $14n$ muffins for the homecoming. **The rule = $14n$**

QUESTION 2: Use your rule to determine how many muffins he will contribute if he makes 7 batches.

ANSWER 2: The number of batches, n, equals 7. Therefore, he will produce $14 \times 7 = 98$ muffins with 7 batches.

QUESTION 3: Determine how many batches he must bake in order to contribute at least 150 muffins.

ANSWER 3: Ten batches will produce $10 \times 14 = 140$ muffins. Eleven batches will produce $11 \times 14 = 154$ muffins. To produce at least 150 muffins, he must bake at least 11 batches.

QUESTION 4: Determine how many muffins he would actually bake in order to contribute 150 muffins.

ANSWER 4: Since Eric actually bakes 18 muffins per batch, 11 batches would result in Eric baking $11 \times 18 = 198$ muffins.

Carefully read and solve the problems below. Show your work.

Tito is building a picket fence along both sides of the driveway leading up to his house. He will have to place posts at both ends and at every 10 feet along the way because the rails come in prefabricated ten-foot sections.

1.   How many posts will he need for a 180 foot driveway?

2.   Write and explain a rule for determining the number of posts needed for n ten-foot sections.

3.   How long of a driveway can he fence along with 32 posts?

Linda is working as a bricklayer this summer. She lays the bricks for a walkway in *sections* according to the pattern depicted below.

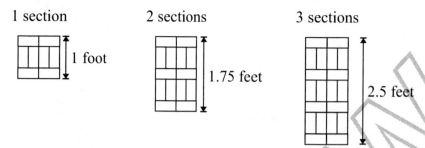

1 section   2 sections   3 sections

1 foot   1.75 feet   2.5 feet

4.   Write a formula for the number of bricks needed to lay *n* sections.

5.   Write a formula for the number of feet covered by *n* sections.

6.   How many bricks would it take to lay a walk that is 10 feet long?

Dakota's beginning pay at his new job is $300 per week. For every three months he continues to work there, he will get a $10 per week raise.

7.   Write a formula for Dakota's weekly pay after *n* three-month periods.

8.   After *n* years?

9.   How long will he have to work before his pay gets to $400 a week?

Amanda is selling shoes this summer. In addition to her hourly wages, Amanda got a $100 bonus just for accepting the position, and she gets a $2 bonus for each pair of shoes she sells.

10. Write and explain a rule that tells how much she will make in bonuses if she sells *n* pairs of shoes.

11. How many pairs of shoes must she sell in order to make $200 in bonuses?

A certain teen telephone chat line, which sells itself as a benefit to teens but which is actually a money-making scheme, is a 900 telephone number that charges $2.00 for the first minute and $0.95 for each additional minute.

12. Write a formula for the cost of speaking *n* minutes on this line.

13. How many minutes does it take to accumulate charges of more than $50.00?

Laura's (unsharpened) pencil was initially 8 inches long. After the first sharpening, it was 7 inches long. Each sharpening thereafter, Laura noticed the pencil would be ½ inch shorter after sharpening than before.

14. Write and explain a rule that tells how long Laura's pencil will be after the nth sharpening.

15. How many sharpenings will it take to get Laura's pencil to only 3 inches long?

Ritchie's dad is tired of not being able to use his own phone whenever he wants, so he started measuring time on the phone and devised a plan for encouraging Ritchie to talk less. Ritchie will receive his ordinary allowance of $20 each week, but for each minute over two hours that Ritchie was on the phone that week, his dad deducts $0.25 from the allowance.

16. Write a rule for the allowance Ritchie receives if he talks on the phone for n minutes a week. (Hint: You actually have two rules: One for less than 120 minutes and one for 120 minutes or more.)

Every time Bob (the used car salesman) sells a car he makes a $150 commission. However, he must pay the owner of the car lot (Mike) a $32 "membership fee" for each car sold. Bob wishes to earn $1,235 for a new riding lawn mower.

17. Write a rule for the net pay that Bob earns after n sales.

18. How many cars will he have to sell in order to purchase his new mower?

Roberta works at the Oakwood movie theater. The first row in the theater has 14 seats, and each row (except the first) has four more seats than the row before it.

19. Write and explain a rule for the number of seats in the nth row at Roberta's theater.

20. Which is the first row to have more than 100 seats?

The table below displays data relating temperature in degrees Farenheit to the number of chirps per minute for a cricket.

Temp (°F)	50	52	55	58	60	64	68
Chirps/min.	40	48	60	72	80	96	112

21. Write a formula or rule that predicts the number of chirps per minute when the temperature is n degrees.

PROPORTIONAL REASONING

Proportional reasoning can be used when a selected number of individuals are tagged in a population in order to estimate the total population.

EXAMPLE: A team of scientists capture, tag, and release 50 deer in a particular national forest. One week later, they capture another 50 deer, and 2 of the deer are ones that were tagged previously. What is the approximate deer population in the national forest?

Solution: Use proportional reasoning to determine the total deer population. You know that 50 deer out of the total deer population in the forest were tagged. You also know that 2 out of those 50 were recaptured. These two ratios should be equal because they both represent a fraction of the total deer population.

$$\frac{50 \text{ deer tagged}}{x \text{ deer total}} = \frac{2 \text{ tagged deer}}{50 \text{ deer captured}}$$

$2x = 2{,}500$ so $x = 1{,}250$ total deer

Use proportional reasoning to solve the following problems.

1. Dr. Wolf, the biologist, captures 20 fish out of a small lake behind his college. He fastens a marker onto each of these and throws them back into the lake. A week later, he again captures 20 fish. Of these, 2 have markers. How many fish could Dr. Wolf estimate are in the pond?

2. Tawanda drew 20 cards from a box. She marked each one, returned them to the box, and shook the box vigorously. She then drew 20 more cards and found that 5 of them were marked. Estimate how many cards were in the box.

3. Maureen pulls 100 pennies out of her money jar, which contains only pennies. She marks each of these, puts them back in the bank, shakes vigorously, and again pulls 100 pennies. She discovers that 2 of them are marked. Estimate how many pennies are in her money jar.

4. Mr. Kizer has a ten acre wooded lot. He catches 20 squirrels, tags them, and releases them. Several days later, he catches another 20 squirrels. One of the 20 squirrels had a tag. Estimate the number of squirrels living on Mr. Kizer's ten acres.

MATHEMATICAL REASONING/LOGIC

The ability to use reasoning and logic is an important skill for solving math problems, but it can also be helpful in real-life situations. For example, if you need to get to Park Street, and the Park Street bus always comes to the bus stop at 3 pm, then you know that you need to get to the bus stop at least by 3 pm. This is a real-life example of using logic, which many people would call "common sense."

There are many different types of statements which are commonly used to describe mathematical principles. However, using the rules of logic, the truth of any mathematical statement must be evaluated. Below are a list of tools used in logic to evaluate mathematical statements.

Logic is the discipline that studies valid reasoning. There are many forms of valid arguments, but we will just review a few here.

A **proposition** is usually a declarative sentence which may be true or false.

An **argument** is a set of two or more related propositions, called **premises**, that provide support for another proposition, called the **conclusion**.

Deductive reasoning is an argument which begins with general premises and proceeds to a more specific conclusion. Most elementary mathematical problems use deductive reasoning.

Inductive reasoning is an argument in which the truth of its premises make it likely or probable that its conclusion is true.

ARGUMENTS

Most of logic deals with the evaluation of the validity of arguments. An argument is a group of statements that includes a conclusion and at least one premise. A premise is a statement that you know is true or at least you assume to be true. Then, you draw a conclusion based on what you know or believe is true in the premise(s). Consider the following example:

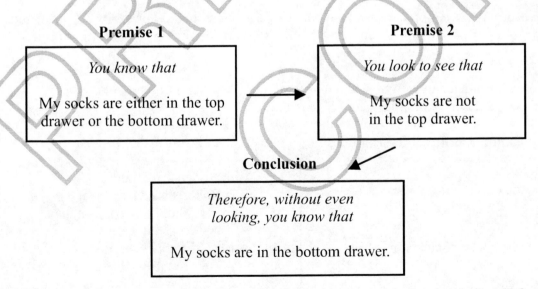

Premise 1

You know that

My socks are either in the top drawer or the bottom drawer.

Premise 2

You look to see that

My socks are not in the top drawer.

Conclusion

Therefore, without even looking, you know that

My socks are in the bottom drawer.

This argument is an example of deductive reasoning, where the conclusion is "deduced" from the premises and nothing else. In other words, if Premise 1 and Premise 2 are true, you don't even need to look in the bottom drawer to know that the conclusion is true.

DEDUCTIVE AND INDUCTIVE ARGUMENTS

In general, there are two types of logical arguments: **deductive** and **inductive.** Deductive arguments tend to move from general statements or theories to more specific conclusions. Inductive arguments tend to move from specific observations to general theories.

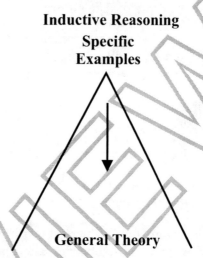

Deductive Reasoning

General Theory

Specific Examples

Inductive Reasoning

Specific Examples

General Theory

Compare the two examples below:

Deductive Argument

Premise 1	All men are mortal.
Premise 2	Socrates is a man.
Conclusion	Socrates is mortal.

Inductive Argument

Premise 1	The sun rose this morning.
Premise 2	The sun rose yesterday morning.
Premise 3	The sun rose two days ago.
Premise 4	The sun rose three days ago.
Conclusion	The sun will rise tomorrow.

An inductive argument cannot be proved beyond a shadow of a doubt. For example, it's a pretty good bet that the sun will come up tomorrow, but the sun not coming up presents no logical contradiction.

On the other hand, a deductive argument can have logical certainty, but it must be properly constructed. Consider the examples below.

True Conclusion for an Invalid Argument

All men are mortal.
Socrates is mortal.
Therefore, Socrates is a man.

Even though the above conclusion is true, the argument is based on invalid logic. Both men and women are mortal. Therefore, Socrates could be a woman.

False Conclusion from a Valid Argument

All astronauts are men.
Julia Roberts is an astronaut.
Therefore, Julia Roberts is a man.

In this case, the conclusion is false because the premises are false. However, the logic of the argument is valid because *if* the premises were true, then the conclusion would be true.

EXAMPLE 1: Which argument is valid?

If you speed on Hill Street, you will get a ticket.
If you get a ticket, you will pay a fine.

A. I paid a fine, so I was speeding on Hill Street.
B. I got a ticket, so I was speeding on Hill Street.
C. I exceeded the speed limit on Hill Street, so I paid a fine.
D. I did not speed on Hill Street, so I did not pay a fine.

Answer: C is valid.
A is incorrect. I could have paid a fine for another violation.
B is incorrect. I could have gotten a ticket for some other violation.
D is incorrect. I could have paid a fine for speeding somewhere else.

EXAMPLE 2: Assume the given proposition is true. Determine if each statement is true or false.

Given: If a dog is thirsty, he will drink.

A. If a dog drinks, then he is thirsty.   T or F
B. If a dog is not thirsty, he will not drink.   T or F
C. If a dog will not drink, he is not thirsty.   T or F

Answer: A is false. He is not necessarily thirsty; he could just drink because other dogs are drinking or drink to show others his control of the water. This statement is the converse of the original.
B is false. The reasoning from A applies. This statement is the inverse of the original.
C is true. It is the **contrapositive** or the complete opposite of the original.

For numbers 1-5, what conclusion can be drawn from each proposition?

1. All squirrels are rodents. All rodents are mammals. Therefore,

2. All fractions are rational numbers. All rational numbers are real numbers. Therefore,

3. All squares are rectangles. All rectangles are parallelograms. All parallelograms are quadrilaterals. Therefore,

4. All Chevrolets are made by General Motors. All Luminas are Chevrolets. Therefore,

5. If a number is even and divisible by three, then it is divisible by six. Eighteen is divisible by six. Therefore,

For numbers 6-9, assume the given proposition is true. Then determine if the statements following it are true or false.

All squares are rectangles.

6. All rectangles are squares.   T or F
7. All non-squares are non-rectangles.   T or F
8. No squares are non-rectangles.   T or F
9. All non-rectangles are non-squares.   T or F

BASE 2 ARITHMETIC

Base 2 (or binary) arithmetic is the system of numbers which computers use to represent numbers with only 0's or 1's. We normally work in the base 10 system, which has ten different digits to represent numbers (0, 1, 2, 3, 4, 5, 6, 7, 8, 9).

When a number is written in base 10, each digit is represented by that place number times its power of ten. For example, 475 is a base 10 number. Let's expand it into its powers of 10:

$$475 = (4 \times 100) + (7 \times 10) + 5$$

$$= (4 \times 10^2) + (7 \times 10^1) + (5 \times 10^0) \quad \textbf{Note:} \text{ Remember that } 10^0 = 1$$

In this example, the powers of ten for each digit are added together to get the base 10 number 475. To get a number in base 2, we will need to find the powers of 2 which add together to give us the number we are looking for.

EXAMPLE 1: What is the base 2 representation for the base 10 number 20?

Step 1:   First we need to find the largest power of 2 that is less than 20.

$$2^4 = 16 \text{ is less than 20} \qquad 2^5 = 32, \text{ which is larger than 20.}$$

Since 2^5 is larger than 20, we must begin with 2^4.

Step 2:   Next we need to find the other powers of 2 that will add together with 2^4 to get 20.

$$20 - 16 = 4 \implies 2^2 = 4$$

Step 3:   Starting with the highest power of two we used for the sum, put together a string of 0's and 1's to represent the number in base 2. If the power of 2 is used, write down a 1 for that place. If it is not, write down a 0 for that place.

> **Very Important:** Because 2^0 represents the 1's place and 2^1 represents the tens place and so on, there will always be one more digit in the base two number than the highest power of two used in the sum. 2^4 is the largest power of 2 needed to represent the number 20 in base 2. Therefore the base 2 number will have 5 digits.

20 in base 2 equals 10100

The first 0 is included because we do not have a 2^3 term. The last two 0's are there because we did not have a 2^1 or a 2^0 term.

EXAMPLE 2: What is the base 2 representation of the base 10 number 475?

Step 1:   Find the largest power of 2 that is less than 475. Subtract it from 475. Find the largest power of 2 less than this difference. Keep going until the difference is 1 or 2.

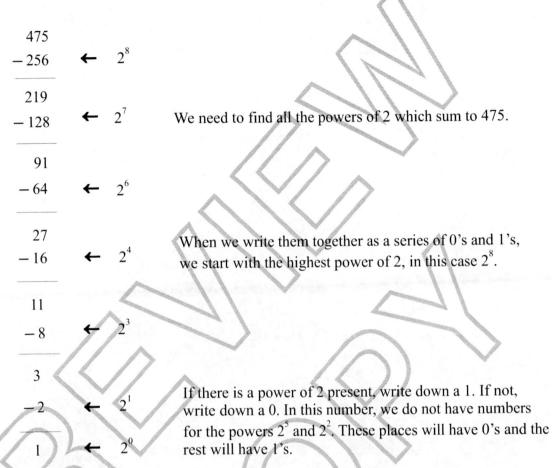

$$\begin{array}{r} 475 \\ -256 \\ \hline 219 \\ -128 \\ \hline 91 \\ -64 \\ \hline 27 \\ -16 \\ \hline 11 \\ -8 \\ \hline 3 \\ -2 \\ \hline 1 \end{array}$$

$\leftarrow 2^8$

$\leftarrow 2^7$ We need to find all the powers of 2 which sum to 475.

$\leftarrow 2^6$

$\leftarrow 2^4$ When we write them together as a series of 0's and 1's, we start with the highest power of 2, in this case 2^8.

$\leftarrow 2^3$

$\leftarrow 2^1$ If there is a power of 2 present, write down a 1. If not, write down a 0. In this number, we do not have numbers for the powers 2^5 and 2^2. These places will have 0's and the rest will have 1's.

$\leftarrow 2^0$

Step 2:

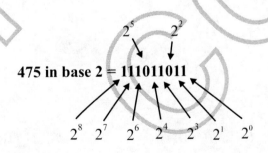

475 in base 2 = 111011011

Please note that the 0's are in place of 2^5 and 2^2, which are not needed to represent 475 in base 2.

Remember: There will always be one more digit in the base 2 number than the highest power of two used. In this problem, 2^8 is the highest power of two found for 475, so the base 2 number will have 9 digits.

Convert the following from base 10 to base 2:

1. 14	11. 402	21. 1602
2. 225	12. 881	22. 2495
3. 33	13. 2048	23. 5195
4. 171	14. 513	24. 7847
5. 1091	15. 4312	25. 8106
6. 4962	16. 7251	26. 9
7. 520	17. 3388	27. 65
8. 999	18. 11,202	28. 131,072
9. 6611	19. 1407	29. 17
10. 10,240	20. 32,768	30. 1029

Convert the following from base 2 to base 10:

1. 10010	11. 101	21. 110100
2. 11101	12. 111	22. 10011111111
3. 1010	13. 10000	23. 100100100100
4. 1101011	14. 10000001	24. 1100110011001
5. 1010101	15. 11111110	25. 10
6. 11100110	16. 10101111	26. 111011101110111
7. 101110011	17. 1011000001	27. 1000001000000
8. 11011000101	18. 1010001100110	28. 11001000100100
9. 111011001	19. 110110110110	29. 101010101010
10. 111111111	20. 1000000000000	30. 111111011111

CHAPTER 22 REVIEW

Find the pattern for the following number sequences, and then find the *n*th number requested.

1. 0, 1, 2, 3, 4   pattern _____

2. 0, 1, 2, 3, 4   20th number _____

3. 1, 3, 5, 7, 9   pattern _____

4. 1, 3, 5, 7, 9   25th number _____

5. 3, 6, 9, 12, 15   pattern _____

6. 3, 6, 9, 12, 15   30th number _____

7. Andrea spent $1.24 for toothpaste. She had quarters, dimes, nickels, and pennies. How many dimes did she use if she used a total of 14 coins?

8. Cody spent 59¢ on a hotdog. He had quarters, dimes, nickels, and pennies in his pocket. He gave the cashier 9 coins for exact payment. How many quarters did he give the cashier?

9. Vince, Hal, Weng, and Carl raced on roller blades down a hill. Vince beat Carl. Hal finished before Vince but after Weng. Who won the race?

10. A veterinarian's office has a weight scale in the lobby to weigh pets. The scale will weigh up to 250 pounds. The following dogs are in the lobby:

    > Pepper ... 23 pounds
    > Jack ... 75 pounds
    > Trooper ... 45 pounds
    > Precious ... 25 pounds
    > Coco ... 120 pounds

    Which dogs could you put on the scale to get as close to 250 pounds as possible, without going over? How much would they weigh?

11. Felix has set a goal to increase his running speed by 1 minute per mile every 5 weeks. He starts out running 1 mile in 11 minutes. If he can accomplish his goal, how many weeks will it take him to run a mile in 8 minutes?

Jessica Bloodsoe and Katie Turick are climbing Mt. Fuji in Japan which is 12,388 ft. high. The higher they go, the slower they climb due to lack of oxygen. The chart below shows their progress.

Days Ascending	Altitude
End of day 1	4,000 feet
End of day 2	7,200 feet
End of day 3	9,600 feet

12. If the weather holds, what will be their altitude on day 5? _____

13. If they can keep the same rate, how many days would it take them to get to the top? _____

In a large city of 200,000, there was an outbreak of tuberculosis. Immediately, health care workers began an immunization campaign. The chart below records their results.

	No. of People Immunized	No. of TB Cases
Year 1	20,000	60
Year 2	60,000	45
Year 3	100,000	30

14. About how many cases of TB would you predict for year 4? _____

Exotic goldfish are kept in different size containers of water. The larger the container, the bigger the size the goldfish can grow. The chart on the right shows how big one goldfish can grow in different size containers.

Fish Size	Container Size
1 inch	20 gallon or less
$2\frac{1}{2}$ inches	50 gallon
5 inches	100 gallon

15. Based on the chart, how large would you predict a goldfish could grow in a 140 gallon container? _____

Olivia starts, maintains, and sells ant farms as a hobby. She had 500 ants in 1996, 2,000 in 1997, and 8,000 in 1998.

16. If her hobby continues to grow as it has since 1996, how many ants will Olivia have in 2000?

17. How many in the *n*th year after 1996?

18. In what year would she have more than 100,000 ants?

---

19. Sean is studying bacteria and antibiotics. Using standard measurement and estimation techniques, he records a reading of about 100,000 bacteria on a lab dish. He then applies a drop of antibiotic and does another bacteria count every 30 minutes. He finds 90,000 after 30 minutes, 81,000 after 60 minutes, 72,900 after 90 minutes, and 65,610 after 120 minutes. How many do you predict he will find 150 minutes after applying the antibiotic?

Justin has just got a bill from his Internet Service Provider. The first four months of charges for his service are recorded in the table below.

	January	February	March	April
Hours	0	10	5	25
Charge	$4.95	$14.45	$9.70	$28.70

20. Write a formula for the cost of *n* hours of Internet service.

21. What is the greatest number of hours he can get on the Internet and still keep his bill under $20.00?

Lisa is baking cookies for the Fall Festival. She bakes 27 cookies with each batch of batter. However, she has a defective oven, which results in 5 cookies in each batch being burnt.

22. Write a formula for the number of cookies available for the festival as a result of Lisa baking *n* batches of cookies.

23. How many batches does she need in order to produce 300 cookies for the festival?

24. How many cookies (counting burnt ones) will she actually bake?

---

25. Jamal wonders how many ants are in his ant farm. He puts a stick in the container, and when he pulls it out, there are 15 ants on it. He gently sprays these ants with a mixture of water and food coloring, then puts them back into the container. The next day his stick draws 20 ants, 1 of which is green. Estimate how many ants Jamal has.

For numbers 26-29, assume the given proposition is true. Then determine if the statements following it are true or false.

All whales are mammals.

26. All non-whales are non-mammals.   T or F
27. If a mammal lives in the sea, it is a whale.   T or F
28. All mammals are whales.   T or F
29. All non-mammals are non-whales.   T or F

For 30-32, chose which argument is valid.

30. If I oversleep, I miss breakfast. If I miss breakfast, I cannot concentrate in class. If I do not concentrate in class, I make bad grades.

    A.   I made bad grades today, so I missed breakfast.
    B.   I made good grades today, so I got up on time.
    C.   I could not concentrate in class today, so I overslept.
    D.   I had no breakfast today, so I overslept.

31. If I do not maintain my car regularly, it will develop problems. If my car develops problems, it will not be safe to drive. If my car is not safe to drive, I cannot take a trip in it.

    A.   If my car develops problems, I did not maintain it regularly.
    B.   I took a trip in my car, so I maintained it regularly.
    C.   If I maintain my car regularly, it will not develop problems.
    D.   If my car is safe to drive, it will not develop problems.

32. If two triangles have all corresponding sides and all corresponding angles congruent, then they are congruent triangles. If two triangles are congruent, then they are similar triangles.

    A.   Similar triangles have all sides and all angles congruent.
    B.   If two triangles are similar, then they are congruent.
    C.   If two triangles are not congruent, then they are not similar.
    D.   If two triangles have all corresponding sides and angles congruent, then they are similar triangles.

358

Chapter 23 | Transformations and Plotted Shapes

Transformations are geometric figures that have been changed by **reflection**, **rotation**, and **translation**, and/or **dilation**..

REFLECTIONS

A **reflection** of a geometric figure is a mirror image of the object. Placing a mirror on the **line of reflection** will give you the position of the reflected image. On paper, folding an image across the line of reflection will give you the position of the reflected image.

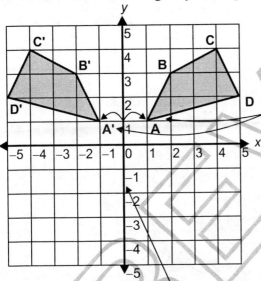

line of reflection: *y*-axis

Quadrilateral ABCD is reflected across the y-axis to form quadrilateral A'B'C'D'. The *y*-axis is the line of reflection. Point A' (read as A prime) is the reflection of point A, point B' corresponds to point B, C' to C, and D' to D.

Point A is +1 space from the *y*-axis. Point A's mirror image, point A', is −1 space from the *y*-axis.

Point B is +2 spaces from the *y*-axis. Point B' is −2 spaces from the *y*-axis.

Point C is +4 spaces from the *y*-axis and point C' is −4 spaces from the *y*-axis.

Point D is +5 spaces from the *y*-axis and point D' is −5 spaces from the *y*-axis.

Triangle FGH is reflected across the *x*-axis to form triangle F'G'H'. The *x*-axis is the line of reflection. Point F' reflects point F. Point G' corresponds to point G, and H' mirrors H.

Point F is +3 space from the *x*-axis. Likewise, point F' is −3 space from the *x*-axis.

Point G is +1 spaces from the *x*-axis, and point G' is −1 spaces from the *x*-axis.

Point H is 0 spaces from the *x*-axis, so point H' is also 0 spaces from the *x*-axis.

line of reflection: *x*-axis

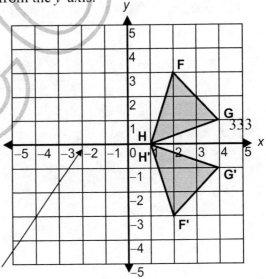

8.4.4

359

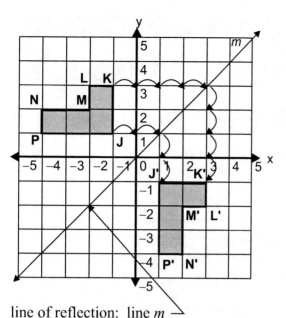

line of reflection: line *m*

Figure JKLMNP is reflected across line *m* to form figure J'K'L'M'N'P'. Line *m* is at a 45° angle. Point J corresponds to J', K to K', L to L', M to M', N to N' and P to P'. Line *m* is the line of reflection. **Pay close attention to how to determine the mirror image of figure JKLMNP across line *m* described below. This method only works when the line of reflection is at a 45° angle.**

Point J is 2 spaces over from line *m*, so J' must be 2 spaces down from line *m*.

Point K is 4 spaces over from line *m*, so K' is 4 spaces down from line *m*, and so on.

Draw the following reflections, and record the new coordinates of the reflection. The first problem is done for you.

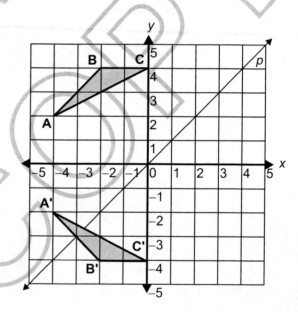

1. Reflect figure ABC across the x-axis. Label vertices A'B'C' so that point A' is the reflection of point A, B' is the reflection of B, and C' is the reflection of C.

    A' = _____   B' = _____   C' = _____
        (–4, –2)   (–2, –4)   (0, –4)

2. Reflect figure ABC across the y-axis. Label vertices A"B"C" so that point A" is the reflection of point A, B" is the reflection of B, and C" is the reflection of C.

    A" = _____   B" = _____   C" = _____

3. Reflect figure ABC across line *p*. Label vertices A'''B'''C''' so that point A''' is the reflection of point A, B''' is the reflection of B, and C''' is the reflection of C.

    A''' = _____   B''' = _____   C''' = _____

8.4.4

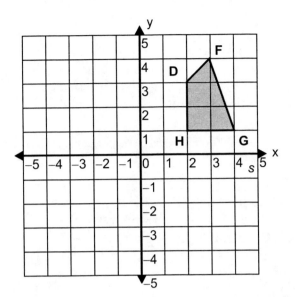

4. Reflect figure DFGH across the *y*-axis. Label vertices D'F'G'H' so that point D' is the reflection of point D, F' is the reflection of F, G' is the reflection of G, and H' is the reflection of H.

   D' = _____   G' = _____

   F' = _____   H' = _____

5. Reflect figure DFGH across the *x*-axis. Label vertices D", F", G", and H" so that point D" is the reflection of D, F" is the reflection of F, G" is the reflection of G, and H" is the reflection of H.

   D" = _____   G" = _____

   F" = _____   H" = _____

ROTATIONS

A **rotation** of a geometric figure shows motion around a point.

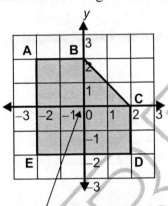

The origin is the point of rotation.

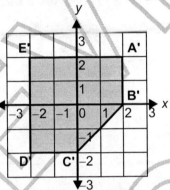

Figure ABCDE has been rotated $\frac{1}{4}$ of a turn clockwise around the origin to form A'B'C'D'E'.

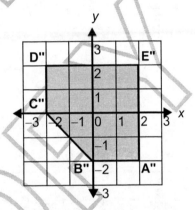

Figure ABCDE has been rotated $\frac{1}{2}$ of a turn around the origin to form A"B"C"D"E".

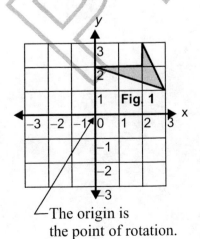

The origin is the point of rotation.

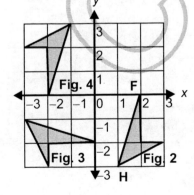

Figure 1 is rotated in $\frac{1}{4}$ turns around the origin. Figure 2 is a $\frac{1}{4}$ clockwise rotation of Figure 1. Figure 3 is a $\frac{1}{2}$ rotation of Figure 1. Figure 4 is a $\frac{3}{4}$ clockwise rotation of Figure 1.

8.4.4

Draw the following rotations, and record the new coordinates of the rotation. The figure for the first problem is drawn for you.

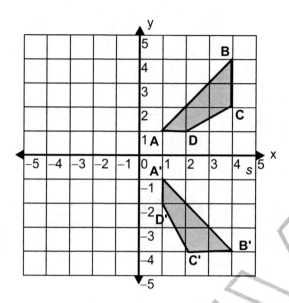

1. Rotate figure ABCD around the origin clockwise $\frac{1}{4}$ turn. Label the vertices A', B', C', and D' so that point A' corresponds to the rotation of point A, B' corresponds to B, C' to C, and D' to D.

    A' = _____   C' = _____

    B' = _____   D' = _____

2. Rotate figure ABCD around the origin clockwise $\frac{1}{2}$ turn. Label the vertices A", B", C", and D" so that point A" corresponds to the rotation of point A, B" corresponds to B, C" to C, and D" to D.

    A" = _____   C" = _____

    B" = _____   D" = _____

3. Rotate figure ABCD around the origin clockwise $\frac{3}{4}$ turn. Label the vertices A''', B''', C''', and D''' so that point A''' corresponds to the rotation of point A, B''' corresponds to B, C''' to C, and D''' to D.

    A''' = _____   C''' = _____

    B''' = _____   D''' = _____

---

4. Rotate figure MNO around point O clockwise $\frac{1}{4}$ turn. Label the vertices M', N', and O so that point M' corresponds to the rotation of point M and N' corresponds to N.

    M' = _____   N' = _____

5. Rotate figure MNO around point O clockwise $\frac{1}{2}$ turn. Label the vertices M", N", and O so that point M" corresponds to the rotation of point M, and N" corresponds to N.

    M" = _____   N" = _____

6. Rotate figure MNO around point O clockwise $\frac{3}{4}$ turn. Label the vertices M''', N''', and O so that point M''' corresponds to the rotation of point M, and N''' corresponds to N.

    M''' = _____   N''' = _____

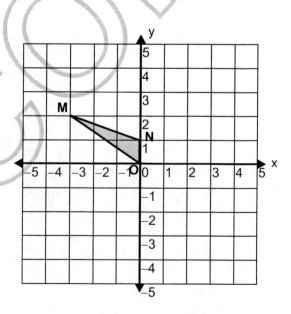

8.4.4

TRANSLATIONS

To make a **translation** of a geometric figure, first duplicate the figure. Then, slide it along a path.

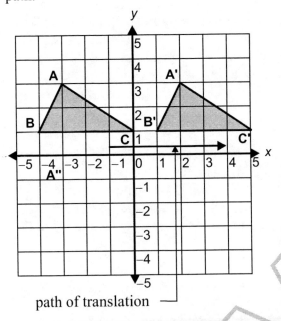

path of translation

Triangle A'B'C' is a translation of triangle ABC. Each point is translated 5 spaces to the right. In other words, the triangle slid 5 spaces to the right. Look at the path of translation. It gives the same information as above. Count the number of spaces across given by the path of translation, and you will see it represents a move 5 spaces to the right. Each new point is found at $(x + 5, y)$.

Point A is at $(-3, 3)$. Therefore, A' is found at $(-3 + 5, 3)$ or $(2, 3)$.

B is at $(-4, 1)$, so B' is at $(-4 + 5, 1)$ or $(1, 1)$.

C is at $(0, 1)$, so C' is at $(0 + 5, 1)$ or $(5, 1)$.

Quadrilateral FGHI is translated 5 spaces to the right and 3 spaces down. The path of translation shows the same information. It points right 5 spaces and down 3 spaces. Each new point is found at $(x + 5, y - 3)$.

Point F is located at $(-4, 3)$. Point F' is located at $(-4 + 5, 3 - 3)$ or $(1, 0)$.

Point G is at $(-2, 5)$. Point G' is at $(-2 + 5, 5 - 3)$ or $(3, 2)$.

Point H is at $(-1, 4)$. Point H' is at $(-1 + 5, 4 - 3)$ or $(4, 1)$.

Point I is at $(-1, 2)$. Point I' is at $(-1 + 5, 2 - 3)$ or $(4, -1)$.

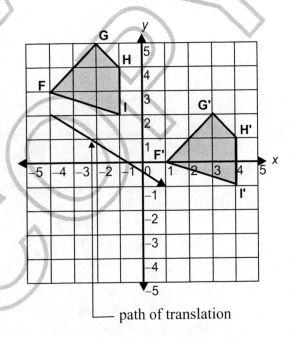

path of translation

8.4.4

Draw the following translations, and record the new coordinates of the translation. The figure for the first problem is drawn for you.

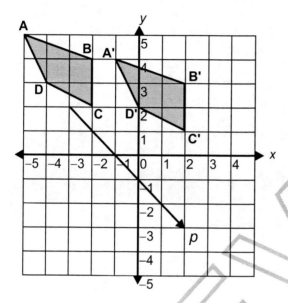

1. Translate figure ABCD 4 spaces to the right and 1 space down. Label the vertices of the translated figure A', B', C', and D' so that point A' corresponds to the translation of point A, B' corresponds to B, C' to C, and D' to D.

   A' = _____   C' = _____

   B' = _____   D' = _____

2. Translate figure ABCD 5 spaces down. Label the vertices of the translated figure A", B", C", and D" so that point A" corresponds to the translation of point A, B" corresponds to B, C" to C, and D" to D.

   A" = _____   C" = _____

   B" = _____   D" = _____

3. Translate figure ABCD along the path of translation, *p*. Label the vertices of the translated figure A''', B''', C''', and D''' so that point A''' corresponds to the translation of point A, B''' corresponds to B, C''' to C, and D''' to D.

   A''' = _____   C''' = _____

   B''' = _____   D''' = _____

---

4. Translate triangle FGH 6 spaces to the left and 3 spaces up. Label the vertices of the translated figure F', G', and H' so that point F' corresponds to the translation of point F, G' corresponds to G, and H' to H.

   F' = _____   G' = _____   H' = _____

5. Translate triangle FGH 4 spaces up and 1 space to the left. Label the vertices of the translated triangle F"G"H" so that point F" corresponds to the translation of point F, G" corresponds to G, and H" to H.

   F" = _____   G" = _____   H" = _____

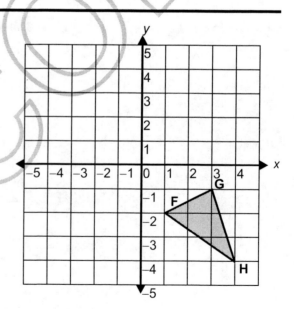

8.4.4

A point can undergo a translation as well.

EXAMPLE: A point at (1,3) is moved to (0, 0). If a point at (−1, 2) is moved in the same way, what will its new coordinates be?

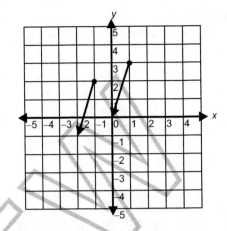

Step 1: Count how many lines down and over the original point is moving. In this problem, the point is moving down three places and then to the left, one place.

Step 2: Find where point (−1, 2) is, and see where it would be when you move it in the same way–down 3 places and to the left 1 place. The new coordinates will be (−2, −1).

Use the grid to answer the questions that follow.

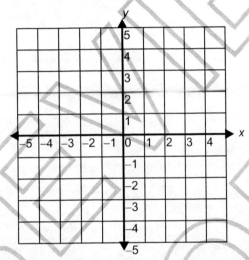

1. A point at (3, 4) is moved to (0, 0). If a point at (2, 2) is moved in the same way, what will its new coordinates be?

2. A point at (−3, 2) is moved to (0, 0). If a point at (−1, 2) is moved in the same way, what will its new coordinates be?

3. A point at (−4, 2) is moved to (0, 0). If a point at (−1, 1) is moved in the same way, what will its new coordinates be?

4. A point at (2, −4) is moved to (0, 0). If a point at (3, 1) is moved in the same way, what will its new coordinates be?

5. A point at (−2, −4) is moved to (0, 0). If a point at (−2, 1) is moved in the same way, what will its new coordinates be?

6. A point at (−3, 1) is moved to (0, 0). If a point at (1, 2) is moved in the same way, what will its new coordinates be?

8.4.4

365

TRANSFORMATION PRACTICE

Answer the following questions regarding transformations.

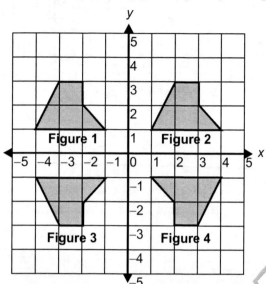

1. Which figure is a rotation of Figure 1? _____

   How far is it rotated? _____

2. Which figure is a translation of Figure 1? _____

   How far and in which direction(s) is it translated?

   _____

3. Which figure is a reflection of Figure 1? _____

---

4. Translate quadrilateral ABCD so that point A' which corresponds to point A is located at coordinates (–4, 3). Label the other vertices B' to correspond to B, C' to C, and D' to D. What are the coordinates of B', C', and D'?

   A' = _____   C' = _____

   B' = _____   D' = _____

5. Reflect quadrilateral ABCD across line m. Label the coordinates A", B", C", D" so that point A" corresponds to the reflection of point A, B" corresponds to the reflection of B, and C" corresponds to the reflection of C. What are the coordinates of A", B", C", and D"?

   A" = _____   B" = _____   C" = _____   D" = _____

6. Rotate quadrilateral ABCD $\frac{1}{4}$ turn counterclockwise around point D. Label the points A'''B'''C'''D''' so that A''' corresponds to the rotation of point A, B''' corresponds to B, C''' to C, and D''' to D. What are the coordinates of A''', B''', C''' and D'''?

   A''' = _____   B''' = _____   C''' = _____   D''' = _____

7.4.2

FINDING LENGTHS OF PLOTTED SHAPES
ON CARTESIAN PLANES

Once pairs of values are labeled as points and lines are drawn to connect these points, shapes are drawn. Use the Cartesian number lines to find the lengths of lines, perimeters of shapes, and area. On some problems, you may be given only the data points and asked to draw lines connecting the points. In this case, you must first plot the points. Then, draw the lines on a separate sheet of paper.

EXAMPLE 1: **What are the lengths of the rectangle *ABCD*?**
Data points: (−4, 2), (3, 2) (3, −2) (−4,−2)
Examine the plotted rectangle to your right. You can find the lengths of each side by counting the number of blocks used. For instance, line segment $\overline{AB}$ on the Cartesian plane is 7 blocks in length. Line segment $\overline{BC}$ is 4 blocks in length, and so on.

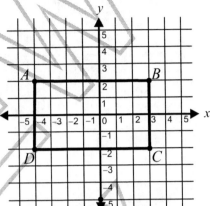

EXAMPLE 2: **What is the length of the hypotenuse of triangle *ABC*?**
Data points: (−4, 4) (−4,−2) (4,−2)
Examine the plotted triangle to your right. First, find the length of the two legs. By counting the blocks, you will find side $\overline{AB}$ has a length of 6 and side $\overline{BC}$ has a length of 8. Next, use the Pythagorean Theorem to find the length of the diagonal side, $\overline{AC}$. $6^2 + 8^2 = 100$. $\sqrt{100} = 10$ Therefore, the length of the hypotenuse, side $\overline{AC}$, = 10.

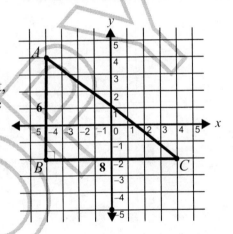

EXAMPLE 3: **What is the radius of the plotted circle?**
Finding the radius and the diameter of a circle on a Cartesian plane is simple. First, examine the circle. Next, draw an imaginary line through the center of the circle. Third, count the number of blocks which span the length of the line segment to find the diameter. To find the radius, simply divide this number by 2. In this example, the diameter = 5, so the radius = 2.5.

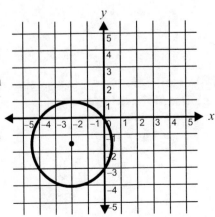

7.4.1

367

Read each of the following graphs carefully. Using the methods taught on the previous page, find the length of each side and the area of each shape.

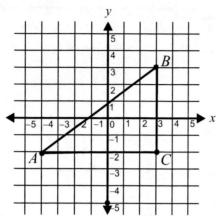

1. $\overline{AB}$ = _____

2. $\overline{BC}$ = _____

3. $\overline{CA}$ = _____

4. Area = _____

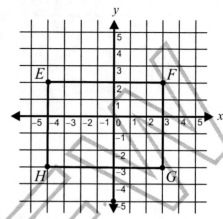

7. $\overline{EF}$ = _____   10. $\overline{GH}$ = _____

8. $\overline{FG}$ = _____   11. $\overline{HE}$ = _____

9. Area = _____

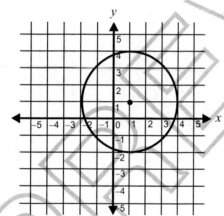

5. radius = _____

6. Area = _____

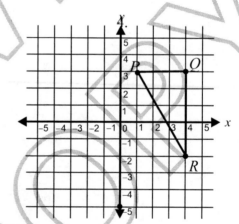

12. $\overline{PQ}$ = _____

13. $\overline{QR}$ = _____

14. $\overline{RP}$ = _____

15. Area = _____

For questions 16-21, plot the points given. Next, calculate the perimeter and the area of each plotted figure.

16. (2,1)(-3,1) (-3,4)

17. (−3,4) (2,4) (−3,−2) (2,−2)

18. (3,3) (3,−1) (−2,−1)

19. (0,0) (−3, 0) (0,4) (−3,4)

20. (4,4) (−2,4) (−2, −3) (4,−3)

21. Circle: center at point (2,1).
    Point (2,−1) on circle. Find radius and area.

7.4.1

368

DILATIONS

A **dilation** of a geometric figure is either an enlargement or a reduction of the figure. The point at which the figure is either reduced or enlarged is called the **center of dilation**. The dilation of a figure is always the product of the original and a **scale factor**. The scale factor is always a positive number that is multiplied by the coordinates of a shape's vertices, which is usually illustrated in a coordinate plane. If the scale factor is greater than one, then the resulting dilated figure will be an enlargement of the original figure. If the scale factor is less than one, then the resulting dilated figure will be a reduction of the original figure.

EXAMPLE: The triangle ABC has been dilated by a scale factor of $\frac{1}{4}$.

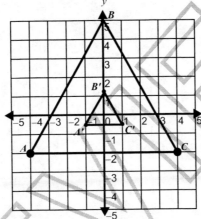

The first step in finding the dilated object is to list all the vertices of the original object, ABC. The next step is to multiply the coordinates of the vertices of ABC by the scale factor, $\frac{1}{4}$ to find the coordinates of the dilated figure. Lastly, draw the dilated object on the coordinate plane as shown above.

A: $(-4, -2)$ A': $(-1, -\frac{1}{2})$

B: $(0, 5)$ B': $(0, \frac{5}{4})$

C: $(4, -2)$ C': $(1, -\frac{1}{2})$

NOTE: Since the scale factor is less than one, the dilated figure $A'B'C'$ is a reduction of original triangle, ABC.

Circle the coordinate plane that contains the shape that has been dilated.

A.
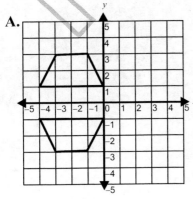

B.
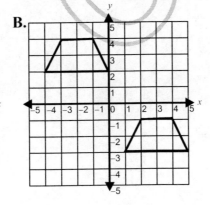

C.
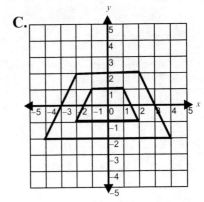

8.4.4

369

On your own graph paper sketch the dilated and original figures.

For questions 1–6, find the coordinates of the vertices of the dilated figure.

1. A: $(-3, 1)$
   B: $(-1, 4)$
   C: $(1, 4)$
   D: $(3, 1)$
   Scale factor: 4

2. A: $(-6, 5)$
   B: $(3, 5)$
   C: $(3, -4)$
   D: $(-6, -4)$
   Scale factor: $\frac{1}{3}$

3. A: $(-10, 0)$
   B: $(0, 10)$
   C: $(8, 5)$
   Scale factor: $\frac{4}{5}$

4. A: $(-1, 7)$
   B: $(1, 7)$
   C: $(5, 5)$
   D: $(5, \frac{1}{2})$
   E: $(1, -3)$
   F: $(-1, -3)$
   G: $(-5, \frac{1}{2})$
   H: $(-5, 5)$
   Scale factor: 2

5. A: $(-8, 7)$
   B: $(-4, 7)$
   C: $(-2, 3)$
   D: $(-6, 3)$
   Scale factor: $\frac{3}{2}$

6. A: $(-4, 12)$
   B: $(6, -2)$
   C: $(-14, -2)$
   Scale factor: $\frac{1}{2}$

For questions 7–10, find the scale factor.

7. A: $(-3, 2)$ A': $(-10.5, 7)$
   B: $(1, 2)$ B': $(3.5, 7)$
   C: $(1, -3)$ C': $(3.5, -10.5)$
   D: $(-3, -3)$ D': $(-10.5, -10.5)$

8. A: $(-6, 9)$ A': $(-2, 3)$
   B: $(3, 12)$ B': $(1, 4)$
   C: $(6, 3)$ C': $(2, 1)$
   D: $(-9, 0)$ D': $(-3, 0)$

9. A: $(0, -3)$ A': $(0, -2)$
   B: $(6, 0)$ B': $(4, 0)$
   C: $(0, 3)$ C': $(0, -2)$

10. A: $(-2, 6)$ A': $(-10, 30)$
    B: $(2, 6)$ B': $(10, 30)$
    C: $(3, 3)$ C': $(15, 15)$
    D: $(2, 0)$ D': $(10, 0)$
    E: $(-2, 0)$ E': $(-10, 0)$
    F: $(-3, 3)$ F': $(-15, 15)$

For questions 11 and 12, determine whether or not $A'B'C'D'$ is a dilation of $ABCD$.

11. A: $(-2, 5)$ A': $(-1, 2)$
    B: $(8, 8)$ B': $(4, 4)$
    C: $(12, 0)$ C': $(6, 0)$
    D: $(2, -6)$ D': $(1, -3)$

12. A: $(0, 8)$ A': $(-2, 6)$
    B: $(5, 8)$ B': $(3, 6)$
    C: $(5, -3)$ C': $(3, -1)$
    D: $(0, -3)$ D': $(-2, -1)$

8.4.4

CHAPTER 23 REVIEW

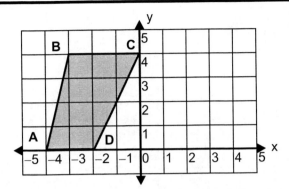

1. Draw the reflection of image ABCD over the y-axis. Label the points A', B', C', and D'. List the coordinates of these points below.

2. A' _____   4. C' _____

3. B' _____   5. D' _____

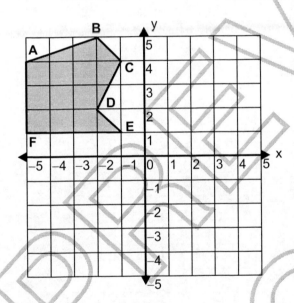

6. Rotate the figure above a $\frac{1}{2}$ turn about the origin, 0. Label the points A', B', C', D', E', and F'. List the coordinates of these points below.

7. A' _____   10. D' _____

8. B' _____   11. E' _____

9. C' _____   12. F' _____

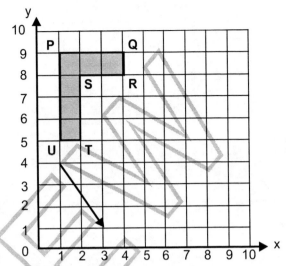

13. Use the translation described by the arrow to translate the polygon above. Label the points P', Q', R', S', T', and U'. List the coordinates of each.

14. P' _____   17. S' _____

15. Q' _____   18. T' _____

16. R' _____   19. U' _____

Use the grid to answer the question that follows.

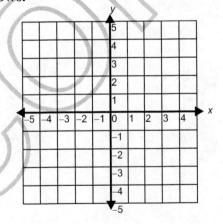

20. A point at (−3, 2) is moved to (0, 0). If a point at (1, 1) is moved in the same way, what will its new coordinates be?

_____

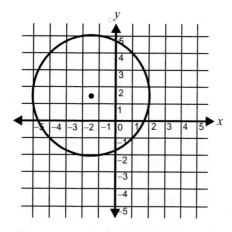

21. What is the length of the radius of the circle above?

22. What is the area of the circle?

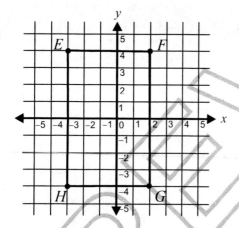

23. What is the length of $\overline{EF}$?
24. What is the length of $\overline{FG}$?
25. What is the perimeter of the above figure?
26. What is the area of the above figure?

27.

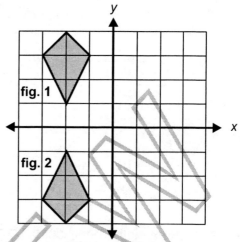

Figure 1 goes through a transformation to form figure 2. Which of the following descriptions fits the transformation shown?

A. reflection across the x-axis
B. reflection across the y-axis
C. $\frac{3}{4}$ clockwise rotation around the origin
D. translation down 2 units

28.

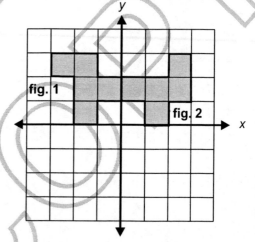

Figure 1 goes through a transformation to form figure 2. Which of the following descriptions fits the transformation shown?

A. reflection across the x-axis
B. reflection across the y-axis
C. $\frac{1}{4}$ clockwise rotation around the origin
D. translation right 3 units

ISTEP+ Grades 9 and 10 Mathematics Review Sheet

Shape	Formulas for Area (*A*) and Circumference (*C*)
Triangle:	$A = \frac{1}{2}bh = \frac{1}{2} \times$ base $\times$ height
Rectangle:	$A = lw =$ length $\times$ width
Trapezoid:	$A = \frac{1}{2}(b_1 + b_2)h = \frac{1}{2} \times$ sum of bases $\times$ height
Parallelogram:	$A = bh =$ base $\times$ height
Square:	$A = s^2 =$ side $\times$ side
Circle:	$A = \pi r^2 = \pi \times$ square of radius $C = 2\pi r = 2 \times \pi \times$ radius $\pi =$ pi ≈ 3.14 or $\frac{22}{7}$

Figure	Formulas for Volume (*V*) and Surface Area (*SA*)	
Rectangular Prism:	$V = lwh =$ length $\times$ width $\times$ height $SA = 2lw + 2hw + 2lh$ $\quad = 2(\text{length} \times \text{width}) + 2(\text{width} \times \text{height}) + 2(\text{length} \times \text{height})$	
General Prisms:	$V = Bh =$ area of base $\times$ height $SA =$ sum of the areas of the faces	
Cylinder:	$V = \pi r^2 h = \pi \times$ square of radius $\times$ height $SA = 2\pi r(r + h) = 2 \times \pi \times$ radius (radius + height) $\pi =$ pi ≈ 3.14 or $\frac{22}{7}$	$\pi \cong 3.14$
Sphere:	$V = \frac{4}{3}\pi r^3 = \frac{4}{3} \times \pi \times$ cube of radius $SA = 4\pi r^2 = 4 \times \pi \times$ square of radius	or
Right Cylinder Cone:	$V = \frac{1}{3}\pi r^2 h = \frac{1}{3} \times \pi \times$ square of radius $\times$ height	$\pi \cong \frac{22}{7}$
Regular Pyramid:	$V = \frac{1}{3}Bh = \frac{1}{3} \times$ area of base $\times$ height	

Equation of a Line

Slope-Intercept Form:

$y = mx + b$
where m = slope and b = y-intercept

Point-Slope Form:

Where m = slope and (x_1, y_1)
and (x_2, y_2) are two points in the plane

Slope of a Line

Let (x_1, y_1) and (x_2, y_2) be two points
in the plane

$\text{slope} = \dfrac{\text{change in } y}{\text{change in } x} = \dfrac{y_2 - y_1}{x_2 - x_1}$ where $x_2 \neq x_1$

Pythagorean Theorem

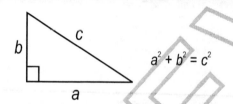

$a^2 + b^2 = c^2$

Distance Formula

$d = rt$
where d = distance, r = rate, and t = time

Temperature Formulas

$°C = \dfrac{5}{9}(F - 32)$
$°Celsius = \dfrac{5}{9} \times (°Fahrenheit - 32)$

$°F = \dfrac{9}{5}C + 32$
$°Fahrenheit = \dfrac{9}{5} \times °Celsius + 32$

Simple Interest Formula

$i = prt$
where i = interest, p = principle,
r = rate, and t = time

Quadratic Formula

$x = \dfrac{-b \pm \sqrt{b_2 - 4ac}}{2}$

where $ax^2 + bx + c = 0$, $a \neq 0$,
and $b^2 - 4ac \geq 0$

Conversions:

1 yard = 3 feet = 36 inches
1 mile = 1,760 yards = 5,280 feet
1 acre = 43,560 square feet
1 hour = 60 minutes
1 minute = 60 seconds

1 cup = 8 fluid ounces
1 pint = 2 cups
1 quart = 2 pints
1 gallon = 4 quarts

1 liter = 1,000 millimeters = 1,000 cubic centimeters
1 meter = 100 centimeters = 1,000 millimeters
1 kilometer = 1,000 meters
1 gram = 1,000 milligrams
1 kilogram = 1,000 grams

1 pound = 16 ounces
1 ton = 2,000 pounds

374

Practice Test 1

1. Simplify: $\dfrac{(3a^2)^3}{a^3}$

   A. $27a^3$ C. $9a^3$

   B. $\dfrac{9a^6}{a^3}$ D. $\dfrac{3a^6}{a^3}$

A1.1.2

2. Simplify the expression shown below.

$$3x^{-4}$$

   A. $\dfrac{81}{x^4}$

   B. $(3x)^{-1}(3x)^{-1}(3x)^{-1}(3x)^{-1}$

   C. $\dfrac{3}{x^4}$

   D. $\dfrac{1}{3x^4}$

A1.1.4

3. Find b: $4b - 7\,(b+2) = 10 + 3b$

   A. $b = \dfrac{{}^+4}{3}$
   B. $b = -4$
   C. $b = 4$
   D. $b = 8$

A1.2.1

4. The steel company employed 512 people last year. This year there are only 448 people working there. What percent decrease is that?

   A.   6%
   B.   8%
   C.   10%
   D. $12\frac{1}{2}\%$

7.2.2

5. Multiply and simplify:
$(3x + 2)(x - 4)$

   A. $3x^2 - 10x - 8$

   B. $3x^2 + 5x - 8$

   C. $3x^2 + 5x - 6$

   D. $8x^2 - 2$

A1.6.4

6. Solve the equations $-2x - 4y = -14$ and $5x + y = -1$ by substitution.

   A. $(-1, 4)$
   B. $(5, 1)$
   C. $(10, -11)$
   D. $(3, 2)$

A1.5.3

7. Which of the following is a graph of the inequality $-y \geq 2$?

A. 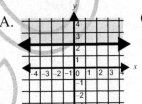   C.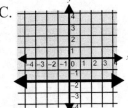

B. 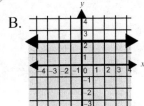   D.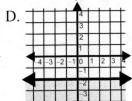

A1.4.6
8.7.9

8. Marla found a dress she liked for $80. The next week it was on the sale rack for $44. What percent discount is that?

   A. 44%   C. 55%
   B. 36%   D. 45%

   7.2.3

9. Solve the equation $\sqrt{6w - 8} = w$

   A. $w = 3, 4$
   B. $w = 2, 4$
   C. $w = 3\sqrt{2}, 4$
   D. $w = 2\sqrt{3}, -2\sqrt{3}$

   A1.8.8

10. Translate the following into an algebraic expression and find the number.

    "The sum of five times a number and eleven is four less than the product of eight and the number."

    A. $\frac{7}{13}$

    B. -5

    C. $-\frac{7}{13}$

    D. 5

    7.3.1

11. Which of the following is the prime factorization of 68?

    A. $2^2 \times 17$

    B. 2×34

    C. 4×17

    D. $12 \times 5 + 8$

    7.1.5

12. Find: $(4y^3 - 8y^2 - 5y) - (2y^3 + 5y - 6)$

    A. $2y^3 - 8y^2 - 6$
    B. $2y^3 - 8y^2 - 10y + 6$
    C. $6y^3 - 3y^2 - 10y - 6$
    D. $2y^3 - 8y^2 + 6$

    A1.6.1

13. The posted speed limit is 704 inches/second. What is the speed limit in miles per hour?

    A. 30
    B. 40
    C. 55
    D. 70

    A1.1.5

14. Solve $V = lwh$ for h

    A. $h = \frac{V}{lw}$
    B. $h = V + 1 + w$
    C. $h = V - lw$
    D. $h = Vlw$

    A1.2.2
    7.3.5

15. Which of the following tables of values does not represent a linear function?

    A. MONTHLY RENT

Bedrooms	Rent
1	$550
2	$625
3	$700

    B.

x	y
−4	−7
0	1
1	4
6	13

    C.

Time	Odometer Reading
7 AM	20825
11 AM	20965
3 PM	21105

    D. COST OF REPAIR

Hours	Cost
1	$72.50
2	$95.00
3	$117.50
4	$160.00

    7.3.1
    7.3.6

16. Amy has 4 pictures on her shelf. How many ways can she arrange the four pictures?

    A. 4
    B. 8
    C. 16
    D. 24

    7.6.7

376

17. Find the equation of the dotted line perpendicular to the line shown below passing through the point (2,1).

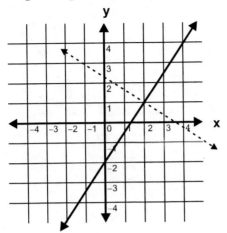

A. $y = \frac{2}{3}x + \frac{7}{3}$

B. $y = \frac{3}{2}x + \frac{4}{3}$

C. $y = -\frac{2}{3}x + \frac{7}{3}$

D. $y = -\frac{3}{2}x + \frac{5}{2}$

7.3.7

18.

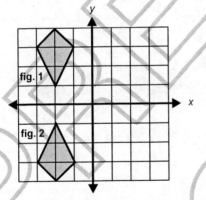

Figure 1 goes through a transformation to form figure 2. Which of the following descriptions correctly describes the transformation shown?

A. reflection across the x-axis
B. reflection across the y-axis
C. $\frac{3}{4}$ clockwise rotation around the origin
D. translation down 2 units

7.4.1

19. Which of the following graphs shows a line with a slope of $-\frac{1}{3}$ that passes through the point $(0, -1)$?

A.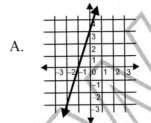

B.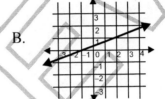

C.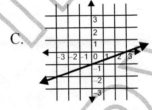

D.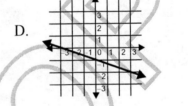

A1.4.2
7.3.8

20. A utility pole is to be fitted with a guy wire. One end of the wire will be attached at ground level, 8 feet from the base of the pole. The other end will be attached to the pole, 15 feet above ground level. What is the length of the wire?

A. 17 feet
B. 18 feet
C. 19 feet
D. 20 feet

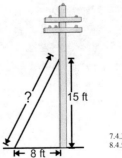

7.4.3
8.4.5

21. Simplify: 1 week 8 days 50 hours

    A.   2 weeks 2 days 2 hours
    B.   2 weeks 3 days 2 hours
    C.   2 weeks 5 days 2 hours
    D.   3 weeks 2 days

7.5.1
8.5.1

22. $\triangle ABC$ is similar to $\triangle DEF$. What is the value of x?

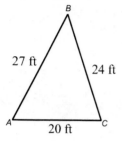
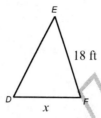

    A.   12 feet
    B.   13 feet
    C.   14 feet
    D.   15 feet

7.5.2

23. On a map drawn to scale, 2 centimeters represent 300 kilometers. How long would a line measure between two cities that are 500 kilometers apart?

    A. $1\frac{1}{5}$ centimeters
    B. $3\frac{1}{3}$ centimeters
    C.   5   centimeters
    D. $\frac{3}{5}$ centimeters

8.5.4

24. Solve the equation $d^2 - 4d + 1 = 0$ by completing the square.

    A. $D = -3, -1$
    B. $D = \sqrt{3}, 2\sqrt{3}$
    C. $D = 2 - \sqrt{3}, \sqrt{3} + 2$
    D. $D = 2i, -2i$

A1.8.4

25. What is the slope of the line in the graph below?

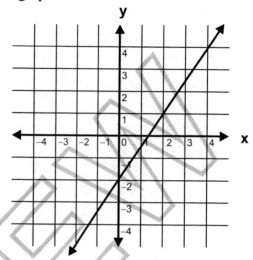

    A.   3
    B.   -3
    C. $\frac{2}{3}$
    D. $\frac{3}{2}$

A1.4.2

26. The figure below is a circle inscribed in a square. What is the area of the shaded region? The formula for the area of a Circle is $A = \pi r^2$. Use $\pi = \frac{22}{7}$.

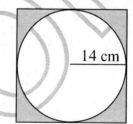

    A.   44 cm^2
    B.   168 cm^2
    C.   196 cm^2
    D.   616 cm^2

7.5.5

27. Find the volume of the figure below. Each edge of each cube measures 4 feet.

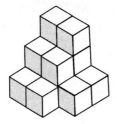

   A.   56 ft^3
   B.   224 ft^3
   C.   896 ft^3
   D.   3584 ft^3

7.5.6

28. Solve the equation $(x - 3)^2 = 1$

   A. $x = 3, -3$
   B. $x = 1, 3$
   C. $x = 2, 4$
   D. $x = 1, -1$

A1.8.3

29. If the figure below were reflected across the y-axis, how would the new figure relate to the original figure?

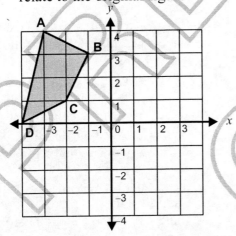

   A.   The figures are similar.
   B.   The figures are congruent.
   C.   The new figure is larger than the original.
   D.   The new figure is smaller than the Original.

7.4.2
8.4.4

30. Which figure below represents a cylinder?

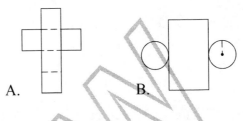

A.

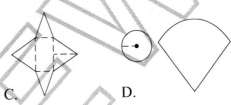

B.

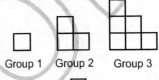

C.   D.

7.4.4

31. Which members of the set $\{0, 1, 2, 3, 4\}$ are solutions for the inequality $5x + 2 < 13$?

   A. $\{0, 1, 2, 3\}$
   B. $\{0, 1, 2\}$
   C. $\{-1, 0, 1, 2\}$
   D. $\{3, 4\}$

A1.2.3

32. Consider the pattern of block designs shown below.

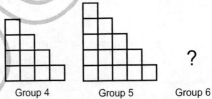

Group 1   Group 2   Group 3

Group 4   Group 5   Group 6

How many squares will be in Group 6?

   A.   20

   B.   21

   C.   22

   D.   23

7.6.2

33. Greg has test scores of 77, 90, 92, and 77 in English class. If he gets a 0 on his next paper, how will it affect the mean, median and mode of his scores?

   A. The mean will go down.
      The median will go down.
      The mode will remain the same.
   B. The mean will remain the same.
      The median will remain the same.
      The mode will remain the same.
   C. The mean, median and mode will all go down.
   D. The mean will go down.
      The mode and median will remain the same.

   7.6.3

34. Which of the following computations will result in an irrational number?

   A. $1\frac{1}{8} \times \frac{3}{4}$
   B. $7\sqrt{2}$
   C. $7.2 - 3.1$
   D. 6×3.25

   8.1.3

35. Write 5^{-3} as a fraction.

   A. $\frac{1}{15}$
   B. $\frac{3}{5}$
   C. $\frac{5}{3}$
   D. $\frac{1}{125}$

   8.1.4

36. Find the point of intersection of the two equations by adding and/or subtracting.

   $$x + y = 4$$
   $$2x - y = 5$$

   A. $(3,1)$
   B. $(-3, 1)$
   C. $(1, 3)$
   D. $(-1, -3)$

   A1.5.4

37. Last year, the Blossom Antique Fair required its vendors to pay $75 to rent a booth and then give 20% of their earnings to a local charity. The cost of participating in the fair is represented in the graph below.

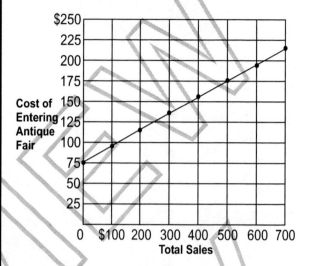

This year, the Blossom Antique Fair is charging only $50 to rent a booth, but is requiring its vendors to give 25% of their earnings to a local charity. How will the graph of the cost of participating in the fair this year compare to the graph of last year's cost?

   A. The graph for this year's cost will have a greater y-intercept and a steeper slope.
   B. The graph for this year's cost will have a greater y-intercept, but there will be no change in slope.
   C. The graph for this year's cost will have a smaller y-intercept and a steeper slope.
   D. The graph for this year's cost will have a smaller y-intercept, but there will be no change in slope.

   A1.4.5

380

38. What is the range of the function $y = x^2 + 2$ as represented in the graph?

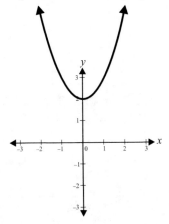

A. $-3 \leq R \leq 3$
B. $R \geq 0$
C. All real numbers ≥ 2
D. All real numbers

A1.3.4

39. $\sqrt{5}$ is between

A. 3 and 4
B. 2 and 3
C. 1 and 2
D. 4 and 5

8.1.7

40. Find the point of intersection of the two equations by adding and/or subtracting.

$$4x + 5y = \tfrac{2}{3}$$
$$7x - 3y = 9$$

A. $(\tfrac{2}{3}, -1)$
B. $(-2, \tfrac{3}{2})$
C. $(-1, 0)$
D. $(1, -\tfrac{2}{3})$

A1.5.4

41. Pat wanted to divide 7.86 by 3.9, but he forgot to enter the decimal points when he put the numbers into the calculator. Using estimation, where should Pat put the decimal point?

A. 0.2015386
B. 2.015386
C. 20.15386
D. 201.5386

8.2.3

42. 3^3 is equal to

A. 3
B. 9
C. 27
D. 81

8.2.4

43. Matt needs to earn at least $1000 this month. He is paid $400 per month plus $50 for each new account he establishes. How many new accounts (a) must he establish in order to earn at least $1000?

A. $a = 8$
B. $a = 12$
C. $a = 20$
D. $a = 24$

8.3.1

44. What is the solution to the following system of equations?

$$y = 4x - 8$$
$$y = 2x$$

A. $(-4, -8)$
B. $(4, 8)$
C. $(-1, -2)$
D. $(1, 2)$

8.3.2

45. What is the slope of the equation

$$3x - 3y = 5$$

A. -1
B. 3
C. -3
D. 1

8.3.6

46. Josh is on a diet. He presently weighs 150 pounds and hopes to continue to lose two pounds per week. Which graph best models Josh's projected weight loss over The next twelve weeks?

A.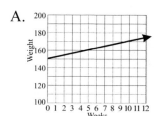

C.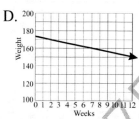

B.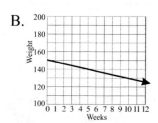

D.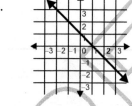

A1.3.2

47. Which is the graph of $2x - y = 1$?

A.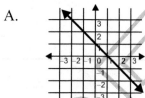

B.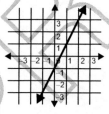

C.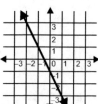

D.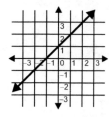

A1.4.1
8.3.5

48. Identify the graph of the following function:

$$y = x^2 - 2$$

A. 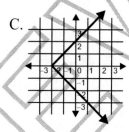 B.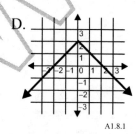

C. 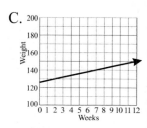 D.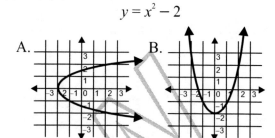

A1.8.1

49. Nihar jogged at 7 miles per hour for forty-five minutes and 3 miles per hour for thirty minutes. How far did he jog?

A. $6\frac{3}{4}$ miles

B. $5\frac{3}{4}$ miles

C. $6\frac{1}{4}$ miles

D. $7\frac{1}{2}$ miles

8.5.2

50. Solve the following quadratic equation by factoring.

$$6x^2 - 16x - 6 = 0$$

A. $2, -3$

B. $-3, \frac{1}{3}$

C. $-1, 1$

D. $3, -\frac{1}{3}$

A1.8.2

382

51. Veronica is comparing the cost of two floor waxing services. Acme Co. charges $50 per 100 square feet. Best, Inc. charges $25 per 100 square feet plus a fixed $75 fee. Two equations representing price as a function of square footage are graphed below.

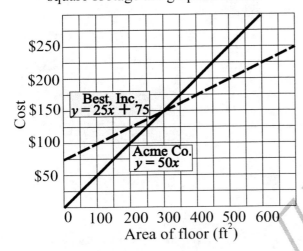

Best, Inc. is less expensive than Acme Co. for

   A.   all size floors.
   B.   exactly 300 ft^2.
   C.   areas less than 300 ft^2.
   D.   areas more than 300 ft^2.

A1.5.6

52. Jack is going to paint the ceiling and four walls of a room that is 10 feet wide, 12 feet long, and 10 feet from floor to ceiling. How many square feet will he paint?

   A.   120 square feet
   B.   560 square feet
   C.   680 square feet
   D.   1,200 square feet

8.5.4

53. Solve the inequality: $2 \leq 2x - 2 \leq -5$

   A. $2 \leq x \leq \frac{3}{2}$
   B. $2 \leq x \geq -\frac{3}{2}$
   C. $-2 \leq x \leq \frac{3}{2}$
   D. $2 \leq x \leq -\frac{3}{2}$

A1.2.5

54. Nick has $10.50 in his pocket. The cab company charges $2.40 for the first mile and $0.70 per mile after that. Which inequality shows how many miles (m) Nick can go in the cab?

   A. $0 \leq m \leq 12$
   B. $0 \leq m \leq 11$
   C. $0 \leq m \geq 12$
   D. $0 \leq m \geq 10$

A1.2.6

55. Which of these is the equation of the line that generalizes the pattern of the data in the table?

   A. $y = -4x - 7$
   B. $y = x + 2$
   C. $y = 2x + 1$
   D. $y = -7x - 4$

x	y
−4	−7
0	1
1	3
6	13

8.3.9

56. What is the mode of the following data set?

   42, 44, 38, 37, 44, 39, 38, 38

   A.   37
   B.   38
   C.   39
   D.   44

8.6.3

57. Consider the following set of data:

28	32	42	37
30	25	57	39
24	32	33	44
38	34	30	44
31	28	31	29

If you made a stem-and-leaf plot for the data, what numbers would you use for the stems?

   A.   2, 3, 4
   B.   1, 2, 3, 4
   C.   20, 30, 40
   D.   2, 3, 4, 5

8.6.4

58. Matthew has 16 fish in an aquarium. The fish are the following colors: 4 blue, 6 orange, 2 black and white striped, and 4 pink. Matthew also has a trouble-making cat that has snared a fish. What is the probability that the cat snared an orange fish if all the fish are equally capable of avoiding the cat?

A. $\frac{1}{4}$

B. $\frac{3}{8}$

C. $\frac{1}{8}$

D. $\frac{1}{6}$

8.6.6

59. Jane has 5 different playing cards in her hand. In how many different orders can the cards be arranged?

A. 5

B. 25

C. 120

D. 240

8.6.7

60. For this graph of a quadratic function in the form of $y = ax^2 + c$, what are the values of a and c?

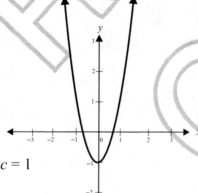

A. $a = 1, c = 1$

B. $a = -1, c = -1$

C. $a = 2, c = -1$

D. $a = 2, c = 1$

A1.3.4

61. Patty has carefully weighed and measured the length of a licorice stick before taking the first bite and again after each bite. From her data shown in the table below, she has concluded that the weight of the remaining licorice stick is proportional to the length.

Bite Number	Length	Weight
0	304mm	28.6 grams
1	280mm	26.3 grams
2	250mm	23.5 grams
3	239mm	22.5 grams
4	202mm	? grams

After the 4th bite, the licorice stick was 202 mm long. Approximately how many grams should the licorice stick have weighed?

A. 17 grams

B. 18 grams

C. 19 grams

D. 20 grams

8.7.1

62. If $3\Delta 5 = -2$, what is $20\Delta 27$?

A. 7
B. −7
C. 5
D. 2

8.7.4

Use grid sheets to record your answers to the following problems.

63.

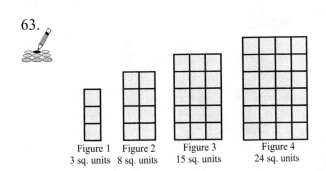

Figure 1 Figure 2 Figure 3 Figure 4
3 sq. units 8 sq. units 15 sq. units 24 sq. units

What would be the area, in square units, of the 12th figure in this pattern?

8.7.1
7.7.5

64. Kevin is predicting the length in centimeters of a particular flowering vine. He is using the formula $L = 0.75(400 + 8w)$, where L is the length of the vine and w is the number of weeks. What will be the vine's length in centimeters after 6 weeks?

7.6.2

65.

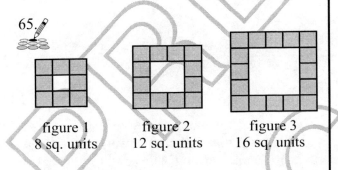

figure 1 figure 2 figure 3
8 sq. units 12 sq. units 16 sq. units

What would be the area, in square units, of the 100th figure in this pattern?

8.7.2
7.7.5

66. What is the missing number in this sequence?

0.04, 0.12, 0.36, ____?____, 3.24

8.7.2

---

67. What is the value of the expression below?

$$\sqrt{4^3} - (-2)^3$$

8.3.4

68.

$$4^4 =$$

7.1.4

69. Sean earns a weekly base salary of $150 per week plus a commission of 20% of his total sales for the week. Last week Sean earned a total of $510. What was the total, in dollars, of Sean's sales last week?

8.3.1

70. The Franklin High School cheerleaders had a car wash to raise money to buy new uniforms. The bar graph below shows the number of cars washed during five one-hour periods.

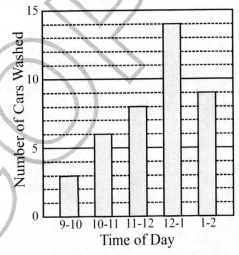

What percentage of the cars washed were washed between 11:00 and 12:00?

7.6.1
8.6.4

71. The bar graph shows the speeds of vehicles on Shore View Boulevard as recorded by a radar speed detector.

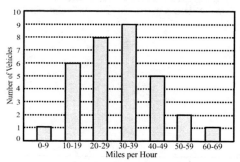

The speed limit is 30 miles per hour. What percentage of the vehicles exceeded the speed limit by 10 miles per hour or more?

7.6.1
8.6.4

72. Betty is planning an irrigation sprinkler system for her yard as shown below.

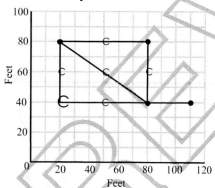

What total length of pipe will she need to complete the system as planned? Answer to the nearest foot.

7.1.6

73. What is the length of line segment $\overline{AD}$ in $\triangle ADE$?

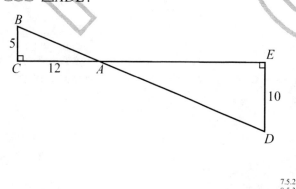

7.5.2
8.5.3

74. Maybelle wants to calculate the temperature in degrees Celsius of a ceramic kiln which has been heating for 30 minutes. She is using the formula $T = I + 12m$ where T is the temperature of the heated kiln, I is the initial, unheated temperature of the kiln, and m is the number of minutes the kiln has been heating. What is the temperature of the kiln in degrees Celcius after 30 minutes of heating if the initial room temperature was 25° Celsius?

7.3.2
8.3.1

75. The number of jelly beans in each of 10 bags is listed below.

61, 63, 59, 57, 57, 63, 64, 64, 63, 59

What is the mode of these numbers?

7.6.3

76. What is the probability that the spinner will land on a shaded section or the number 4? Grid your answer in fraction form.

7.6.6

77. An automatic pitching machine releases balls with an initial upward velocity of 72 feet per second. The height of the ball at any time can be described by the function $h = 72t - 16t^2$, where t is the time in seconds and h is the height in feet. How many seconds does it take for a ball to return to the ground?

A1.8.7

Practice Test 1 - Session 1

Directions: Write out your response to each of the following problems on your own paper. Show all of your work. You may NOT use a calculator on this portion of the practice test.

78. Simplify: $\dfrac{36y^2 - 25}{24y^2 + 2y - 15}$

A1.7.1

79. Mrs.Tucker, a freshman algebra teacher, divided her class into two teams, rewarding the team with the highest average with a pizza party. Each team had 6 students. The scores on the first test for Team 1 were: 84,75,95,76,84 and 72. Then Jamal was transferred into Mrs. Tucker's algebra class and joined Team 1. Jamal got 88 on his first test. How did his score affect the mean, the median, and the mode of the scores for Team 1?

7.6.3

80. Use the subtraction method to solve the equations:

$4x + 2y = 24, \ 4x + 4y = 28$

7.4.1
8.4.5

81. Using the quadratic formula, find solutions to the equation $x^2 - 3x - 8 = 0$. Show your work.

A1.8.6

82. Write the equation of the line with points (2, 5) and (4, 9). Then find an equation of the line through point (2, 3) perpendicular to the first line.

A1.4.4

83. Write 12,300,000,000,000 using decimals in scientific notation.

7.1.1
8.1.1

84. If the triangle in the graph on the right is reflected across the y-axis, what would be the coordinates of the vertices of the new triangle?

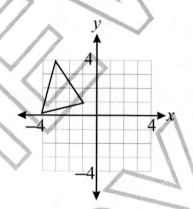

7.4.1

85. Factor: $x^2 - 7x - 18$

A1.6.7

86. Write an equation of a line with slope -5 and y-intercept 2.

8.3.6

87. Graph the equation $x = \frac{1}{2}y$ and $x + 2y = 6$ on the same graph to find where the lines intersect. Use the graphs to estimate the solution of the pair of equations.

A1.5.1

Practice Test 1 - Session 2

Directions: Write out your response to each of the following problems on your own paper. Show all of your work. You MAY use a calculator on this portion of the practice test.

88. The scale of a map is $\frac{1}{2}$ inch = 60 miles. If two towns are $4\frac{1}{2}$ inches apart, how many miles apart are they? Show all of your work.

7.5.3
8.5.3

89. Given the equation: $2(5x - 3) - 6x = 2$, solve the equation for x.

7.3.2
8.3.1

90. Tom earns $300 per week plus 10% of all his sales. Last week he earned $450. What was the amount of his sales last week to the nearest dollar?

7.7.7
8.7.7

91. Sarah sells musical instruments. She is paid $80 per day plus $30 for each instrument that she sells. Write an equation to show how many instruments Sarah must sell to earn at least $250 in one day. Then solve the equation to show that it works.

7.7.6
8.7.6

92. A golf course manager needs to find the distance, c, across the pond from a golf tee A to hole B. Use the Pythagorean Theorem to help him find the answer. Show all of your work.

$a = 105$ yd

$c = ?$

$b = 130$ yd

7.4.3
8.4.5

93. Find the probability of rolling an even number or 7 with two number cubes. Show all of your work.

8.6.6

94. You have 26 cards, each with a different letter of the alphabet, face down in front of you so you can't see the letters. The cards were mixed up before they were placed on the table. What is the probability that you will pick the first letter of your last name with the first try?

7.6.6

95. Compute: $44 + 7.9 \times 4.5 - 6\frac{1}{4} \times \frac{1}{5} - 17 =$

7.2.1
8.2.1

Practice Test 2

1. $\sqrt{96} =$

   A. $4\sqrt{6}$
   B. $2^5 \times 3$
   C. $\sqrt{90} + \sqrt{6}$
   D. $96\frac{1}{2}$

   A1.1.2

2. Simplify $4^{\frac{3}{2}}$

   A. $\frac{1}{4^{\frac{2}{3}}}$

   B. $\sqrt[3]{4^2}$

   A1.1.4

   C. 8

   D. $3(4^2)$

3. Simplify: $(5x - 4)(x + 2)$

   A. $5x^2 - 2$

   B. $5x^2 - 8$

   C. $5x^2 + 6x - 8$

   D. $5x^2 - 18x - 8$

   A1.1.3
   A1.6.4

4. Bonnie makes toy wagons to sell to gift shops. She buys wheels for the wagons at $12.72 per dozen. Which formula will calculate the cost of 4 wheels per wagon?

   A. $12x = 12.72$

   B. $\frac{12}{4}x = 12.72$

   C. $\frac{12.72}{12} = x$

   D. $4x = 12.72$

   7.3.1

5. Solve the inequality $8x - 5 < 12$ for x in the set $\{-2, -1, 0, 1, 2, 3, 4\}$

   A. $x = \{-2, -1, 0, 1, 2\}$
   B. $x = \{-2, -1, 0, 1\}$
   C. $x = \{0, 1, 2, 3, 4\}$
   D. $x = \{-2, -1, 0, 1, 2, 3, 4\}$

   A1.2.3

6. A set of table and chairs that normally sells for $450.00 is on sale this week for 30% off the regular price. How much would I save if I bought it on sale?

   A. $ 30.00
   B. $ 31.50
   C. $ 135.00
   D. $ 315.00

   7.2.3

7. For the following pair of equations, find the point of intersection (common solution) using the substitution method.

   $$3x + 3y = 9$$
   $$9y - 3x = 6$$

   A. $(1, 2)$

   B. $(\frac{7}{4}, \frac{5}{4})$

   C. $(1, 1)$

   D. $(\frac{1}{3}, \frac{1}{6})$

   A1.5.3

390

8. Solve for x; $4x + 7 = 35$

   A. 5
   B. 6
   C. 7
   D. 11

7.3.2

9. Which of the following is the prime factorization of 90?

   A. $2 \times 3^2 \times 5$
   B. 30×3
   C. 15×6
   D. $2 \times 3 \times 15$

7.1.5

10. Simplify:

   $$(3b^3 - 5b^2 - 7) + (2b^2 - 6b + 4)$$

   A. $3b^3 - 3b^2 - 6b - 3$
   B. $3b^3 - b^2 - 6b - 3$
   C. $3b^3 + 3b^2 - 6b - 3$
   D. $5b^2 - 11b - 3$

A1.1.3

11. Which order of operations should be used to simplify the following expression:

   $$6(7 - 2) \div 2 + 5$$

   A. subtract, multiply, divide, add
   B. subtract, add, multiply, divide
   C. add, subtract, multiply, divide
   D. multiply, subtract, divide, add

7.3.4

12. Solve $ab + cd = 20$ for b
   A. $b = 20 - cda$

   B. $b = \dfrac{20 + cd}{a}$

   C. $b = \dfrac{20 - cd}{a}$

   D. $b = 20cda$

A1.1.2

7.3.5

13. Which of the following table of values below represents a linear function?

A.
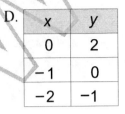

Capacity of the Coffee Urn	Number of Scoops of Coffee
4 quarts	18 scoops
6 quarts	26 scoops
10 quarts	42 scoops

B.

x	y
-3	-2
0	-1
3	1

C.

x	y
-1	-5
1	-3
3	0

D.

x	y
0	2
-1	0
-2	-1

A1.3.3

14. Find the equation of the line perpendicular to the line graphed below with the same y-intercept.

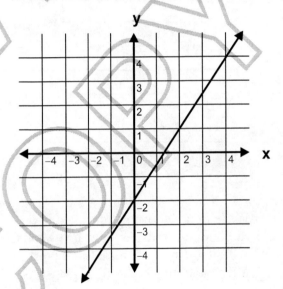

   A. $y = -2x - 2$
   B. $y = -\frac{2}{3}x - 2$
   C. $y = -\frac{3}{2}x + 2$
   D. $y = \frac{2}{3}x - 2$

A1.4.4
7.3.7

15. Which graph shows a line with slope $\frac{5}{2}$, passing through the point (1, −3)?

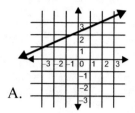
A.

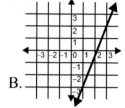
B.

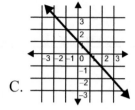
C.

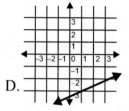
D.

7.3.8
8.3.5

16.

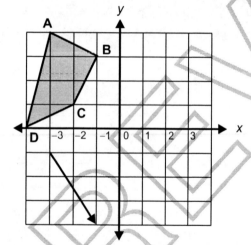

If the figure above is translated in the direction described by the arrow, what will be the new coordinates of point B after the transformation?

A. (2, 1)   C. (1, 0)
B. (0, 1)   D. (1, 1)

7.4.1

17. Find the length of the missing side of the triangle below.

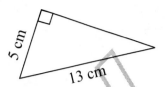

A. 10 cm   C. 12 cm
B. 11 cm   D. 15 cm

7.4.3
8.4.5

18. Simplify:
2 yd 2 ft 3 in + 3 yd 1 ft 10 in

A. 5 yd 1 ft 1 in
B. 5 yd 2 ft 1 in
C. 6 yd 1 ft 1 in
D. 6 yd 1 in

7.5.1
8.5.1

19. Victoria recorded how far her pet snail crawled. After 10 minutes the snail had crawled 20 centimeters. After 15 minutes it had crawled 30 centimeters. Assuming the snail crawls at a constant rate, which of these graphs shows the distance traveled by the snail as a function of time?

A.
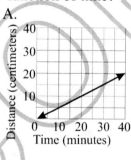

C.
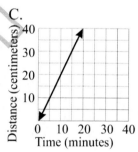

B.
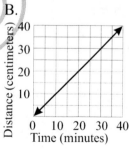

D.
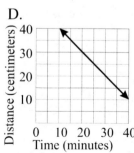

A1.3.2

20. Madison is reading the floor plans of her new house. What is the perimeter of the room shown below?

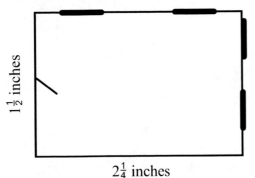

$1\frac{1}{2}$ inches

$2\frac{1}{4}$ inches

Scale: $\frac{1}{8}$ inch = 1 foot

A. 20 feet   C. 40 feet
B. 60 feet   D. 30 feet

7.5.3

21. The following graph depicts the weight of an individual as a function of age.

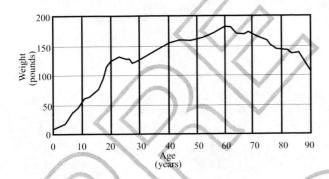

What is the range (R) of this function?

A. 0 pounds ≤ R ≤ 90 years

B. 0 years ≤ R ≤ 90 years

C. 8 pounds ≤ R ≤ 110 pounds

D. 8 pounds ≤ R ≤ 180 pounds

A1.3.4

22. Solve: $6a2 + 11a - 10 = 0$, using the quadratic formula.

A. $(-\frac{2}{5}, \frac{3}{2})$

B. $(\frac{2}{5}, \frac{2}{3})$

C. $(-\frac{5}{2}, \frac{2}{3})$

D. $(\frac{5}{2}, \frac{2}{3})$

A1.8.6

23. What is the volume of the figure below?

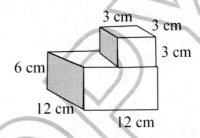

3 cm 3 cm
3 cm
6 cm
12 cm
12 cm

A. 21 cm
B. 39 cm
C. 216 cm
D. 891 cm

7.5.6

24. Which equation is non-linear?

A. $y = \frac{1}{4}x + 2$
B. $y = -x^2$
C. $x + 2y = -4$
D. $2x - 4 = 0$

7.3.9

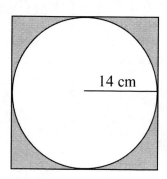

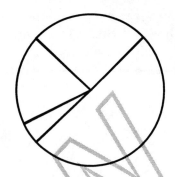

25. The figure above is a circle inscribed in a square. What is the area of the shaded region?

    A.   44 square centimeters
    B.   168 square centimeters
    C.   196 square centimeters
    D.   616 square centimeters

7.4.2
8.4.4

26. Which of the following figures represents a pyramid?

A.

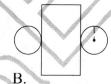

B.

C.   D.

7.4.4

The circle graph most accurately represents which of the situations below.

    A.   John's after school-activities between 4:00 and 6:00 pm are divided between watching TV, 30%; playing video games, 20%; playing sports, 30%; and surfing the internet, 20%.

    B.   The Johnson household family budget is divided into the following categories: housing, 40%; food, 25% ; clothing, 20%; charities, 15%.

    C.   The basketball team spends their budget in the following categories: uniforms, 50%; equipment, 10%; travel, 30%; and snacks, 10%.

    D.   Tina spends her $80.00 monthly allowance in the following categories: make-up, 25%; clothes, 50%; snacks, 5%; and music Cds, 20%.

7.6.1
8.6.4

28. Solve the equation $(x + 9)^2 = 49$

    A. $x = -9, 9$
    B. $x = -9, 7$
    C. $x = -16, -2$
    D. $x = -7, 7$

A1.8.3

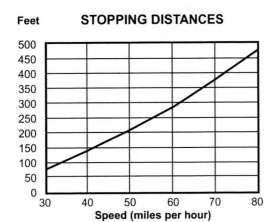

Feet   **STOPPING DISTANCES**

29. According to the chart above, how many more feet does it take to stop a car traveling at 70 miles per hour than at 55 miles per hour?

   A.   175 feet
   B.   150 feet
   C.   125 feet
   D.   100 feet

7.6.1
8.6.4

30. Shannon has grades of 60, 70, 60, 55, and 68 on her Spanish quizzes. What will happen to the mean, median and mode of her grades if she gets a 100% on her next quiz?

   A.   The mean will go up.
         The median will go up.
         The mode will remain the same.
   B.   The mean will remain the same.
         The median will remain the same.
         The mode will remain the same.
   C.   The mean, median and mode will all go up.
   D.   The mean and median will go up.
         The mode will remain the same.

7.6.3

31. Find the difference:
$$(x^2 + 4x - 7) - (2x^2 + x - 1)$$

   A. $-3x^2 - 3x - 8$

   B. $-x^2 - 3x - 8$

   C. $-x^2 + 3x - 6$

   D. $-2x^2 + 4x + 7$

A1.6.1

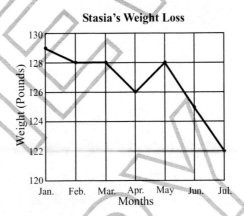

32. Why is the graph shown above misleading?

   A.   The graph is misleading because it Does not account for how much Stacia is eating.
   B.   The graph is misleading because it does not include August - December.
   C.   The graph is misleading because the weight does not go down to 0.
   D.   The graph is misleading because the graph shows most of the weight loss over a 2 month period.

7.6.4

33. Solve: $1 + 3x - 9 = 4x - 7$

    A. -15

    B. $-\frac{15}{7}$

    C. -1

    D. $-\frac{7}{15}$

A1.2.1

34. Brad has 5 trophies on his shelf. How many ways can he arrange the 5 trophies?

    A.   5
    B.   25
    C.   120
    D.   200

7.6.7

35. Which of the following computations will result in an irrational number?

    A. $8\sqrt{5}$
    B. $1\frac{1}{2} \div \frac{2}{3}$
    C. $2.4 \div 3.1$
    D. $7 + 3.7$

7.1.3
8.1.3

36. Write 2^{-5} as a fraction.

    A. $\frac{1}{32}$

    B. $\frac{2}{5}$

    C. $\frac{5}{2}$

    D. $\frac{1}{10}$

8.1.4

37. Solve the equation $y = \sqrt{16 - 6y}$

    A. $y = 4, 2\sqrt{3}$
    B. $y = 3 -\sqrt{2}, 2 -\sqrt{3}$
    C. $y = 2$
    D. $y = 2 + 3i, 2 - 3i$

A1.8.8

38.

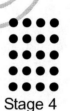 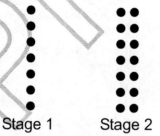

Stage 1   Stage 2   Stage 3   Stage 4   ?   Stage 5

What will be the next stage (Stage 5) in the pattern?

  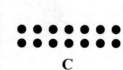 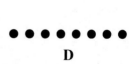

    A.   B   C   D

8.2.1

39. Simplify $(\sqrt{5})^2$

    A.   10

    B.   15

    C.   25

    D.   5

8.1.6

40. $\sqrt{13}$ is between

    A.   3 and 4
    B.   4 and 5
    C.   2 and 3
    D.   12 and 13

8.1.7

41. Hanna earns 12% commission on any jewelry sales she makes. About how much is her commission on a $45 sale?

    A. $1.00
    B. $4.00
    C. $5.00
    D. $12.00

7.2.3
8.2.3

42. Emily had $100.00 to spend. She spent $37.85 on clothes. About how much change should she receive from $100.00?

    A. $30.00
    B. $40.00
    C. $60.00
    D. $70.00

7.2.1
8.2.4

43. Solve: $3x - 2 = 17$

    A. $6\frac{1}{3}$

    B.   5

    C. $12\frac{2}{3}$

    D. $16\frac{1}{4}$

7.3.2
8.3.1

44. Which ordered pair is a solution for the following system of equations?

$$-3x + 7y = 25$$
$$3x + 3y = -15$$

    A. $(-13, -2)$
    B. $(-6, 1)$
    C. $(-3, -2)$
    D. $(-20, -5)$

8.3.2

45. Solve the equation $c^2 + 3c - 9 = 0$ by completing the square.

    A. $c = 3, -3$
    B. $c = \frac{3}{2}\sqrt{5} - \frac{3}{2}, -\frac{3}{2}\sqrt{5} - \frac{3}{2}$
    C. $c = \pm\sqrt{3}$
    D. $c = 3i, -3i$

A1.8.4

46. What is the slope of the equation

$$-2x^2 + 4y = 7$$

    A. $\frac{1}{2}$
    B.   2
    C. -2
    D. $-\frac{1}{2}$

8.3.6

47.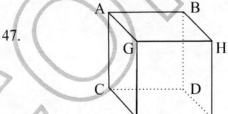

Which of the following pairs of line segments are perpendicular?

    A. $\overline{EF}$ and $\overline{GH}$
    B. $\overline{EF}$ and $\overline{BH}$
    C. $\overline{EF}$ and $\overline{AB}$
    D. $\overline{EF}$ and $\overline{DF}$

8.4.1

48. Lupe has an accident while cooking. She wants to add a dash of pepper to two liters of boiling water for a soup she is making. Instead, the cap falls off and three hundred grams, the entire bottle of pepper, falls into her pot of water. What is the density of the pepper/boiling water mixture?

    A.   15 grams/liter
    B.   30 grams/liter
    C.   150 grams/liter
    D.   300 grams/liter

8.5.2

49. How many of the smaller cubes will fit inside the larger box? (Figures are not drawn to scale.)

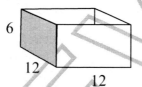

    A.   6
    B.   16
    C.   20
    D.   32

7.5.3
8.5.3

50. Find the area of the rectangle below.

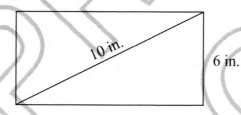

10 in.

6 in.

    A.   42 sq. in.   C.   54 sq. in.
    B.   48 sq. in.   D.   60 sq. in.

7.4.3
8.5.4

51.

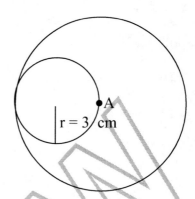

Point A is in the center of the larger circle. The radius of the smaller circle is 3 cm. What is the area of the larger circle? Use 3.14 for π.

    A.   9.42 cm^2
    B.   18.84 cm^2
    C.   28.26 cm^2
    D.   113.04 cm^2

8.5.5

52. Solve the following quadratic equation by factoring.

$$6a^2 - 5a - 6 = 0$$

    A. $\frac{3}{2}, -\frac{2}{3}$

    B. $-\frac{3}{2}, -\frac{2}{3}$

    C. $-\frac{2}{3}, \frac{2}{3}$

    D. $-\frac{3}{2}, \frac{2}{3}$

A1.8.2

53. Examine the following two data sets:

   Set #1: 49, 55, 68, 72, 98
   Set #2: 20, 36, 47, 68, 75, 82, 89

   Which of the following statements is true?

   A. They have the same mode.
   B. They have the same median.
   C. They have the same mean.
   D. None of the above.

   8.6.3

54. Sally, Janet, and Nancy are helping with a reforestation project. Sally plants between 50 and 90 seedlings per hour. Janet and Nancy each plant between 40 and 75 trees per hour. Which of these inequalities represents the possible range of total number of trees the three workers can plant in one hour?

   A. $90 \leq x \leq 165$

   B. $130 \leq x \leq 240$

   C. $150 \leq x \leq 280$

   D. $180 \leq x \leq 330$

   A1.2.6

55. David owns a dog named Wishes. David reached into his box of 8 lamb-flavored, 12 liver-flavored, 6 chicken-flavored, and 20 milk-flavored dog biscuits and gave one to Wishes without looking. What is the probability of Wishes getting a liver treat?

   A. $\frac{1}{12}$ B. $\frac{6}{17}$

   C. $\frac{6}{23}$ D. $\frac{1}{23}$

   8.6.6

56. Which of the following is a graph of the inequality $x + y \geq 4$?

A.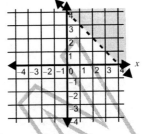

B.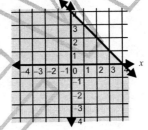

C.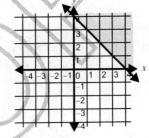

D.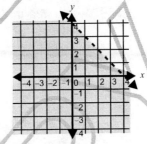

A1.4.6
8.7.9

57. Consider the pattern of square-block designs shown below.

Stage 1 Stage 2 Stage 3

Area = 5   Area = 12   Area = 21

Stage 4   Stage 5

Area = 32   Area = ?

What will be the area in Stage 5?

A. 45

B. 46

C. 47

D. 48

8.7.2

58. District Court judges are elected from a two county area. Here are the results from this year's election.

	PERCENTAGE OF TOTAL VOTE IN	
Candidate	County A	County B
Lou Smith	58%	10%
John McDermitt	30%	25%
Patrick Mann	12%	65%
Total Votes	10,500	4,800

Which candidate had the overall majority vote?

A. Lou Smith
B. John McDermitt
C. Patrick Mann
D. Not enough information is given.

8.7.3

59. The cost of operating a hospital's emergency electrical generator on a particular day is calculated using the formula $c = .05x + 30$, where c is the cost in dollars, and x is the number of kilowatt hours generated on that day.

If the cost of operation for a particular day is known, which of the following formulas may be used to calculate the number of kilowatt hours generated on that day?

A. $x = .05c - 1.5$
B. $x = 20c - 600$
C. $x = 15c - 2000$
D. $x = 500c - 1500$

8.3.8

60. Gena has 4 trinkets to put on her shelf. How many ways can she arrange them?

A. 24
B. 12
C. 6
D. 4

8.6.7

61. Julie has a part-time business selling cosmetics out of her home. This week she spent 5 hours on her business, and cosmetics sales totaled $147.50. To find out how much Julie made per hour, you also need to know

A. how many samples she gave away.
B. how much she spent at the grocery store.
C. how many customers she had.
D. how much she paid for the cosmetics and supplies.

8.7.1

Use the provided grid sheets to record your answers to the following Problems.

62. Jill is flying a kite on 100 feet of string. She holds the end of the kite string to the ground while Jack measures the distance to a point directly under the kite.

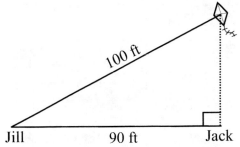

If the distance from Jill to Jack is 90 feet, how high is the kite above the ground? Answer to the nearest whole foot.

7.1.6
8.4.5

63. Acme Alarm Co. charges $85 per month plus $175 per response ($R$) to an alarm according to the formula $C = 85 + 175R$, where C represents total monthly charge. Wilson Pharmaceutical Supply received a monthly bill of $1310. How many times did Acme Alarm Co. respond that month?

8.3.1
A1.2.6

64. The diagram below represents two animal pens.

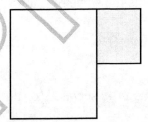

The shaded area represents an old pen which can contain approximately 100 animals.
The unshaded area represents a new pen. Approximately how many animals can the new pen contain?

7.5.5
8.5.5

65. Katelyn is a pitcher on the softball team. Out of 130 pitches, 91 were strikes. What percent of Katelyn's pitches were not strikes?

7.2.3
8.2.4

66. The table shows values of x and y for the equation.

$3x^4 \downarrow y = 0$

What is the value of y when $x = 4$?

x	y
-3	243
-2	48
-1	3
0	0
1	3
2	48
3	243
4	?

8.3.4

67. There are 52 cards in a standard deck of cards: 13 clubs, 13 diamonds, 13 hearts, and 13 spades. Dawn draws two cards at random from the deck, both are hearts. If she draws again, without replacing the first two cards, what is the probability that the third card she draws will also be a heart?

8.6.6

68. Jeanette is reading a circle graph with an angle of 50°. What fractional part of the circle does the 50° angle represent?

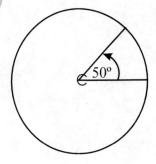

8.6.4

69. The scatter plot below shows the weight of worms as a function of the length of the worm.

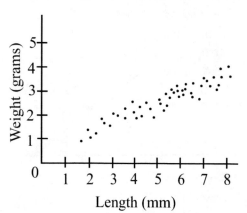

Length (mm)

A worm which is 6 mm long is most likely to weigh, to the nearest gram, how many grams?

8.6.5

70. The heights of 10 corn plants are listed below.

34 cm	41 cm
29 cm	52 cm
36 cm	24 cm
48 cm	37 cm
30 cm	32 cm

What is the median height in centimeters?

8.6.3

71. The population of Belleview Island increased from 2,057 to 2,607 last year. What was the percent of increase in Belleview Island's population last year? Round to the nearest whole percent.

7.2.3
8.2.4

72.

$4^3 =$

7.1.4

73. The bar graph below shows the number of members of the chess club by grade level.

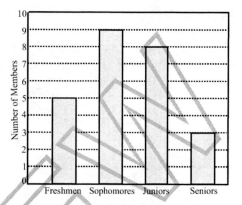

What percentage of the chess club members are NOT seniors?

7.6.1
8.6.4

74.

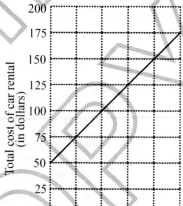

Miles Driven

What is the slope of the line representing the cost of a car rental as a function of miles driven?

7.3.7

75. If it takes 40 seconds for the shadow of a helicopter to cross the length of a 100 yard long football field, how fast is the helicopter flying in miles per hour? (Round answer to the nearest whole mile per hour.)

8.5.2

76. Joyce dropped a nickel and a dime on the ground. What is the probability that both coins will come up tails?

7.6.6
8.6.6

Practice Test 2 - Session 1

Directions: Write out your response to each of the following problems on your own paper. Show all of your work. You may NOT use a calculator on this portion of the test.

77.   Find the slope and *y*-intercept of the equation $3x + 2y = 11$.

A1.4.2

78.   Write 34,500,000 in scientific notation.

7.1.1

79.   Factor: $4x^2 + x - 3$

A1.6.7

80.   A candy store in the mall sells candy separately and prepackaged. The clerks must count each prepackaged bag of candy to find the number of chocolates in each one. They currently have 13 bags of prepackaged candy. After counting the number of chocolates in each bag, their data was as follows:

17, 14, 15, 20, 12, 17, 19, 21, 14, 17, 17, 18, 21,

Find the minimum and maximum value, lower and upper quartiles, and median of this data.

8.6.3

81.   Find the square root of $x^6 y^{10}$.

7.6.3

82.   Graph the inequalities $2x + 4y \leq 8$ and $x > 1$ and shade the portion of the graph where both inequalities are true.

A1.5.2

83.   Using the coordinates below, graph the original shape and its dilation on a coordinate plane.

A: $(-10, 0)$
B: $(0, 10)$
C: $(5, 5)$
Scale factor: $\frac{4}{5}$

8.4.4

84.   What would be the coordinates of the parallelogram below if it were reflected over the y-axis then translated down three units?

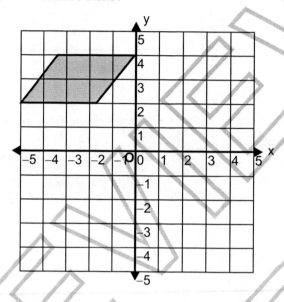

7.4.1

85.   Using the quadratic formula, find the solutions of the equation $x^2 - 8x = -7$. Show your work.

A1.8.6

86.   The area of a farmer's corn field is 2.5 square miles. How large is the farmer's corn field in square feet? Use dimensional analysis.

A1.1.5

87.   Divide: $\dfrac{64x^6y^8 + 44x^2y^6 - 36x^6y^8}{4x^2y^2}$

A1.6.5

Directions: Write out your response to each of the following problems on your own paper. Show all of your work. You MAY use a calculator on this portion of the test.

88. Mr. Dumple wants to open up a savings account. He has looked at two different banks. Bank 1 is offering a rate of 5% compounded daily. Bank 2 is offering an account that has a rate of 8%, but is only compounded semi-yearly. Mr. Dumple puts $5000 and wants to take it out for his retirement in 10 years. Which bank will give him the most money back? Show all your work.

    8.2.2

89. Sandra is going shopping for a new couch for her living room this weekend. She has had her eye on one that costs $749.00. She has already saved $400, and her job is allowing her to work extra hours if she needs to. If she gets an hourly pay of $8.00, how many hours will she have to work this week to buy the new couch?

    8.3.1
    7.3.2

90. David invested $2400, part of it at 5% annual interest and the rest at 6% annual interest. He received a total of $138 in interest at the end of one year.

    Part A Write a system of two equations that could be used to find the amount of money David invested at each interest rate. Let x represent the amount of interest at 5% and let y represent the amount invested at 6%.

    Part B Then, solve the system of equations for x and y to determine how much money David invested in each account.

    A1.5.6

91. Determine the shape of the net below, then calculate the volume of the shape.

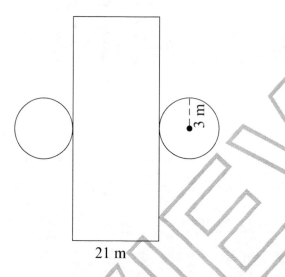

3 m

21 m

7.4.4
7.5.4

92. Write the number 26 in base 2 notation. Show all your work.

8.7.3

93. Solve the inequality $-5 < 2x + 4 < 13$

A1.2.5

94. Using the diagrams below, find the next pattern in the sequence. How many blocks would you need to construct the 6th group?

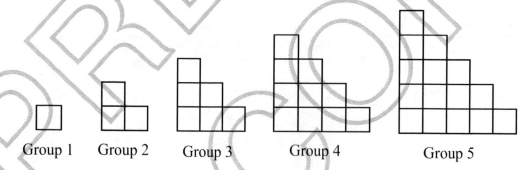

Group 1   Group 2   Group 3   Group 4   Group 5

7.7.1
7.7.3

95. Write the prime factorization of 70.

7.1.5

96. Graph the equations $x - 2y = 6$ and $-2x = y$ on the same graph to find where the lines intersect. Use the graphs to estimate the solution of the pair of equations.

A1.5.1

97. Write the equation $9x + 3y = 18$ in slope intercept form. What is the slope of this line?

A1.4.3

American Book Company

Meeting Standards, Exceeding Expectations

Product Order Form

Please fill this form out completely and fax it to 1-866-827-3240.

Purchase Order #: _____

Date: _____

Contact Person: _____

School Name (and District, if any): _____

Billing Address: _____

Street Address: ☐ same as billing

_____

Attn: _____

Attn: _____

_____

_____

Phone: _____

E-Mail: _____

Credit Card #: _____

Exp Date: _____

Authorized Signature: _____

Order Number	Product Title	Pricing* 5 books	Qty	Pricing 30 books	Qty	Total Cost
IN-ES0904	Passing the ISTEP+ 7 & 8 in Mathematics	$59.75 (1 set of 5 books)		$268.50 (1 set of 30 books)		
IN9-M0905	Passing the New ISTEP+ 9 in Mathematics	$59.75 (1 set of 5 books)		$268.50 (1 set of 30 books)		
IN-M0904	Passing the ISTEP+ GQE in Mathematics	$84.95 (1 set of 5 books)		$419.70 (1 set of 30 books)		
IN-L0904	Passing the ISTEP+ GQE in Language Arts	$84.95 (1 set of 5 books)		$419.70 (1 set of 30 books)		
INS-OLL0806	Passing the ISTEP+ GQE Language Arts On-Line Testing**	$399.00 (1 year subscription)				
INS-OLM0806	Passing the ISTEP+ GQE Mathematics On-Line Testing**	$399.00 (1 year subscription)				

10-1-07

*Minimum order is 1 set of 5 books of the same subject.
**Each subscription is per subject. Only $299 for customers who have previously purchased a site license! If you qualify, please call us today to secure your lower price!

Subtotal

Shipping & Handling 12%

Total

American Book Company ● PO Box 2638 ● Woodstock, GA 30188-1383
Toll Free: 1-888-264-5877 ● Fax: 1-866-827-3240 ● Web Site: www.americanbookcompany.com

ALASKA'S
PERFECT MOUNTAIN

STORIES FROM GIRDWOOD AND THE ALYESKA RESORT

Text By

Lana Johnson

Photography by

Randy Brandon

EPICENTER PRESS

Epicenter Press is a regional press founded in Alaska whose interests include but are not limited to the arts, history, environment, and diverse cultures and lifestyles of the Pacific Northwest and high latitudes. The press seeks both the traditional and innovative in publishing nonfiction books and contemporary art and photography gift books.

Publisher: Kent Sturgis
Acquisitions Editor: Lael Morgan
Cover and book design: Elizabeth Watson
Proofreader: Susan Ohrberg, The Copyright
Production Coordination: Todd Communications Inc.
Printer: Samhwa Printing Co., Ltd

ISBN 0-9745014-1-7

PRINTED IN THE REPUBLIC OF KOREA

First Printing, January 2004

10 9 8 7 6 5 4 3 2 1

To order single copies of ALASKA'S PERFECT MOUNTAIN, mail $19.95 plus $5.95 for shipping (WA residents add $2.25 state sales tax) to: Epicenter Press, PO Box 82368, Kenmore, WA 98028.

Discover exciting ALASKA BOOK ADVENTURES™! Visit our online Alaska bookstore at www.EpicenterPress.com, or call our 24-hour, toll-free hotline at 800-950-6663. Visit our online gallery featuring dog mushing's favorite artist, Jon Van Zyle at www.JonVanZyle.com.

All photographs in this book are taken by Randy Brandon except where indicated.

Cover Photographs: A skier looks down Mount Alyeska, through Glacier Valley and out onto the waters of Turnagain Arm. *Conde Nast Traveler* **magazine calls this view the best at any ski area in North America. At left, from top: the community that stretches from the mountain to the arm is named for James Girdwood, a turn-of-the-century gold miner.** [CITY OF GIRDWOOD] **Downhill racer Kjersti Bjorn-Roli is one of the many world-ranked speed skiers who grew up on the mountain.** [BJORN-ROLI FAMILY] **The 307-room Alyeska Prince Hotel is the only Alaska hotel to earn an AAA Four Diamond Rating. Back cover: Nina and Chris von Imhof pose for an early Alaska Airlines brochure about the Alyeska Resort.**

In memory of Nina von Imhof,

a vivacious redhead who filled Glacier Valley

with laughter, beauty, and dance.

Contents